EARTH

A Tenant's Manual

EARTH

A Tenant's Manual

FRANK H. T. RHODES

Cornell University Press
Ithaca and London

First published 2012 by Cornell University Press
First printing, Cornell Paperbacks, 2012

Printed in the United States of America

Design by Scott Levine

Library of Congress Cataloging-in-Publication Data

Rhodes, Frank Harold Trevor.
   Earth: a tenant's manual / Frank H.T. Rhodes.
      p. cm.
   Includes bibliographical references and index.
   ISBN 978-0-8014-7823-9 (pbk: alk. paper)
1. Earth—Popular works.   2. Earth sciences—Popular works.   I. Title.
   QB631.2.R46   2012
   550—dc23   2012002190

The author and publisher have made a good-faith effort to reach all rights holders for figures and tables. If the current rights holder is unknown or has not responded to multiple inquiries, the original source is noted.

Cornell University Press strives to use environmentally responsible suppliers and materials to the fullest extent possible in the publishing of its books. Such materials include vegetable-based, low-VOC inks and acid-free papers that are recycled, totally chlorine-free, or partly composed of nonwood fibers. For further information, visit our website at www.cornellpress.cornell.edu.

Paperback printing   10   9   8   7   6   5   4   3   2   1

Contents

Part III. Earth Future: The Sustainable Planet

Preface

This book consists of three parts, each of which is, to some degree, self-contained. The first—Earth Present: The Third Planet—sets the Earth in its larger context. It explains how Earth's location and composition determine what it is, how it works, why it is so bountiful, why life is so exquisitely suited to it, or, depending on one's viewpoint, why the Earth is so well suited to life. And it explores why, for all its marvelous richness, Earth's limits are both real and constraining.

The second part—Earth Past: The Changing Planet—explores Earth's long and continuing development and its particular significance to our present pressing challenges. How real is climate change? How serious a threat to us is it, and how predictable a process? How big a problem is pollution? Should we—can we—reduce it? How many people can Earth support, and under what conditions?

The third part—Earth Future: The Sustainable Planet—deals with our human prospects on a finite but changing planet. The question is not whether the planet is sustainable, which it clearly is. The pertinent question is rather whether our burgeoning, high-consumption, human population is sustainable. And the answer is no, our human population is not sustainable, at least not under existing conditions. The essential commodities on which our day-to-day existence depends—fresh water, clean air, fertile soil, adequate food, available energy, and essential metals—are themselves now constrained.

Only new resources, new priorities, new policies, and most of all, new knowledge can offer prospects for a sustainable human future. And a sustainable human future will require a sense of responsible stewardship, for we are not owners of this planet. We are tenants.

Acknowledgments

I have a substantial debt of gratitude to many friends who have readily given me their help and advice in preparing this volume, and I am happy to acknowledge their kindness and support.

Joy Wagner, my executive assistant for more years than either of us care to count, has given me endless technical help and has patiently nudged and encouraged me to see the manuscript through to publication. I am deeply grateful to her.

Rachel Parks, dedicated environmentalist and lawyer-to-be, typed the entire manuscript with consummate patience and skill and helped me at every turn. She chased down text, figures, and references, unearthed data, and played the role of constructive critic. My debt to her is great and I am profoundly thankful for all her support. Without her skillful help I could not have completed the manuscript.

Two Cornell colleagues, Jack Bird and Larry Cathles, have had the fortitude and kindness to read the entire manuscript. Their comments and suggestions for improvements have been most helpful and I am grateful to both of them, though I do not imply, of course, that they would necessarily agree with my conclusions.

Erin Haemling and Amara Fatima Pinnock corrected later versions of the manuscript with cheerfulness and precision. I am grateful to each of them.

A number of colleagues and friends have given me the benefit of their specialized professional help in reviewing particular chapters. These reviewers include Norman Scott, who read the whole of parts 2 and 3, Yervant Terzian (chapter 1), David Pimentel (chapter 15), Frank Di Salvo (chapters 16, 23, and 24), the late Walter Lynn (chapter 17), Harold Mathijs Van Es (chapter 19), Malden Nesheim (chapter 20), Jeffrey Tester (chapter 21), and Steven Sass (chapter 22). I am deeply grateful for their help. The errors that remain are, of course, my own.

John Ackerman, director of Cornell University Press, has been consistently helpful, supportive, and creative. I am most grateful for his enthusiastic advice and support. I am also most grateful for the generous help of Ange Romeo-Hall, Assistant Managing Editor, and for her consummate editorial skills. Sincere thanks are due as well to Scott Levine for expert design and Sarah Grossman for skilled assistance with permissions.

A Note on Related Reading

At the end of the book is a list of books and articles, organized by chapter, for further reading. Although I have included a few introductory-level textbooks, most of the references listed are written for the general reader. In a few controversial areas, such as nuclear power, climate change, and the volume of future petroleum supplies, I have deliberately included references to authors of varying viewpoints.

I have generally not included specialized scientific research articles, except where these are quoted in the text. These are listed in the footnotes.

PART I

EARTH PRESENT

The Third Planet

CHAPTER 1

The Third Planet

Suppose, just suppose, you were somewhere out in space, beyond the confines of the solar system, and that you were in radio contact with an alien, a friendly alien, from another part of the galaxy. And suppose, just suppose, that in chatting about things in general, you wanted to describe your home, the place where you live or had grown up. Not that small town in New Hampshire, or that particular suburb of Los Angeles or Cape Town or London, but the planet, Earth—the home planet. How would you identify it? How would you give some sense of where it is, what it is like, how it is to live there?

Well, you'd have to locate the Sun, our parent star, of course, and then relate the Earth to that. So, how do we locate, how do we describe the Sun? The Sun, as astronomers describe it, is a medium-sized, main sequence G2 star, one of more than a billion, perhaps as many as 250 billion, stars in our galaxy. There are other galaxies, too, of course. Perhaps hundreds of billions. And Earth is one of the dense, rocky, inner planets, the third planet out from the Sun, between the pale, cloud-draped Venus and that rather reddish planet, Mars. "But what's it like to live on a tiny planet like that?" asks the alien. "If I were to go and live there, if I could travel there, what would I find? What kind of place is it?"

That could be a long conversation, because it's a difficult question. It's difficult because no two of us see Earth, or anything else for that matter, in quite the same way. Think of all the artists, poets, writers, composers, explorers, cartographers, landscape architects, urban planners, and more, and the endless richness and variety of their views, perceptions, and descriptions of Earth. And it's difficult, too, because it's impossible to see Earth "whole," even from space. "My view of our planet was a glimpse of divinity," declared the astronaut Edgar Mitchell. But most of us can't see that. Scale and perspective limit us. Even if we took all the descriptions that have ever been written, or painted, or composed, thousands upon thousands of them, we'd still have trouble giving a truly comprehensive description, however well we know our own planet. Let me try to explain what I mean.

There is something comforting, something reassuring, about Earth. It is where we are from, the place where we grew up: our backyard, our street, our hometown, our state. It is the place we know, the place where our family and our friends live. It is the place where we work. Even the word *Earth* is somehow warm and familiar. It is the summer evening barbecue on the lawn; it is fall, with the glorious colors of the maple near the porch; it is Thanksgiving night with the family around the table, enjoying turkey and pumpkin pie; it is spring rain, with trillium under the new-leafed branches. It's Earth. It's home.

And it is not just my local neighborhood. It is the other places that we know so well. It's the Merritt Parkway, it's Fifth Avenue, it's Yankee Stadium, it's Rockefeller Plaza at Christmas, it's summer on Nantucket, and it's days on the beach. It's taking the kids to camp by the lake in Ontario; it's flying to Denver to visit the in-laws. It's all that and more.

And just then the astronomer—a terrestrial astronomer—strolls by. "Earth, ah, yes, Earth," he muses. "The fifth-largest planet of the solar system: third in order from the Sun, between Venus and Mars. It has a mean distance of about 93 million miles from the Sun. That's about 150 million kilometers, by the way. It rotates from west to east, completing one rotation every day. Ah, but you know that, don't you? Its diameter at the equator is 7,926 miles, and its circumference is about 24,830 miles. It's a rocky planet, not gaseous like . . ."

"Rocky it certainly is," interrupts the geologist. "Its mean density is about 5.5 grams/cm^3, but here's the remarkable thing. The average density of all the rocks we can measure at the surface is only 2.8. Quite remarkable, isn't it? How do you account for that? Well, the deeper rocks must be much denser than the ones at the surface. In fact, the crust of the Earth, its outer shell, is very thin—well, comparatively thin. It ranges in thickness from five to twenty-five miles, and the continental areas are thicker, as well as higher, than the oceanic areas. Below the crust, the mantle consists . . ."

"What you omit," interrupts the biologist, "is that Earth is the only planet known to support life. Lewis Thomas described it well: 'Viewed from the distance of the moon, the astonishing thing about the earth, catching the breath, is that it is alive. . . . Aloft, floating free beneath the moist, gleaming membrane of bright blue sky, is the rising earth, the only exuberant thing in this part of the cosmos. . . . It has the organized, self-contained look of a live creature, full of information, marvelously skilled in handling the sun.' "[1]

"What you mechanistic merchants overlook is the splendor and beauty of this remarkable place," interrupts the poet. "Remember what Wordsworth wrote:

There was a time when meadow, grove, and stream,
The earth, and every common sight,
To me did seem
Apparelled in celestial light,
The glory and the freshness of a dream."[2]

"Wait, wait," demands the environmental activist. "Measure it, analyze it, eulogize it if you will. But the most important thing people need to know about Earth is that we are abusing it, plundering it, polluting it, poisoning it, depleting it, destroying it. We're squandering our heritage, defaulting on our tenancy, robbing our children. Every year thousands of acres of tropical forest are destroyed; every hour another species becomes extinct."

"Now let's not dramatize this. Let's not overdo this doomsday scenario," replies the solid citizen. "Climate change is nothing new. It's been going on since the end of the last Ice Age. And a good thing, too. Without it, this would be a bleak and inhospitable place to live. As for depleting resources, global warming, rising sea level, and all that doomsday stuff . . ."

So each blind man, each specialist, touches the elephant Earth. We all do. We must. And we desperately need the knowledge that each one brings. The problem is that it is impossible to comprehend it all: it's impossible to grasp the whole planet, to throw our intellectual arms around it or integrate all the descriptions of it.

But because we live here, we have to try. This is not just an artistic compulsion or an existential yearning, still less an academic exercise. It's a survival issue. This is the only planet we have. There is no escape to Mars for us. We're stuck here, and we don't own the place—it would be the height of arrogance to assume that we do. We're tenants here, not owners, but we're tenants with hope for a long-term tenancy. We want to extend our lease just as far as we can.

That doesn't mean we are intruders, or interlopers, of course. Our species has been here a long time, at least in human terms: about 130,000 years, in fact. Early tool-making hominids appeared about 2 million years ago. And as tenants we have rights, rights to peaceable occupancy, just as every lease explains. But tenants also

---

1. Lewis Thomas, *The Lives of a Cell* (New York: Bantam Books, 1975), 170.
2. William Wordsworth, "Ode: Intimations of Immortality from Recollections of Early Childhood," *The Complete Poetical Works of William Wordsworth* (New York: Houghton Mifflin, 1919), 5:54.

have obligations as well as rights. It is our job to look after the property, to care for it, to safeguard it. It's our job as tenants to be responsible occupants, careful custodians, good neighbors with the 7.0 billion other people with whom we share our tenancy, not to mention some 2 million known species of other animals and plants who share the property.

So we need some understanding about this remarkable planet we share. We need to be clear about the terms of our tenancy. We need to know something about the property itself, its location, its layout, and its boundaries. We need to know a little bit about its history, and we need as much information as we can possibly gather on how to take care of the place: how the plumbing works, how the heating system operates, and the longer-term outlook for our residence and our well-being here, as well as the prospects for our families and children.

And that's an urgent discussion. The old Native American proverb is right: "We don't inherit the Earth from our parents; we borrow it from our children."

This book is intended to provide a kind of manual that will help us become knowledgeable and responsible tenants. It is a book that seeks to explore the property and get a glimpse of the wider neighborhood. It is an attempt to create a kind of mental map, not unlike those beautiful medieval navigation maps, which charted intricate coastlines and rivers, mountains and settlements, relationships and distances, winds and hazards, and guided our forebears across unknown seas. If we can create that kind of chart, we can have an informed discussion about our roles and responsibilities. For there are relationships that are subtle, trends that are complex, and hazards that are more threatening than medieval sea dragons.

Every one of the countless species that ever lived has contributed to the building of our planet. And every living species—bacteria, insects, flowers, trees, worms, fish, mammals—every living creature, influences this planet that we share together. But we, our species, influence it more than all the rest at the present time; we know that we influence it, and we begin to understand a little how we influence it. We can even choose, actually select, the way in which we wish to continue to influence it, for good or ill. So we are not just tenants, we are something more. We are stewards, custodians, caretakers, and "it is required of stewards that they be found faithful."

Now suppose we grant all that. Suppose we agree we are not owners but tenants, with needs and rights as well as obligations. And let's agree, too, that we have neighbors we care for, not only next door, but also beyond, to whom we have obligations. And we have families, too—parents, siblings, children, grandchildren—about whom we care, about whose future well-being we are deeply concerned. Grant all that. But still, we live on a small planet with finite resources, a burgeoning population, and stark inequalities in the distribution of wealth and resources. People of the developed world have an insatiable appetite for goods, services, food, energy, and raw materials, while more than 3.7 billion humans suffer from malnutrition.[3] We live in a world where our long-term exploitation of the planet's capital—soils, fossil fuels, metals, forests, and fisheries—may threaten not only to deplete but to exhaust some of our nonrenewable treasures, and certainly threatens to reduce the quality of the air we breathe, the soil we till, and the water we drink. We live in an age where much of the once widespread pastoral landscape, that "green and pleasant land," has been lost to development and urbanization. We live at a time when the environmental impact of relentless population growth and industrialization of the past two centuries in the developed world is about to be reinforced by the far more rapid industrialization of the less developed world.

One doesn't need to be an environmental zealot, or even a Green Party member, to see the widespread effects of global warming, of climate change, of rising sea level, of deforestation, of loss of arable land, of atmospheric pollution, of resource depletion. Continuing population growth multiplies all these problems. And the growing industrialization of the developing world multiplies and magnifies them even more.

All this sounds a little like Woody Allen: "More than any other time in history, mankind faces a crossroads. One path leads to despair and utter hopelessness. The other, to total extinction. Let us pray we have the wisdom to choose correctly." And we might have to concede his point unless—and it's a big unless—we can have a realistic assessment of where we stand and begin to think and talk and act together as tenants. No coherent planning, still less any global strategy or UN program, is likely to emerge until we can begin this informed discussion. That's the starting place for this book.

---

3. D. Pimentel, "Ecological Footprint, World Population, and the Environment" (keynote paper presented at the International Ecological Footprint Conference, "Stepping Up the Pace: New Developments in Ecological Footprint Methodology, Policy and Practice," BRASS Centre, Cardiff University, Wales, http://www.brass.cf.ac.uk/events/International_Ecological_Footprint_Conference-Keynote_Speeches.html).

CHAPTER 2

The Home Planet

Beating the Bounds

Though we know our neighborhood like the back of our hand, the Earth itself, the larger place that includes all our neighborhoods, is a strange place. This homestead, at once so stable, is spinning, we're told, and spinning fast at about 1,000 miles per hour. This secure dwelling, this sturdy foundation, is also hurtling around the Sun at 66,600 miles per hour; that's 18.5 miles per second.

It's unsettling, contradictory, bewildering, even intimidating. So let's step back and get a sense of our position. Just where are we? How can we get a useful sense of the place we call home?

There is, in some parts of Europe, an ancient ceremony known as "beating the bounds." The ceremony has existed in Britain, for example, for well over two thousand years, and in earlier days it involved the inhabitants of a manor house, village, or parish marking property lines by walking the boundaries and beating certain boundary markers—trees, boulders, fences, hedges, and so on—with sticks.

The original purpose of this regular beating was perhaps to drive out evil spirits, but by Anglo-Saxon times it was used chiefly to assert the limits of the property and especially, in days before maps, to acquaint all the local inhabitants with the nature and extent of their land. At certain points, young boys were held upside down and beaten with wands or had their heads bumped on particular markers, presumably to emphasize the importance of the boundary line.

Beating the bounds defined the boundaries of the dwelling place. It also emphasized one more thing: it acknowledged the bounty of the land. It reminded tenants and serfs (and almost all the participants were either tenants or serfs) of their obligations and confirmed their rights. It united everyone in a sense of community, place, and dependence.

How do we define the boundaries and explore the limits of our global village today? How do we encourage that sense of place, of community, of responsibility? Well, perhaps by beating the bounds, just as our forebears did, marking the extent of this wonderful homeland of ours.

But, of course, there is more to the home planet than the Earth alone. Earth's surface doesn't mark our boundaries. There's the atmosphere above it, extending some 350 miles above Earth's surface. There are neighbors to consider: Venus, Mars, and others. There is the heating plant, our energy source for the whole property, our grandparent: the effulgent Sun. There's our terrestrial daughter, the Moon, which creates the tides. There are "lights for signs" as the psalmist called them, stars for navigation and direction. There are bodies beyond the terrestrial boundary that influence our every turn and on whose benevolence our lives depend. We need to look around not only the Earth but also the larger neighborhood.

Even so, we are Earth creatures, and likely to remain so. So let's start with Earth.

Measuring the Earth

On April 7, 2006, a British traveler was interrupted as he tried to cross the Russian border, where he was detained by immigration officials because his passport did not have the required visa stamp. What was remarkable was that the border was the one across the Bering Strait and that the traveler, Karl Bushby, an ex-paratrooper, was seeking to walk around the world. His journey had started in Chile and continued along the length of South America and North America to Alaska and then over the Bering Strait. He'd already walked eighteen thousand miles, about half the total distance of his planned journey, and had been walking since 1998. He hopes to complete the walk in 2014, so his walk around the Earth will have taken sixteen years. That's how large the Earth really is.

Our ancient ancestors regarded the Earth as an object of veneration, but various cultures suggested several views as to its form. The Homeric-age Greeks regarded it as a flat disk, surrounded by a world ocean; the Egyptians as a broad surface, with the sky stretched as a canopy over supporting mountain ranges; the Hindu Vedas as a tree of knowledge; and the Brahmins as a mountainous land, resting on elephants, who stood on the shell of a giant turtle, in the middle of an encompassing sea. It was Pythagoras and later Plato in about 500 BC who suggested, on aesthetic grounds, that the Earth must be spherical. Aristotle, in the third century BC, confirmed this by arguing that it was demonstrated by the hulls of ships disappearing over the horizon being lost to view before the masts and also by the Earth's curved shadow on the Moon during a lunar eclipse.

But it was left to another Greek, Eratosthenes, some years later, to measure the Earth. Eratosthenes was born in Cyrene, in what is now Libya, in about 276 BC. He was a man of extraordinary talent: poet, philosopher, administrator, mathematician, and astronomer; he headed the great library at Alexandria and invented the "sieve of Eratosthenes" for finding prime numbers. Eratosthenes was a prolific author, but he is best remembered as the one who made the first measurements of the circumference of the Earth.

Like several of his contemporaries, Eratosthenes reasoned that, because both the Sun and Moon were spherical, it was probable that the Earth, too, was spherical rather than flat. He had noted that at noon on Midsummer Day, the Sun shone vertically down a well at Syene (now Aswân in Egypt). He argued that, if the Sun were assumed to be so remote that its rays could be regarded as essentially parallel to one another, measurement of its angle of inclination (by means of the shadow of a vertical object) at another locality (Alexandria) and measurement of the distance between the two points could be used to calculate the circumference of the Earth. This he estimated (using reports from caravaneers who took fifty days to make the journey at a rate of one hundred stadia a day) to be 250,000 stadia, and, though there is some scholarly debate as to exactly how long a stadium was, this is interpreted as about 28,700 miles, a remarkably accurate result. Eratosthenes also calculated the inclination of the Earth's axis with similar accuracy, as well as the distances to the Moon and the Sun.

There the matter rested until the sixteenth century, when Galileo's observations with the telescope opened up a new era of astronomical interest and encouraged precise observation. The Royal Academy of Sciences in Paris assigned Jean-Felix Picard to measure the radius of the Earth, which he did by accurately measuring the distance between two points on the same meridian whose latitudes differed by 1°. He derived a figure for the Earth's radius corresponding to 3,963 miles, giving a circumference of 24,900 miles, about 13 percent less than Eratosthenes' estimate 1,700 years earlier, which he published in 1671.

Refinements continued over the centuries and the figure presently accepted for the Earth's radius is 3,963.44 miles, with slight flattening at the poles. Remote sensing satellites have added continuing refinements to these measurements.

Now measuring the Earth is one thing; weighing it is another. In 1665 Sir Isaac Newton, then twenty-two years old, proposed his theory of gravitation, which stated that the attractive force between two objects is equal to the universal force of gravity, multiplied by the product of their masses and divided by the square of the distance between them.

Newton did not measure the gravitational constant. In fact, it is exceedingly difficult to do so because the force is so weak. The force of attraction between one-foot lead balls, for example, each weighing four hundred pounds and placed two feet apart, is only two one-hundred-thousandths of an ounce. It was a century after Newton that Sir Henry Cavendish in 1798 performed a remarkable experiment to measure the density of the Earth using two lead spheres on the ends of a horizontal bar, suspended from a long thin metal wire. When another, heavier pair of lead spheres were moved toward them, the slight attraction between the objects twisted the wire, and Cavendish used this deflection as a measure of the force between them. His measurement of the density (5.448 times greater than that of water) was subsequently used to obtain the mass of the Earth and, much later, to calculate the force of gravity. The figure for the mass of the Earth—6.6 billion tons—is about twice as much as we would expect Earth to weigh if it consisted only of rocks of similar density to those known at the surface, suggesting that the deeper parts of the Earth consist of much denser materials than those of its crust.

Beyond Our Reach: The Larger Universe

Large as the Earth is, it is small when viewed as part of the solar system: all the planets together make up only 0.13 percent of the mass of the solar system. And that system, vast as it is, would be invisible even from the nearest stars, using our present observational techniques.

Observations of distant galaxies indicate the immensity of this enveloping universe. Five qualities of the universe stand out: its size, its density, its expansion, its age, and its temperature. The universe is *vast:* some 10^{20} times larger than the Earth. It is *light:* Earth's density is 5.5 grams per cubic centimeter; the comparable density of the universe is about 1×10^{-31}. That's equivalent to one hydrogen atom for every cubic meter of space. It is *ancient:* perhaps some 13.7 billion years old. That's about three times the age of the Earth. And it is *cold:* away from the stars, its average temperature is only 2.726 degrees Kelvin (°K). That is only 2.726 degrees above absolute zero, which is −273.15 degrees Centigrade (°C). And it is *exploding* apart, with the most distant galaxies receding at speeds approaching the velocity of light.

This expansion of the universe implies that it once existed in a smaller, denser form. Studies in 1964 detected weak background radiation that seems to represent the remnant of the "big bang," in which this earliest, ultradense phase of the universe exploded in a fireball of unimaginable intensity.

The history of the first few seconds of this maelstrom is still debated by cosmologists, but there is agreement that after a few minutes it had already assumed its present gross composition—about 75 percent hydrogen by mass and 25 percent helium, with traces of other light elements such as deuterium, lithium, and beryllium—and that its violent turbulence gave rise to islands of matter, which formed galaxies, all rotating and continuing to hurtle apart.

It is these galaxies—giant islands of millions of stars—that are the basic components of the universe. Some are spiral in shape, some elliptical, some disklike, and some irregular. Our planetary system is part of a great spiral galaxy—the Milky Way—a whirling, spiral disk of perhaps some 100 billion or more stars, all rotating around the center of the galaxy, visible on a clear night as a faint, diffused band of starlight across the width of the heavens. So vast are these galaxies that their dimensions and distances are measured in light years: the distance light travels in one year at its velocity of 186,000 miles per second. That's a distance of about 6 trillion miles. Our galaxy is about 100,000 light years in diameter. The nearest galaxies beyond our own are 180,000 light years away, while the nearest spiral galaxy, Andromeda, is 2 million light years away. The most distant galaxies are some 10 billion light years away.

How many galaxies are there? We do not know; perhaps 200 billion. We are surrounded by an uncounted and perhaps uncountable number of neighbors. To the universe in which we find ourselves, there are no obvious boundaries.

The Third Planet

It took humankind a long time to recognize what a remarkable position we occupy in the universe. The ancient tradition—articulated by Ptolemy in the second century AD, established by reason, confirmed by common sense, and embraced by the church—was that the Earth stood motionless at the center of the universe. The heavenly bodies rotated around it, embedded in concentric, crystalline spheres. The irregular movements of the planets against the fixed positions of the stars were explained by an elaborate system of epicycles, in which the Sun and each planet moved around the Earth, much as we now understand the Moon to rotate around the Earth. This was a complex system, and it became more complex still when Ptolemy was compelled to adjust his theory to accommodate the movements of the outer planets.

It's commonplace today to speak of Earth as a planet. But that designation was a long time coming. Although the Greeks viewed the planets as "wanderers" (*planetai* = "wanderers") because of their distinctive movements against the wholesale east-west travel of the "fixed" stars, the Earth seemed fixed; it was the firm base from which they could watch the grand cavalcade of the Sun, Moon, and planets moving above us.

The theory had great appeal. It placed man, created in the divine image, in his appropriate place at the center of the universe, so it was endorsed by the church; it satisfied our everyday experience of the Sun rising and setting around the Earth;

it confirmed common sense, for surely the ground beneath us was firm, solid, immovable. It accorded with a sense of fitness, for the movements of the planets in circular orbits had a pleasing order and symmetry. For most people, it was a reassuring, comfortable belief.

But not for Nicolaus Copernicus, the gentle revolutionary who brought us from a closed world into an infinite universe, changing the whole tone and scale of human thought, by freeing us from our first intuitive concepts, indeed, by changing our whole notion of common sense. Copernicus was born in 1473 in Torn, an old city of the Hanseatic League on the bank of the River Vistula, now in Poland. The fall of Constantinople in 1453 created an exodus of scholars, bringing with it Islamic and Greek learning and a wealth of ancient manuscripts. The growing influence of the Renaissance, the recent invention of printing, and the stirrings of the Reformation contributed to the yeasty atmosphere of the times. Voyages of discovery brought a flood of new knowledge to the Old World. This was a time of relative peace and affluence; it was also a time of discovery and awakening. Copernicus was a man of extraordinary talent: trained as a lawyer, physician, astronomer, and theologian, he became a canon of Frauenberg Cathedral and proved an able ecclesiastical administrator.

Copernicus's mathematical sense of fitness was offended by the Ptolemaic, geocentric picture of the universe, which he regarded as "inconsistent and un-systematic." Copernicus used few original observations in his argument, but his sense of correctness and unity was a compelling influence. He displaced the Earth from its central position in the solar system and replaced it by the Sun. He knew that Aristarchos of Samos had suggested a similar arrangement. The treatise in which Copernicus proposed this—*De revolutionibus orbium coelestium*—was published as he lay on his deathbed.

With the new Copernican explanation, the planets' wandering now made sense. We now know that the planets speed around the Sun in regular elliptical (though almost circular) orbits, the most distant being the planet Pluto, which is some 3.67 billion miles from the Sun. (We won't argue here, by the way, about whether or not Pluto is "really" a planet.)

Even as the Greeks tracked the paths of the planets, it was the Romans who later named them after their gods. The innermost planets—Mercury, Venus, Earth, and Mars—are rocky bodies. The outer—Jupiter, Saturn, Uranus, and Neptune—are large and gaseous. The most distant planet, Pluto, is made mostly of ice.

So let's consider our neighbors in space.

The Sun: The Home Star as a Crucible of Creation

With clear weather, a sharp eye, and a lot of time to spare, a careful observer might count two thousand stars in the night sky. As the seasons change and bring new stars into view, another four thousand or so become visible. With the aid of the best telescopes, countless others can be seen. But as vast as the numbers of stars are, the spaces between them are still vaster. Most of the universe is space: dark, cold, and all but empty. If the Sun—93 million miles away from us—were reduced to the size of a dot on the top of an "i" in this sentence, the nearest star would still be some ten miles distant.

We now know that the Sun, our grandparent star, with a diameter of 865,000 miles, is large enough to contain more than a million Earth-sized bodies. It is gaseous, composed of hydrogen, which acts as the solar fuel, producing helium in a radioactive solar furnace, and in the process losing a little mass and releasing energy. It is this process that keeps the Sun shining. It is this energy that sustains life. From this nuclear inferno, we ultimately emerged.

Certain qualities of stars are measurable: their distance, their brightness or magnitude, their surface temperature, and their composition, which we can determine from their spectra—the signature of their light—and the dark, distinctive absorption lines that characterize them. Studies conducted over many years reveal that most stars fit into a well-defined sequence when their absolute magnitude (or the energy emitted) is plotted against their surface temperature. Most stars, including our Sun, fall within this "main sequence," as it is called, and this provides a clue to their history.

Stars are being formed continuously within the vastness of the universe, condensing from clouds of interstellar gas and contracting under gravity to form protostars, until at last their fusion fires are lit. At that critical point, hydrostatic equilibrium is established where the inward force due to gravity is balanced by the outward pressure generated at the star's core by the conversion of hydrogen to helium.

Stars vary greatly in mass, and this affects both their history and their life span. The larger stars have higher temperatures and shorter lives, ending in supernova explosions. At some point all stars exhaust their hydrogen fuel, the smaller ones cooling down and collapsing to form dense white dwarfs.

All stars, whatever their type, are the crucibles of creation. They are the factories of the universe, the sites at which new chemical elements are made. All the elements, other than hydrogen and helium, are manufactured in stellar furnaces. Studies of the Sun's spectra show it to consist of about three-quarters hydrogen and one-quarter helium, but with about 1 percent of the heavier elements. Most stars have a similar composition, and it seems reasonable to conclude that this is not unlike the primeval composition of the universe. In fact, recent astrophysical studies suggest the universe reached this composition within the first ten minutes of its existence. But Earth and the other "inner" planets of the Sun are made of heavier elements: silicon, oxygen, iron, and all the rest. It now seems clear that all these elements up to iron are the products of fusion reaction in stars, while the heavier elements are produced by the supernova explosions we have just described. So Earth and all its components, living things included, are products of the life and death of the stars that existed long before us.

The Sun is one of millions of stars that make up our galaxy, the Milky Way. It is already known that at least some of these stars support planetary systems of their own.

Most cosmologists reconstruct the history of the solar system from the time of formation of the proto-Sun as the history of a rotating, collapsing cloud of interstellar dust and gas, with some cloud fragments rotating around it in a flat, disk-like orbit. From this swirling disk of gas and dust protoplanets gradually formed, growing by the accretion of orbiting debris. And from this maelstrom we have emerged.

The Moon is the most familiar, the most celebrated, body in our night sky. Constant yet changing in both appearance and position, hauntingly beautiful but heavily pockmarked, luminous but pale, smaller than Earth but encircling it, the Moon illuminates, defines, and inspires, influencing the behavior not only of humankind but of other creatures too. It is the only other heavenly body that we have, as yet, visited. The Moon is a near neighbor: only about thirty times Earth's diameter away. The Sun, by contrast, is about four hundred times more distant.

Ancient peoples—Babylonians, Greeks, Romans, Arabs, and Jews among them—used the Moon, rather than the Sun, as a timekeeper. The variable date on which Christians celebrate Easter is a reflection of its ancient definition as falling on the first Sunday after the first full Moon of spring and is a reminder of this earlier lunar calendar. But the Moon's gentle influence extends beyond churchmen, poets, lovers, and other animals. It is the chief agent in creating the tides that so influence the lives of the significant portion of creatures who live in the shallow seas or on the coasts. The Moon's mass, though small, is so "near" to Earth that, although its gravitational attraction is negligible, its attraction over the water of the oceans is marked. It draws the water nearest to it into a bulge of high tide, while its lesser attraction of water on the opposite side of Earth produces another corresponding tidal crest. These tidal crests travel around the oceans in the course of a day (a synodic day of twenty-four hours, forty-eight minutes: the time between two culminations of the Moon), so coastal localities in most parts of the Earth experience high and low tide twice a day. The effect becomes less in the polar regions.

The Moon is an ancient satellite, as old as the Earth, in fact. It formed, much as Earth did, some 4.6 billion years ago, by a giant collision, followed by the falling together of debris, whose impact heated up the new body. After its initial heating, the Moon cooled, and was bombarded again by meteorites, whose impact is still evident in the higher areas forming the heavily cratered lunar highlands. The Moon was later flooded with volcanic lavas, which spread over huge areas of its

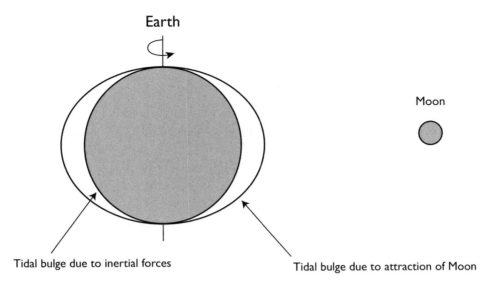

Earth

Moon

Tidal bulge due to inertial forces

Tidal bulge due to attraction of Moon

2.1 Formation of tides. Courtesy of Earth Science Australia.

surface, forming the maria, the seas of Galileo. Although a few fragments of older lunar rocks are known, most lunar rocks are 3.1–3.8 billion years old: about the same age as the oldest known Earth rocks. For the past 3 billion or more years, the Moon has been frozen in time: nothing about it has changed, except for the occasional impact of a meteorite. It is a dead, but beautiful, body. Its surface facing the Sun boils at 200°C. Its dark side is a frigid face: –150°C.

But why are Earth and Moon so different? Although the moon is an unusually large satellite compared to its planet Earth—so big, in fact, that it is often described as a planet—its gravity is only about one-sixth that of Earth; it cannot retain an atmosphere or water. So its ancient crust has been unprotected by an envelope of air, and its surface is pitted and scarred with impact craters from meteorites that have showered upon it throughout its long history. The Moon's surface, unlike Earth's, has not been constantly renewed by subsurface processes; its oldest—and its most densely cratered—rocks are those of its highlands, but after the great volcanic overpourings of some 3 billion years ago, it became inactive, its surface dust stirred only by asteroid impacts.

And beyond the bounds of the Moon and our own neighborhood lies the rest of the familiar solar system: our parent Sun and the rest of its family of five (or six) other, more distant planets; about 140 moons or satellites; thousands of minor "planets" and asteroids; multitudes of meteors; and scores of comets.

Vast as it is, our solar system represents the offspring of just one star near the periphery of a galaxy, the Milky Way, made up of billions of stars. Even so, it's not easy to grasp the immensity of the solar system. From the Sun to the nearest planet, Mercury, is an average distance of 6 million miles, with the planet circling the Sun every three Earth months. The planet Pluto is, as we have seen, some 3.67 billion miles away and takes 250 Earth years to circle the Sun.

And out beyond our solar system, there lie more distant stars. The nearest, Alpha Centauri, is 4.37 light years (25.8 trillion miles) away. Even from that nearest star, it would be impossible to see our planetary system using methods now available to us.

But we are concerned, not with remote regions, but with our corner of space, our neighborhood. We've seen that only the four inner planets are rocky. Mercury, for example, the nearest planet to the Sun, has a cratered surface, rather like that of the Moon. It is rocky and dense, with a relatively thick outer shell of silicate rock and a dense and larger iron core, at least some of which is probably molten.

Our inner neighbor Venus, the second planet, is, in many ways, very like Earth: it's about the same size, the same density, the same composition. It's a beautiful sight, a welcoming wanderer—the morning and evening star—so bright it shines steadily even in the light sky of dawn and sunset. Its brightness reflects not only its nearness to the Sun but also the dense white clouds that perpetually shroud its surface.

But these perpetual clouds hide an inferno. For Venus, though named for the goddess of love, is no hospitable neighbor, no benevolent deity. Space probes have shown that its barren surface is scorchingly hot (480°C/842°F), its cloud cover is formed from droplets of sulfuric acid, and its atmosphere consists of carbon dioxide, torn by violent electric storms. The atmosphere of Venus is one hundred times denser than ours—so dense, in fact, that it produces a runaway greenhouse inferno whose searing temperatures have long ago evaporated off whatever oceans the planet may have had. So hot is the surface of Venus that acid rain, falling continuously through its atmosphere, never reaches its surface, evaporating and recycling before it lands on the barren landscape.

Venus, though close to Earth in size, density, gravity, and orbital position, is therefore quite unlike it in its heavy atmosphere of stifling carbon dioxide, its crushing pressure (ninety times that of Earth), its acid rain, and its scorching temperature (480°C/842°F), which together make it a "truly inhospitable, terrible place." Carl Sagan once wrote that "the average person's view of Hell—sizzling, choking, sulfurous red—is a dead ringer for the surface of Venus." This is global warming gone mad. Beneath its blanketing cloud cover, though it has a rocky crust, with volcanoes and mountain ranges, Venus is starkly different from Earth.

Now consider our other near neighbor, Mars, the fourth planet. Its surface is a dusty red desert. It is only about half the diameter of Earth or Venus, but, like these two bodies, it is a rocky planet with a past. Generations of Mars watchers have considered it as a possible place of life. Percival Lowell, a wealthy Bostonian, devoted his life and his fortune in the late nineteenth and early twentieth centuries

2.2 The solar system. Courtesy of NASA.

to building an observatory in Arizona to study its canals. Fiction writers from H. G. Wells to C. S. Lewis and Robert Heinlein have created Martians, both friendly and menacing.

But two unmanned vehicles ("Mars rovers") that landed on the surface in 2003 confirm it to be a red, cratered landscape with gigantic volcanoes—now cold, dry, and desolate—that bears the scars of catastrophic erosion by ancient floods of water on a gigantic scale. Its poles are marked by glacierlike ice caps of carbon dioxide, frozen out from the thin atmosphere that envelops the planet. But how can what is now a bone-dry desert display this remarkable landscape? Studies suggest that it must have had an earlier history of active erosion and volcanism, with a humid climate and a major river system. Some observers believe there may still be significant subsurface frozen water.

Mars has an atmosphere some 150 times less dense than that of Earth and thus experiences a far greater range of surface temperatures. Water—though it seems once to have existed on Mars in liquid form—would now evaporate or freeze at its surface. Whether life ever could or ever did exist there is a matter of continuing research and debate

Outward from the Sun, beyond Mars, all the planets have gaseous surfaces, with the gas becoming denser with depth. Jupiter and Saturn—separated from Mars by a belt of asteroids, apparently composed of leftover debris from the formation of the solar system—are giants, ten times bigger than Earth. Made of hydrogen and helium, they are intensely turbulent and ringed by their own orbiting systems of satellites, at least one of which has active volcanoes.

Jupiter consists of about 90 percent hydrogen and 10 percent helium (75/25 percent by mass), as well as traces of water, methane, ammonia, and "rock." It probably has a core of rocky material, and its main portion is probably liquid metallic hydrogen. Saturn has a similar composition, and this is thought probably to have some resemblance to the primordial nebula, from which the Sun and all the planets developed.

We have already seen that life as we know it—carbon-based life—can exist only in a narrow range of physical and chemical conditions. Earth's present atmospheric oxygen, for example, would probably have destroyed the large organic molecules from which the oldest forms of life arose had it been in the very early atmosphere. But Earth's present atmosphere and magnetic field shield life from all but the largest incoming meteorites as well as from harmful ultraviolet radiation, and also moderate Earth's climate to the narrow range in which life is possible.

For all its apparent insignificance, Earth's particular place in the solar system, its composition, and its mass have been critical factors in influencing its development, including the development of life. Yet to compare Earth only with the solar system is to fail to comprehend its scale in relation to the rest of the universe. All the planets, taken together, account for only about 0.13 percent of the mass of the solar system, and most of even this combined mass is represented by the large planet Jupiter, which, if it had been just a bit larger, would have been a second Sun. It emits more energy than it receives from the Sun. Small as that proportion is of the mass compared to the solar system, it is even smaller when Earth is compared to the larger universe of which we are a part. The universe is about 10^{28} more massive than the Earth, and their radii differ by twenty powers of ten! There are two other significant differences: Earth, though ancient, is only about one-third as old as the universe, and whereas Earth has a high density (5.5×10^3 kg/m^3) from its rocky composition, the larger universe, as we have seen, has a mean density of only about 1×10^{-26} kg per cubic meter, which represents about one hydrogen atom per cubic meter.

Perhaps in this vastness beyond our solar system there may be other planets with habitable zones—conditions equally hospitable to life. Perhaps, indeed, life in other parts of the vastness of space is not carbon based, so that its description may require a whole new vocabulary. But on this planet Earth, the third planet, there is life, and we are a part of it. We are the products of a long and intricate history—some 3.5 billion years long—of the interaction of living things with this rare and remarkable planet. But, for all our self-awareness and sophistication, we remain Earth's children, wholly dependent on it for every aspect of our existence. It is to this dependence that we now turn.

CHAPTER 3

The Rocky Planet

The Singular Advantages of Being Third

We travel together on this remarkable planet Earth, plunging through the darkness of space at nearly 67,000 miles per hour, as well as moving around our star, the Sun, at 66,000 miles per hour, at just the right distance to allow us to live. It is this distance that makes Earth neither too hot, thus allowing liquid water to remain here, nor too cold, which would freeze it all, with an atmosphere dense enough to allow animals to breathe and plants to respire but not so dense as to filter out sunlight. And our planet's molten nickel-iron core produces an extensive magnetic field, which in the upper atmosphere shields us from harmful extraterrestrial radiation.

By contrast, Venus, Earth's inner neighbor, though close to Earth in size, is an uninhabitable inferno, while Mars, Earth's smaller outer neighbor, has an atmosphere so thin that its huge range of surface temperatures would prohibit life as we know it.

Earth occupies a favorable orbit around the Sun. It's often said that for a home the three most important features are location, location, location, and what's crucial for a home is even more crucial for the home planet. It's our location, our place in space, that makes Earth what it is, that accounts for its distinctive features, its physical properties, its landscape, its character, that makes life possible. Given its location, it has been remarked, the development of life was not just possible, it was inevitable. To a quite remarkable degree, Earth is a Goldilocks planet: "Not too cold, not too hot, but just right" for life to exist.

So Earth's location is "just right." Earth's atmosphere is also "just right," being made up of 77 percent nitrogen and 21 percent oxygen, with trace amounts of argon, carbon dioxide, and water. The oxygen is derived from living organisms, without which it quickly reacts with other elements to form oxides. The early (prehistoric) Earth must have had a much higher atmospheric concentration of carbon dioxide, but that is now largely locked up in "carbon sinks"—carbonate rocks, the oceans, and living plants. The small amount of residual carbon dioxide in the atmosphere plays a vital role, together with water vapor and methane, in maintaining Earth's surface temperature at a comfortable average level of 59°F. Though we tend to label greenhouse gases as harmful, without this beneficial greenhouse effect of carbon dioxide and water vapor in the atmosphere, Earth's temperature would be 91°F colder, and the planet would be uninhabitable.

Apart from its "just right" atmosphere and temperature, Earth also enjoys the benefit of being a rocky planet. Though all four of the inner planets have rocky surfaces, only Earth has one in a state of continuing geologic activity. In contrast, the surfaces of the Moon and Mars are "dead," subject only to the effects of weathering and impact. That's why we can read the early history of asteroid bombardment from their surfaces. Earth's surface, by contrast, shows relatively few impact scars, in part because smaller meteorites burn up in the atmosphere and partly because the Earth's surface undergoes slow, but constant, recycling and change. On the "dead" planets, erosion produces rock debris that accumulates on an essentially stable surface. On Earth, much of the surface itself is slowly reabsorbed into the planet and then melted, reformed, and subsequently released again at the surface in volcanoes and other fissures and vents.

Seventy percent of Earth's surface is covered by water. Earth is the only planet in the solar system capable of retaining liquid surface water, an essential requirement of life. But surface water also plays a major role in shaping Earth's surface and moderating its temperature. Rain and running water sculpt the landscape, depositing sediment that is carried into the oceans, and they play a continuing role in the endless cycle of rainfall and evaporation on which much human sustenance and activity depend.

Earth enjoys not only a "just right" active rocky crust, but also a "just right" almost circular orbit around the Sun and a "just right" rotation period. A highly eccentric elliptical orbit, for example, would lead to extreme temperature variations. A slow rotation speed would give prolonged hours of darkness, and both features would be less hospitable to life. Such "fine tuning" to the needs of life implies, for some, a providential planning, with Earth's position and composition marvelously adapted to the needs of life. Perhaps so, but life has proved to be sufficiently opportunistic to at least to raise the question of whether it may not be the other way around, with life as we know it emerging, perhaps no less providentially, in response to the opportunities and constraints of our given environment.

Rock, air, and water. These are the fundamental "elements" of Earth, the construction materials of both the home planet and of all living things. We'll take a closer look at each of them, but before we do, it's worth remembering that—for all their distinctive "Earthiness"—they have their origin far outside our present planet. They derive, not from our parent, Earth, but from stars that have died and blown apart. "Dust we may be," Maurice Boyd once remarked, "but the dust of a Star, and troubled by dreams."

Is our solar system unique, or are there other planets, other possible sites that may have nurtured the development of living things? Given the vastness of the universe, it is unlikely that our solar system is singular. Discoveries made in 1992 and later have confirmed the existence of extrasolar planetary systems.[1] Perhaps as many as one in every four Sun-like stars have Earth-like planets circling in very close orbits. That could mean that, in our galaxy alone, there could be at least 46 billion Earth-sized planets. To be habitable, such planets would have to have surface liquid water, available organic material, and a reliable source of energy, as well as a congenial balance of atmosphere, orbit, and rotation.

It seems entirely possible, then, that countless other stars, both in our galaxy and beyond it, support planetary systems that would be hospitable to life. That assumes, of course, that whatever extraterrestrial life there is, is similar to the carbon-based life of Earth. Perhaps it is, though it could also, one supposes, be "life" of a kind unlike any we now know, or even can visualize. Even our use of the terms "life" and "living" may require revision, and we may need an entirely new vocabulary to describe it.

Our definition of the unique structure of life had to be revised after the discovery of deep-ocean vent "smoker" fauna in 1977. These previously unknown animals live, not in a temperate world of sunlight and oxygen, but in a realm of scorching 660°F temperatures, total darkness, and hydrogen sulfide. So life forms on other planets could also exist in "unearthly" conditions, with their simpler forms living by chemosynthesis rather than photosynthesis.

Meditations on a Rocky Crust

Earth is the only planet whose name is not derived from ancient mythology. The Greeks called it *Gaia*, "Mother Earth"; the Romans, *Tellus*, "fertile soil." The

---

1. Thomas H. Maugh II, "Many Earth-Like Planets Orbit Sun-Like Stars," *Los Angeles Times*, October 29, 2010, http://www.latimes.com/news/science/la-sci-earthlike-planets20101029,0,5584432.story; NASA, "Many, Perhaps Most, Nearby Sun-Like Stars May Form Rocky Planets," http://www.nasa.gov/mission_pages/spitzer-20080217.html.

3.1 Comparison of the nine major planets

Planet	Distance from Sun (Millions of Kilometers)	Radius (Compared to Radius of Earth = 1)	Mass (Compared to Mass of Earth = 1)	Density (Compared to Density of Water = 1)	Composition of Planet	Density of Atmosphere (Compared to Earth's Atmosphere = 1)	Number of Satellites
				Terrestrial Planets			
Mercury	58	0.38	0.06	5.4	Rocky with metallic core	One billionth	0
Venus	108	0.95	0.82	5.2		90	0
Earth	150	1	1	5.5		1	1
Mars	229	0.53	0.11	3.9		0.01	2
				Jovian Planets			
Jupiter	778	11.2	318	1.3	Liquid hydrogen surface with liquid metallic mantle and solid core	Dense and turbulent	60
Saturn	1420	9.4	94	0.7			31
Uranus	2860	4.0	15	1.3	Hydrogen and helium outer layers with solid core	Similar to Jupiter except that some compounds that are gases on Jupiter are frozen on the outer planets	22
Neptune	4490	3.9	17	1.7			11
				Most Distant Planet			
Pluto	5910	0.17	0.0025	2.0	Rock and ice	0.00001	1

Source: From Graham R. Thompson and Jonathan Turk, *Earth Science and the Environment (with EarthScienceNow and InfoTrac)*, 3rd ed., copyright 2005 Brooks/Cole, a part of Cengage Learning, Inc. Reproduced with permission, www.cengage.com/permissions.

word *Earth* comes from the Middle English *erthe,* which, in turn, comes from the Old English *oerthe,* akin to the Old High German *erda, earth.*

Earth is, as we have seen, a rocky planet, and rock has great benefits for tenants: no rock, no terrestrial life, that's the arrangement, at least as far as we now know.

So we are the beneficiaries of this rockiness. We are here because Earth is a rocky planet. It might not have been. Had it been a little farther out from the Sun, it would have had a gaseous surface, as all the outer planets have. How can we account for these distinctive features of the planets? We are a curious species, and one question that seems to have haunted us since our earliest days concerns our origins. Where did we come from? How did we come into being? And how did Earth develop? Every civilization has its own traditional answers to those questions. And the answers, varied as they have been, have in turn shaped and formed the cultures within which they developed.

Our own age dwells on the scientific answers to these profound questions, abstracting from the totality of our experiences those that can be quantified and measured. To those answers, others, more metaphysical, are complementary.

What suggestions can we make, what models can we provide, for the development of Earth, Sun, and all the rest of our galactic family? Any hypotheses are tentative and speculative, the events are distant, the evidence indirect, and replication and "proof" are impossible. But there is broad agreement on at least the outlines of a hypothesis.

We might suppose that turbulence within the swirling megacloud of cosmic dust and gas that became the protogalaxy would produce local eddies that led to the aggregation of islands of matter, until the attraction of gravity was sufficiently strong to prevent their breakup and dissolution. From these countless clouds, one formed the early Sun, surrounded by a disklike, swirling cloud of gas and dust, whose cooling allowed its separation into rings from which the protoplanets in time emerged, probably by the incessant collision of fragments of cooling stellar material. The original size of the planets may have been very much smaller than their present size, as continuing bombardment added to both their size and their temperature. Later cooling and shrinkage of the Sun and the protoplanets allowed the former to shine and the latter to heat up, with the denser materials melting and sinking to form a molten core, the lighter rising to form a crustlike surface, and the lightest—the hydrogen and helium from which they had been formed—escaping to form a protoatmosphere.

The gaseous outer planets—so unlike the rocky inner planets in size and composition—are thought to have emerged either as the cooler sites of the outer orbits at which gases and ices developed, or, perhaps, as remnants of similar envelopes of gases that also once surrounded the inner rocky planets, but were blown away by the solar wind once the sun's fusion fires were lit.

It seems probable that early in Earth's history, radioactive decay—reinforced by continuing influx of cosmic debris—led to the melting of a significant part of the entire planet, from which the separation of the rocky crust eventually took place. This is also suggested by studies of the oldest lunar rocks, which show that the Moon was very hot at the time of their formation, some 4.6 billion years ago. Those rocks also display all the markings of impact, as does the surface of the innermost planet, Mercury. So we must assume that Earth also underwent this long initial period of bombardment by extraterrestrial debris, reflected now in the cratered surface of its neighbors, but obscured on Earth by countless aeons of crustal activity and weathering. At this early period all the inner terrestrial planets lost their lighter gases. Unlike the Earth, the smaller Moon has less internal radioactive heating, and thus its surface has been inactive for the last 3+ billion years.

The rocks of the Earth's crust, so vital to life as we know it, remain just that—a crust, a thin layer five to twenty-five miles in thickness, of light rock on a denser molten and mobile interior. It is this hot and mobile interior core and outer mantle that give rise to the active "living" crust that distinguishes Earth from its inert planetary neighbors, forming another link in the remarkable chain of features that make it congenial to life.

The Earth's crust is in a slow but continuous cycle of change, of which erosion, deposition, burial, uplift, and subsidence are the observable surface expressions. Living, as many of us do, in the cultivated landscapes of towns and cities or the lush farmlands of the rural areas, we see Earth's rocky surface less than its soils and gravels, which form the surface of the land and support its vegetation and life. But soils are the thinnest veneer, and everywhere, at shallow depths a few feet below them, are the rocks whose weathered remains they are.

The deepest mines in this rocky crust reach a depth of only some two miles; the deepest boreholes penetrate some five miles, and both encounter rocks just like those at the surface, though we do know that temperatures increase rapidly as we move downward. Miners in South Africa's deepest mines, for example, struggle with temperatures of 125°F. So our knowledge of the rocks of the lower crust is indirect, and depends on studies of the travel paths of earthquake waves, which indicate that crustal rocks have a typical thickness of some twenty miles below the continents, twice that below mountain ranges, but only some five miles below the ocean basins. This thin crust—which seems to us to be so solid—is in slow but steady motion, undergoing constant change. Massive as it seems to us, it represents only some 0.5 percent of Earth's radius. (Figure 3.2.)

One remarkable feature of the crust is its composition. Almost half of its mass is made up of oxygen and over a quarter of silicon. Some metals, which are so much a part of our daily lives—tin, lead, copper, and silver, for example—are rare, though others are more common. Iron, aluminum, and magnesium represent

5.80, 8.00, and 2.77 percent of the crust by weight, for example, though gold, by contrast, makes up only 0.0000002 percent. Carbon and hydrogen—of which we and all living things are made—make up less than 0.2 percent of the total composition of the crust. (Figure 3.3.)

Earth's crust is divided into a dozen large, rigid plates, and several smaller ones. From the mobile mantle that separates Earth's crust from its innermost core, forces act on these crustal plates, slowly moving them laterally, jostling one great plate against another, dragging them inch by inch, steadily apart at the mid-ocean ridges, and pulling them downward, melting and consuming them under the great arcs of volcanoes, deep ocean trenches, and earthquake zones that girdle the globe. (Figure 3.4.)

It was similar volcanoes that, after the period of intense asteroid bombardment earlier in Earth's history, produced the atmosphere and oceans of the planet,

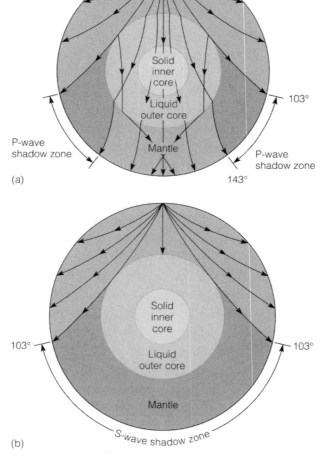

3.2 Seismic paths. (a) P-waves are refracted so that little P-wave energy reaches the surface in the P-wave shadow zone. (b) The presence of an S-wave shadow zone indicates that S-waves are being blocked within Earth. From Reed Wicander and James S. Monroe, *Essentials of Geology (with Geology NOW)*, 4th ed., copyright 2006 Brooks/Cole, a part of Cengage Learning, Inc. Reproduced with permission, www.cengage.com/permissions.

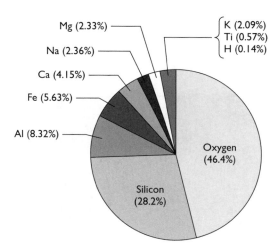

3.3 Common elements in the Earth's crust.

slowly releasing from the planet's interior gases and water vapor that were later to form the atmosphere and oceans, which now cling like a breathing membrane around Earth's rocky surface. And because we are creatures of the air and the oceans, as much as of the land, these too played a vital role, not only in shaping Earth's form and character but also in making it a habitable abode for living things.

All that we know directly about the Earth's rocks comes from examining its thin outermost layer, the crust, on which we live. Even there, our observations are limited. The rocks we can study fall into three great families, reflecting the way in which they were formed. Large areas of Earth's surface, including its most ancient rocks, which form the "cores" of the continents, are the cooled and crystallized products of once molten material (magma): they are the *igneous rocks.* Some igneous rocks, such as granite, are formed deep within the Earth, their coarse crystalline texture reflecting their slow cooling. Others, such as volcanic basalts, are finer-grained, more homogeneous rocks, and are the products of extrusion and rapid cooling at or near Earth's surface. (Figure 3.6.)

Both kinds of igneous rock play a prominent part in human affairs. Some of the world's most important ore bodies are the products of mineral differentiation and concentration within igneous rocks. These include ore deposits such as those of the Bushveld in South Africa, which contain vast platinum and chromite-rich ore bodies, and those of Sudbury, Ontario, which provide about 90 percent of all the world's nickel. Diamonds are the products of a distinctive igneous rock—kimberlite—formed in deep, ancient volcanic pipes. Almost all these minerals and other similar ore bodies are the products of intrusion of once molten magma, deep within Earth's crust.

Although igneous rocks are widely distributed, ore bodies are rare. Platinum, for example, has an average abundance of only 0.0000005 percent by weight in

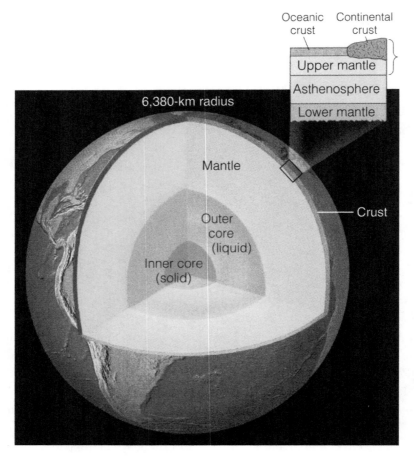

3.4 A cross section of the Earth, illustrating the core, mantle, and crust. The enlarged portion shows the relationship between the lithosphere (composed of the continental crust, oceanic crust, and solid upper mantle) and the underlying asthenosphere and lower mantle. From Reed Wicander and James S. Monroe, *Essentials of Geology (with GeologyNOW)*, 4th ed., copyright 2006 Brooks/Cole, a part of Cengage Learning, Inc. Reproduced with permission, www.cengage.com/permissions.

the igneous rocks of the Earth's crust. It is only by the natural concentration of these widely scattered minerals, by igneous differentiation, and other processes that workable ore deposits have come into existence. The metals and minerals on which our current prosperity depends are products of ancient geological processes and changes, and are typically difficult to find, expensive and often hazardous to mine, and frequently remote from areas of settlement.

Other igneous rocks represent hazards rather than benefits. There are over five hundred known active volcanoes, and though many of these—such as those of Hawaii—are relatively benign, others, such as those of the Pacific Northwest and Asia, are dangerous. Volcanoes have posed a hazard throughout human history. We shall explore them further in chapter 6.

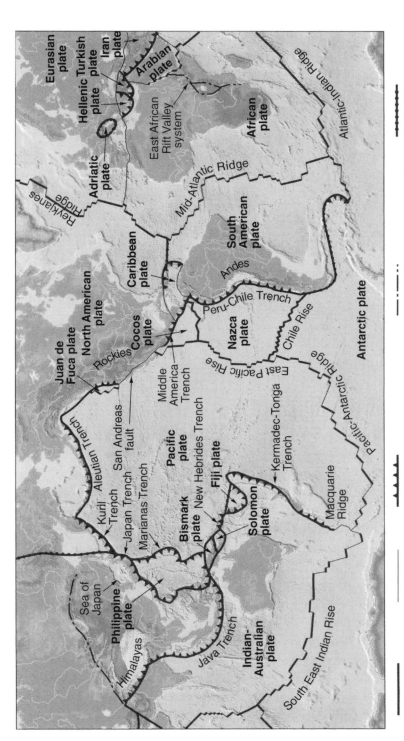

3.5 Major plates. Earth's lithosphere is divided into rigid plates of various sizes that move over the asthenosphere. From Reed Wicander and James S. Monroe, *Essentials of Geology (with GeologyNOW)*, 4th ed., copyright 2006 Brooks/Cole, a part of Cengage Learning, Inc. Reproduced with permission, www.cengage.com/permissions.

Ridge axis — Transform — Subduction zone — Zones of extension within continents — Uncertain plate boundary

Divergent boundary — Convergent boundary

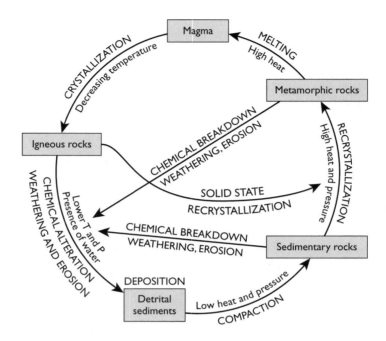

3.6 The rock cycle. From University of Texas at Austin, Texas Natural Science Center. Redrawn with permission.

All Earth's oldest rocks must have been igneous, but as the young Earth cooled, the new-formed atmosphere, with its turbulence and torrential storms, began slowly to erode the planet's new rocky surface, in the process forming a second great family of rocks—*sedimentary rocks*—from the surface debris of the planet. These rocks—created by the erosional effects of running water, ice, wind, and the oceans—are formed from sediment carried by streams, rivers, and glaciers toward the oceans, where many come to rest, to be buried and lithified, and thus to become familiar sandstones and shales. Other sedimentary rocks are formed by precipitation from solution—limestone, for example—and still others are formed on land—the by-products of deposition by glaciers, rivers, lakes, and wind. (Figure 3.6.)

Sedimentary rocks also provide many vital materials—oil, gas, coal, iron, phosphates, and a host of others—which we shall later explore. It is from such rocks that the ancient history of seas and landscapes can be deduced, and it is from them that the entombed remains of once-living organisms (fossils) can be extracted and pieced together to provide an outline of the evolution of life.

We should know little of this history were it not for the fact that Earth's crust is restless, with great areas undergoing uplift, folding and faulting under

3.7 Comparison of concentration of specific elements in Earth's crust with concentration needed to operate a commercial mine

Element	Natural Concentration in Crust (% by Weight)	Concentration Required to Operate a Commercial Mine (% by Weight)	Enrichment Factor
Aluminum	8	24–32	3–4
Iron	5.8	40	6–7
Copper	0.0058	0.46–0.58	80–100
Nickel	0.0072	1.08	150
Zinc	0.0082	2.46	300
Uranium	0.00016	0.19	1,200
Lead	0.00010	0.2	2,000
Gold	0.0000002	0.0008	4,000
Mercury	0.000002	0.2	100,000

Source: From Graham R. Thompson and Jonathan Turk, *Earth Science and the Environment (with EarthScienceNow and InfoTrac)*, 3rd ed., copyright 2005 Brooks/Cole, a part of Cengage Learning, Inc. Reproduced with permission, www.cengage.com/permissions.

the intense pressures of deep-seated Earth movements, of which earthquakes are a surface expression. Such activity leads to the formation of a third family of rocks—*metamorphic*—by the melting and recrystallization of some older rocks, both in mountainous regions—the Alps or the Rockies, for example— and deep within the Earth to produce metamorphic rocks: schists and gneisses, marbles and slates. Rocks of this kind are especially conspicuous among the more ancient rocks, formed as the products of alteration of still older rocks by heat and pressure deep beneath the Earth's surface. Metamorphic rocks also provide some vital materials, such as asbestos, graphite, talc, and other minerals.

Earth's Engine

Earth is a restless planet. Its land, waters, air, and creatures interact endlessly with one another. Much of the movement is slow, almost imperceptible. The stone retaining wall on the hillside orchard bulges outward over time and sags downhill; the gravestone of a century ago leans awkwardly downhill on the sloping cemetery lawn; the river slowly undercuts its crumbling bank. Other movement is more perceptible: torrential rain, moving clouds, rising morning mist, gentle breezes, sand on the beach shifting with the tide. And still other movement is invisible: the endless movement of carbon through the carbon cycle or nitrogen through the nitrogen cycle. All these movements, for all their differences, mark the steady interchanges of Earth's surface: its atmosphere, hydrosphere, lithosphere, and biosphere interacting endlessly one upon another under the influence of solar energy and gravity. Their slow pace and inconspicuous effects make it easy to forget how vital and unceasing these interactions are in the economy of the planet and in the sustenance of life.

There's another kind of movement that is not superficial: a restlessness that comes from Earth's deep interior. This is the movement whose surface manifestation is neither gentle, slow, nor immediately benevolent, but violent, sudden and often destructive—volcanic eruptions and earthquakes. Earthquakes and volcanoes are broadly interconnected: together they reshape the surface of the planet. And they, too, represent an endless cycle of interchange: a deep-seated interaction between Earth's crust and the underlying mantle, on which it rests. This is Earth's most profound cycle.

For much of the world's population, living in areas unaffected by Earth's instability, volcanoes and earthquakes are a short news item at the bottom of page A24 of the *New York Times* or the *Sydney Morning Herald,* or, if sufficiently destructive, a brief item on the evening news. But about a million earthquakes occur every year, as well as some sixty volcanic eruptions on land and many more below the oceans. Most earthquakes are very slight and are recorded only by instruments: stronger ones kill an average of ten thousand people a year, both from their direct damage and their secondary effects, such as tsunamis.

The reason most of us are unaware of these volcanoes and earthquakes is their distribution: they are not scattered randomly or uniformly across the face of the Earth, but occupy two broad zones: the Mid-Ocean Ridge and the boundaries between Earth's major tectonic plates. The first of these is, perhaps, Earth's most majestic but least conspicuous feature: *the Mid-Ocean Ridge* is an extended zone of mid-ocean volcanic mountains, made of basalt, marked along their crest by active volcanism, shallow earthquakes, and a deep, narrow, trenchlike valley. (Figure 3.8.)

Until fifty years ago, those who studied the Earth struggled to understand the broad architecture of its surface. Surely, they supposed, the vast variety of mountain chains, continents, volcanoes, and earthquakes must reflect some underlying pattern. But the quest for an underlying unity remained elusive. It was not until the 1960s that the results of new techniques of remote sensing and underwater exploration—some of them developed for antisubmarine defense in World War II—began to reveal a broader pattern. For all their generally placid surfaces, the oceans proved to be the site of much of Earth's most significant activity. Most volcanic activity takes place, not on land, but beneath the oceans, where its effects are largely hidden from us. But in one notable location, both land and ocean come together. In Iceland, the forces that remake the oceans and create the land come ashore. (Figure 3.9.)

Iceland sits astride the Mid-Atlantic Ridge, which is part of a great mountain range—the Mid-Ocean Ridge—that encircles the globe. Though most of the ridge

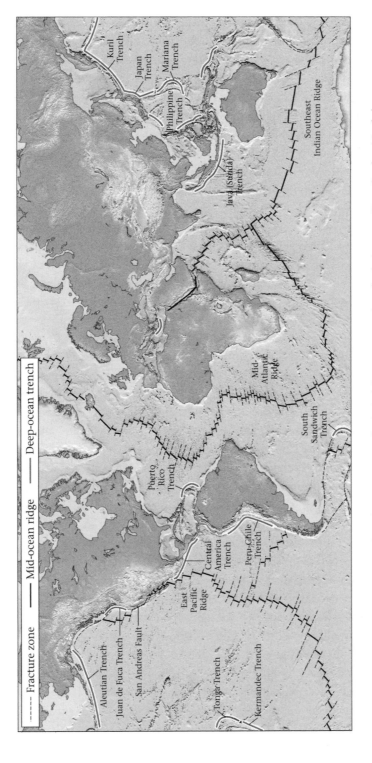

3.8 The mid-ocean ridges, fracture zones, and principal deep-ocean trenches of today's oceans. From Stephen Marshak, *Earth: Portrait of a Planet*, 2nd ed., copyright 2005 W. W. Norton. Reproduced with permission.

can be "seen" only by remote sensing methods, this worldwide mountain range is the most conspicuous feature of the planet. Rising up to 18,000 feet above the ocean floors, and extending up to 1,500 miles in width, the Mid-Ocean Ridge system is the site of major volcanism, shallow earthquakes, and high heat flow. Its crest is marked by a deep, slitlike rift valley, up to 12,000 feet deep and thirty miles wide, in which new volcanic rock is erupted and along which the Earth's surface is slowly but relentlessly being pulled apart. Though hidden from view beneath the oceans over most of the planet, this great rift valley comes ashore in Iceland and the process of Earth being remade can be observed. Hot springs, gas fumaroles, periodic volcanic eruptions, geysers, earthquakes, and the great topographic, trenchlike valley of Thingvellir all testify to the continuing reality of the deep ocean process that reshapes Earth's surface. Surtsey, a small volcano that emerged from the Atlantic in 1963, as well as other intense volcanism, reflect its restless activity.

The Iceland Ridge, and the Mid-Atlantic Ridge of which it is a part, together form a section of the world's greatest mountain range—40,000 miles long—which is marked by the slow eruption of pillow-shaped lavas, high heat flow, and a forest of chimneylike structures, which pour out water, superheated to 600°F, and contain a concentrated mixture of minerals, including iron, zinc, silver, and copper sulfides. It is around these "smokers" that a rich variety of new forms of marine life was recently discovered, living in total darkness, high pressure, and toxic conditions. The molten magma below the ridge periodically erupts in new lava outpourings, which form scarps, parallel to the axis of the rift—displaced here and there by great transform faults—marking the separation of adjacent crustal plates at the rate of an inch or two a year. Here and there, stable "hot spots" below the ocean floor, such as that below the Hawaiian Islands, support more or less continuous plumes of volcanic activity, which penetrate the moving crustal plates lying above them.

If the Earth's crust is pulled apart at the Mid-Ocean Ridge, it must somewhere be carried downward again and resorbed—unless of course, the Earth is expanding, for which there is no evidence. This resorption of the moving crust is exactly what seems to happen in the deep trenches developed along the western coast of South America and along the islands of the western and northern Pacific (Tonga, Java, the Marianas, Japan, and the Aleutians, among them). These trenches are up to 35,000 feet deep, and they are marked by intense volcanic and deep earthquake activity, low heat flow, and strong negative gravity anomalies, all suggesting a down-flexing of the ocean crust below the adjacent island arcs at or below the continental margins. (Figure 3.11.)

This general movement of the crust—plate tectonics—reflects the existence of a dozen crustal slabs (plates) that undergo slow but inexorable lateral movement, apparently driven by giant convection cells moving slowly within the underlying plastic mantle zone of the Earth, which lies between its molten core and its solid crust. These plates, each about eighty miles thick, carry the continents and

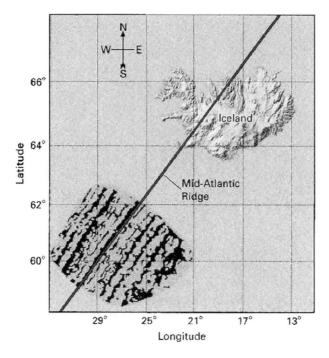

3.9 The Iceland ridge. Magnetic orientations of sea-floor rocks are shown in the lower left portion of the map. The black stripes represent sea-floor rocks with normal magnetic polarity, and the intervening stripes represent rocks with reversed polarity. From Graham R. Thompson and Jonathan Turk, *Earth Science and the Environment (with EarthScienceNow and InfoTrac)*, 3rd ed., copyright 2005 Brooks/Cole, a part of Cengage Learning, Inc. Reproduced with permission, www.cengage.com/permissions.

the neighboring oceans across the face of the globe, diverging at the Mid-Ocean Ridge, colliding at the trenches, and jostling and sliding past one another at the transform faults—such as the San Andreas. Over long periods, this movement has changed the geography and geology of the planet. Although rocks as old as 3.8 billion years are known from the continents, there are no rocks older than about 200 million years known from the ocean floors, presumably because they are being continually resorbed and replaced by "new" lavas at the mid-ocean ridges.

The most conspicuous part of this ridge runs, as we have seen north–south, right down the middle of the Atlantic Ocean. A second arm runs below Africa and Australia across the southern Indian Ocean and thence northward to form the East Pacific Rise. These ridges are major topographic structures, up to 1,500 miles wide and 18,000 feet high, with central rift valleys up to 12,000 feet deep and thirty miles wide, along which new volcanic rocks are erupted. They are the sites of "*smokers*," chimneys of black belching gases, rich in the metallic sulfides characteristic of major ore bodies. The mid-ocean ridges are frequently laterally offset by major fracture zones, known as *transcurrent faults*. Some of these faults are up to one thousand miles in length and produce submarine scarps ten

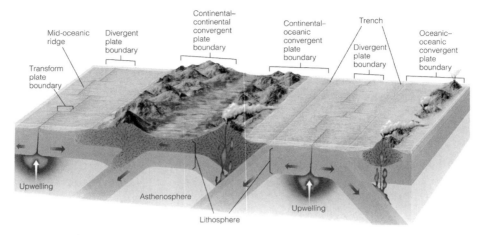

3.10 Major crustal plate downward lateral movement. An idealized cross section illustrating the relationship between the lithosphere and the underlying asthenosphere and the three principal types of plate boundaries: divergent, convergent, and transform. From Reed Wicander and James S. Monroe, *Essentials of Geology (with GeologyNOW)*, 4th ed., copyright 2006 Brooks/Cole, a part of Cengage Learning, Inc. Reproduced with permission, www.cengage.com/permissions.

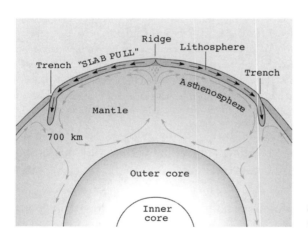

3.11 Crustal convection currents. Courtesy of USGS.

thousand feet high. Detailed surveys show they produce lateral displacements up to six hundred miles. (Figures 3.10–3.12.)

There is one other feature that gives a distinctive character to mid-ocean ridges: they are marked by conspicuous anomalies in the magnetic properties of the rocks that lie beyond their flanks. These iron-rich, volcanic rocks all bear *remnant magnetism*, imprinted by Earth's changing magnetic field characteristic of the time of their eruption, cooling, and formation, and this provides an indication of their geologic age. It is now known that this magnetic field frequently reversed direction, every half million years or so, with the north magnetic pole becoming the south and vice versa.

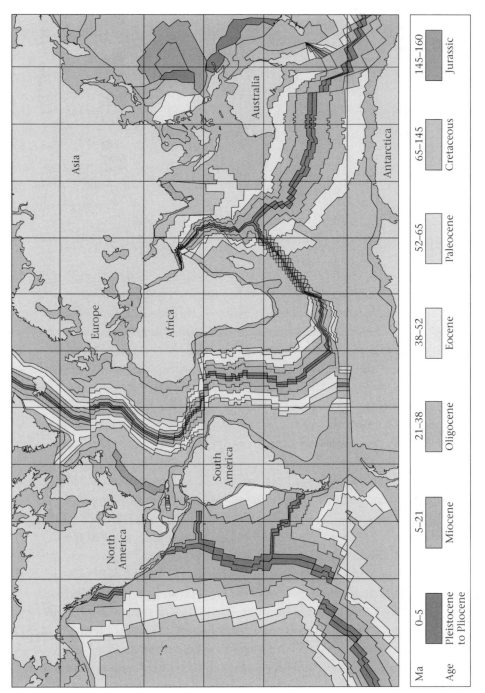

3.12 Seafloor spreading. This map of the world shows the various ages of the sea floor. Note how the sea floor grows older with increasing distance from the ridge axis (Ma = million years ago). From Stephen Marshak, *Earth: Portrait of a Planet*, 2nd ed., copyright 2005 W. W. Norton. Reproduced with permission.

Ma	0–5	5–21	21–38	38–52	52–65	65–145	145–160
Age	Pleistocene to Pliocene	Miocene	Oligocene	Eocene	Paleocene	Cretaceous	Jurassic

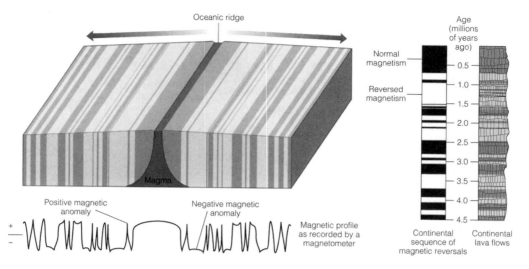

3.13 Remnant magnetism. From A. Cox, "Geomagnetic Reversals," *Science* 163 (January 17, 1969), copyright 1969 American Association for the Advancement of Science. Reprinted with permission as conveyed through Copyright Clearance Center, Inc.

The remarkable parallel pattern produced by plotting these anomalies indicates that the underlying rocks of the ocean floor were erupted successively at the mid-ocean ridges and then carried laterally away from them, so that they become successively older as one moves away from the ridge. Nowhere is this better seen than in Iceland, where the Mid-Atlantic Ridge "comes ashore," and the active volcanism can be studied on land. (Figures 3.13 and 3.14.)

The mid-ocean ridges, in fact, seem to be major architectural elements of the planet, along whose length new ocean crust in the form of basaltic lava is constantly erupted, carried outward away from the ridge, and replaced by newer lavas from new eruptions.

Studies of the remnant magnetism of ancient igneous rocks, imprinted on them by the Earth's particular magnetic field existing at the time of their formation, reveal the pattern of changing continental locations in the distant past. During the later part of the so-called Paleozoic (ancient life) Era, for example, some 250 million years ago, the continents were joined in a single land mass, Pangaea, which later split into separate plates. The familiar mountain ranges of our present time arose when continental plates later collided. The Himalayas, for example, reflect the opposing movement and collision of the Indian and Asian plates that began some 40 million years ago, and whose continuing impact is seen in their slow, but measurable, continuing uprising. The Alps, likewise, reflect the African and Eurasian plates' collision some 80 million years ago. The higher peaks of some of these mountain ranges are formed of rocks containing marine fossils, which testifies to the reality of changing crustal elevations.

Unlike the older rocks of the continents, which are 3.8 billion years old, the oldest rocks of the ocean floor are "only" some 200 million years old, with their

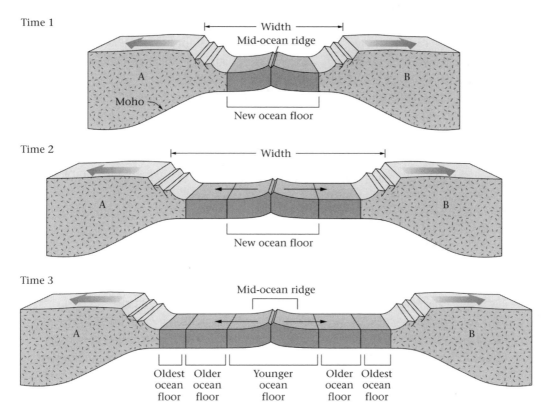

3.14 Three stages of mid-ocean ridge spreading. From Stephen Marshak, *Earth: Portrait of a Planet*, 2nd ed., copyright 2005 W. W. Norton. Reproduced with permission.

age increasing symmetrically on both sides of the mid-ocean ridges. Along the ridges themselves, the rocks are very young, being formed by continuing volcanic activity. Drilling of lava sequences on land has confirmed a series of magnetic reversals and dates that match the mirror images seen across the mid-ocean ridges. Thin sedimentary deposits on the ocean floor, which lie above these lavas, are also young. Here and there on the ocean floor are *submarine seamounts* and volcanoes, some with coral reefs. The oldest are those most distant from the mid-ocean ridges, confirming their probable origin there.

Studies of these submarine rocks indicate that the floors of the oceans are in motion, being newly formed by active volcanoes along the mid-ocean ridges and then carried away from the ridges at rates of a few centimeters a year. This extraordinary discovery was made some thirty-five years ago and has proved to be the key to understanding the major architecture of the Earth. But if ocean crust is carried away laterally from the ocean ridges, where does it go? After all, there is no evidence that Earth's diameter is increasing.

This resorption of Earth's moving crust is exactly what seems to be happening along some of the continental margins. We have already seen that the margins of the oceans are generally characterized by wide continental shelves and slopes, with the deeper oceans forming vast abyssal plains. But around other ocean margins, however, especially those of the Pacific, the Aleutians, and the Caribbean, there are trenches, deep depressions in the ocean floor, marked by zones of great instability, with deep-focus earthquakes, active volcanoes, and ultradeep, narrow, elongated submarine trenches, up to 35,000 feet deep. These margins are also characterized by negative *gravity anomalies*; that is, the difference between the measured reading of the local force of gravity and the anticipated value based on latitude adjusted by some corrections. These negative anomalies, together with a high heat flow on the landward side of volcanic island arcs, and their distinctive, *andesitic volcanic lavas*—which are intermediate in composition between the "acidic" granitic rocks of the continents and the "basic" basaltic rocks of the oceans—suggest that the ocean crust is being carried down below the trenches into the interior of the Earth. The sharply inclined trace of earthquake epicenters seems to confirm this, as does the negative gravity anomaly associated with the trenches, while the volcanoes are explained as arising from the melting of the down-turning slab of oceanic crust.

This, Earth's greatest cycle, is the process that renews the crust and re-creates the planet. It is known as *plate tectonics*, and interprets Earth's surface as consisting of a dozen major elements or *plates* and a number of smaller ones, all of which move slowly in relation to one another. The crustal plates diverge, converge, or slide horizontally in relation to one another. *Diverging plates* are pulled apart at the mid-ocean ridges, as new sea floor is continually formed by volcanism, spreading outward away from the ridge, where young, thin crust rises to fill the gap of separation.

The crust becomes progressively older with greater distance from the ridge at which it was formed. Most areas of diverging plates occur in the ocean, but in a few places, such as East Africa and Iceland, they occur on land, producing rift valleys, volcanoes, and shallow earthquakes. (Figure 3.12.)

Convergent plate boundaries mark the sites where two plates move toward one another, colliding to produce areas of intense geological activity. Three cases exist: (Figure 3.15.)

- *Converging ocean plates* that produce island arcs and deep ocean trenches, marked by down-sinking (subduction) of one plate, deep-focus earthquakes, intense volcanism, and strong deformation and instability. The Japanese islands and the Aleutians are typical examples.
- *Converging plates where one carries a continent and the other an ocean* are marked by mountain ranges and ocean trenches, associated with deep earthquakes, subduction of the oceanic plate, volcanoes, strong deformation, and folding of the rocks involved. The Andes are a classic example.
- *Converging plates in which both bear a continent* produce mountains, with intense deformation and earthquakes. The Himalayas are a striking example.

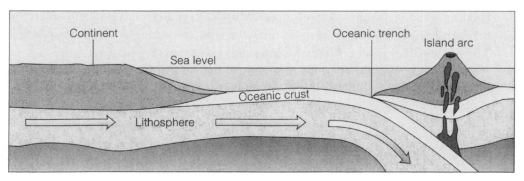

(a) Ocean-ocean convergence

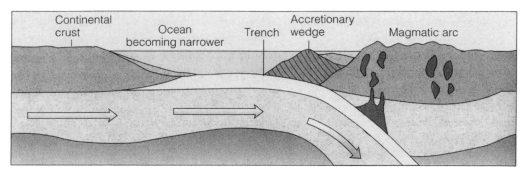

(b) Ocean-continent convergence

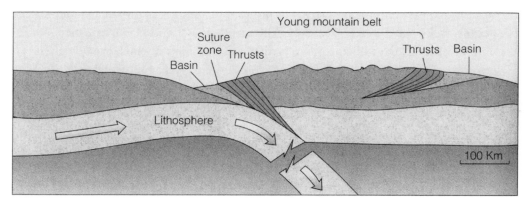

(c) Continent-continent convergence

3.15 Three types of convergent plate boundaries: top: ocean-ocean; middle: ocean-continent; bottom: continent-continent. From Brendan Murphy and Damian Nance, *Earth Science Today* (Pacific Grove, CA: Brooks/Cole, 1999). Used with permission of the authors.

Transform plate boundaries mark the sites of plates that slide laterally against each other in a "sideswiping" fashion. Where they occur in the oceans, they are marked by the major offsets that characterize the mid-ocean ridges, producing small earthquakes. The many offsets along the Mid-Atlantic Ridge or the East

Pacific Rise are examples. Transform boundaries may also exist on land, where they produce earthquakes, minor mountain ranges, and strong rock deformation. The San Andreas Fault of California is a classic example. (Figure 3.8.)

In all these cases the crust forms the upper part of a brittle layer (*lithosphere*) "floating" and moving on the surface of a hot and plastic layer (*asthenosphere*) below it. The speed of the movement of plates, measured by comparison of the ancient magnetic imprint of the rocks and confirmed by satellite laser studies, ranges from 0.4 to 7.1 inches a year.

Studies of the path of earthquake waves, and their deflection and reflection, have been used to study the interior structure of the Earth. They show that the outer part of the Earth—the *lithosphere*—is rigid, brittle, and thin in relation to the size of the planet, though its thickness and composition vary from place to place. Beneath the older oceans, it is some one hundred kilometers (sixty-two miles) in thickness and consists of basalt, a relatively heavy rock. The lithosphere is thicker, averaging perhaps two hundred kilometers (124 miles) below the oldest continents and even more where it forms the "root" below mountain ranges. The upper part of the lithosphere is a chemical differentiate known as the continental crust consisting of granitic rocks, which are lighter than those of the thinner crust under the oceans. The crust and the outermost part of the underlying layer, the mantle, are made up of brittle, rigid, and cool rocks. The crust averages about thirty kilometers thick (18.6 miles) under the continental shields, being about twice that under mountain ranges, but only about a fifth of that (6 km) under the oceans. Below about sixty-two miles, the mantle is hot and becomes plastic, its outermost layer, the asthenosphere, acting as a cushion on which the "floating" crust rests. The thin oceanic crust floats low and the thick continental parts float high. Though it is formed of solid rock, the mantle is so hot that it can flow, much as pitch can flow on a hot day. It is across this layer that the plates move and it is on it that they rest, much as ice floats in water. Earthquakes are restricted to the crust and uppermost parts of the underlying lithosphere, except where the crust is subducted in trenches and is still cold enough to be brittle, giving the two their relative elevation.

The *lower part of the mantle*, which under increasing heat and pressure becomes rigid and brittle, extends to a lower depth of 1,802 miles and consists of metamorphic rocks, whose component minerals become different with increasing depth and pressure. (Figure 3.5.)

At the heart of the Earth, extending from 1,800 miles to its center at 3,958 miles, lies the *core*, consisting of the heavy metals iron and nickel. The high temperature and pressure make the *outer core* molten, but the still higher pressure (over 2×10^6 atmospheres) makes the *inner core* solid, even though its temperature has been estimated to be between 4,000°C (7,232°F) and 7,000°C (12, 632°F).

The Distant Past

All evidence we have reviewed indicates that the crust of the Earth is dynamic, with its component plates in slow but constant movement. If we accept that view of the present operation of the planet, what are we to make of its past? Have continents and oceans changed in shape? Have mountains come and gone? Is Earth's familiar geography a thing only of the present? Can we reconstitute the Earth of ancient times?

The answer to these questions is yes. By careful comparisons of the geology and geography of the present continents, it is possible to develop a coherent and consistent history of the last 600 million years. Consider the following indications of earlier geography:

- *The fit of present coastlines*: Africa and South America fit together almost as well as pieces of an ancient jigsaw puzzle.
- *Geological matching*: Distinctive rock types and tectonic structures match across the two sides of such reconstructed jigsaws.
- *Fossil similarities:* Similar fossil reptiles and plants are found in rocks of the same age on continents that are now widely separated, such as rocks deposited 200 million years ago in South Africa, Antarctica, and South America.
- *Climatic similarities:* Similar climatic conditions are represented by sedimentary rocks of the same age on continents that are now widely separated, such as deposits of what seems to have been a single great continental ice sheet that blanketed much of what is now Antarctica, India, South Africa, and South America, when they were once united in a single, ancient continent, Gondwanaland.

By these and similar methods, including remnant magnetism in iron-bearing rocks, it is possible to reconstruct Earth's changing ancient geographies. There is remarkable consistency between the results obtained by these different methods.

The Mechanism of Moving Plates

Plate tectonics is now accepted as an established theory to interpret these various features of Earth's crust, including the consequences of observed sea floor creation and spreading. Earth's plates move, and their movements can be measured and charted. The crust of the Earth is in constant motion.

But what accounts for this movement? What forces could be capable of dividing the ocean floor and moving continent- and ocean-sized slabs across the face of the planet? And how are plate movements related to the conspicuous geologic activity—volcanoes, earthquakes, uplift, and crustal instability—that defines their boundaries? There is, as yet, no clearly demonstrated single mechanism, but there are several promising indications of possible causes.

First, there are strong indications that Earth's mantle, though most of it is solid rock, flows slowly in great convection cycles, much as a viscous liquid, such as syrup in an open pan, would move, rising from the hottest depths to impinge on the overlying lithosphere, where down-turning, colder material completes the convective loop. This may be a relatively minor driver of plate motion, which seems to be driven by the cold, dense nature of plates leading to their down-sinking (subduction), much as the crust of a lava lake subducts.

Second, it has been suggested that large, upwelling plumes of the hot material—rather like the clumps of a lava lamp—may arise from deep within the mantle, striking the lithosphere and so impinging on or even creating or triggering plate movement. The volcanoes of the Hawaiian Islands are thought to reflect the presence of such a plume.

Third, it has been proposed that, once triggered by an event of this or similar kind, the dense, cold, downward plunging edge of the plate might carry the rest of the plate with it, downward into a subduction zone. (Figure 3.11.)

More work is needed to determine how, or even if, these mechanisms influence plate movement. What is clear is that plates are in constant motion, that their boundaries reflect the elemental architecture of the crust, and that the form of the oceans, the position and shapes of the continents, the distribution and form of mountain ranges, the locations and character of earthquakes and volcanoes all reflect the pattern and form of plate movement. Playing back the film of Earth, rerunning the tape of magnetic anomalies, shows that once, 200 million years ago, the continents coalesced into a single great land mass, Pangaea ("all lands"), with the northern continents being known as Laurasia, and the southern, Gondwana-land. They later split up, moving slowly toward their present locations.

But before Pangaea, before 200 million years ago, what then? What was the geography of pre-Pangaean Earth? Details are sketchy, but there is evidence of earlier continental movement, including the moving together of smaller, fragmentary continental units to assemble the supercontinent of Pangaea. In fact, it has been suggested that the assembly, breakup, and reassembly of continents is itself a cyclic process, requiring perhaps half a billion years or more to complete; so that in the last 2 billion years, Earth may have passed through three such cycles marked by the assembly and breakup of three successive supercontinents. What triggers the cycle—if it is indeed a cycle—what starts the breakup, what drives the process are matters of continuing and intensive study. Whatever the cause, this may be Earth's greatest dance of all: stately in its cadence, global in its embrace, endless in its sequence, Earth-shaping in its impact.

These deep-seated crustal movements have profoundly influenced both our planetary home and the availability of materials that are vital to our everyday pattern of life in the developed world. Downward movements (subduction) of plates have formed some of the world's greatest mineral deposits. The vast copper, silver, tin, and gold deposits of the Andes, for example, were formed by the collision and subduction of the eastern Pacific Nazca plate below the South American plate. The sulfide copper deposits of Cyprus, which have been worked since antiquity, were initially formed and concentrated, not by plate convergence and collision, but by divergence, as smokers along an ancient spreading zone, which was elevated above sea level by the later convergence of the Eurasian and African plates.

Earth's Unity

The theory of plate tectonics was developed in the late 1960s and early '70s and has now become a great integrating theme in Earth sciences. With it, much that was once inexplicable in Earth history—the distribution of mountain ranges and distinctive fossil groups, for example—now makes good sense. The theory even provides glimpses of future development, ranging from metallic mineral prospecting to earthquake studies and prediction.

So the planet that for most of us provides a peaceable dwelling is the product of powerful forces that continue to modify its surface. There is tragic irony in the fact that the volcanoes and earthquakes that can devastate some areas and bring death and suffering to countless people reflect the same continuing forces that have made the planet habitable, and provided the very air, water, and materials on which our lives depend.

The theory of plate tectonics is one of the most elegant and powerful theories ever developed. The austere economy and haunting beauty of it excite the senses and inspire the imagination. It challenges our aesthetic imagination with a sense of wonder and awe. It is also a theory and a process with profoundly practical consequences: it is the process of plate tectonics that renews the planet, that represents the contrast between the dead surface of Mars or the Moon and the moving, heaving, pulsing surface of planet Earth. For these forces not only move continents and elevate mountain ranges; they also account for the concentration and formation of hydrothermal mineral deposits, on which modern technology depends. It may well be that it was along an oceanic ridge that living things first developed. The same processes account also for the positions of the continents, and these, in turn, account for the distribution of animals and plants. We live on a moving crust, whose instability reflects the innermost workings of our planet. It is these same forces that brought life into being and still undergird and renew the remarkable planet on which we dwell.

CHAPTER 4

The Blue Planet

Almost all homes have numbers. A few have names. I've lived in twenty houses over my lifetime, but only two had names. The first was called "Pilgrim Cottage." It was in Wales, close by the sea. We rented it while our own home was being built nearby. We liked the name, it was modest: "cottage" not "hall" or even "house." Still less, "grange" or "manor." And it had a nice ring to it. After all, like it or not, we are all tenants, temporary residents, and therefore pilgrims. Our landlords did not like the name so they changed it, and gave the old name to us. That's how our next home—actually the first we'd ever built and owned—also became "Pilgrim Cottage." The name suited the house somehow. It was a snug home, with big windows, overlooking the sands and sea of Langland Bay and—far across the water on a clear day—the distant coast of Devon. This was the place where our four daughters grew up, and the name—Pilgrim Cottage—seemed to fit it well. So names—house names, home names—are significant: they tell us something about the owners, or the tenants, or the place. They are chosen to fit: to be appropriate, to have meaning.

It's curious then that the planet we have chosen to call "Earth"—the home planet, our planet—is misnamed: surely it should be called planet Ocean. After all, more than two-thirds of its surface is covered by water. Seen from space, it is the ocean—not the earth—that gives it its distinctive blue color. In all the vastness of the solar system, only one planet—Earth—has an ocean. Other planets have "earths": they are rocky. Though other planets may prove to have subsurface water, our watery ocean is distinctive. The "earth" of our planet, the rocky land-scape, the "terra firma" as we are pleased to call it, the "solid earth," is a series of continental islands, encircled in a larger, all-embracing ocean. For even though we talk of oceans—the Atlantic, Pacific, Indian, Arctic, and Antarctic—they are all one, single, great interconnected ocean. Lump all these island continents together, and they would still be smaller in area than the largest of these "sub-oceans": the Pacific. The average depth of the ocean is more than five times the average height of the land above sea level. In fact, if you sliced off the continents at sea level and dropped them into the oceans to make a smooth, uniform sphere, the ocean would still have an average depth of about eight thousand feet over the whole planet. This vast world ocean contains 325 million cubic miles of water, weighing over a million trillion tons.

And that's not all. It is the ocean that gives birth to new land. It is along the mid-ocean ridges that new land is born. It is at the margins of the oceans that older land is consumed and destroyed. It is the water of the oceans, endlessly recycled as rain and snow, that sculpts the surface of the land. Perhaps half of the land sur-face of the Earth is formed of rocks that were deposited as sediments beneath the oceans. The longest, highest, largest mountain range on Earth lies—you guessed it—beneath the oceans. Life—unique, so far as we yet know, to our planet—began in the oceans.

How ironic, then, that we have christened this home planet "Earth." Under-standable, perhaps. After all, we're terrestrial animals. But still, surely, ironic. So let's rename it. Why not Planet Ocean or Planet Oceanus? "Too late," you reply. "Too radical," you add. All right. Perhaps you're right. Let's compromise. Let's call it "The Blue Planet." The exquisitely beautiful, Blue Planet.

That blueness is the thing that catches the breath of astronauts glimpsing the Earth from space. Kathy Sullivan, who in 1984 became the first U.S. woman to walk in space, remarked, "I see the deep black of space and this just brilliantly gorgeous blue and white arc of the earth and totally unconsciously, not at all able to help myself, I said, 'Wow, look at that.'"

Portrait of the Oceans

Now, if Earth is the Blue Planet, the Ocean Planet, where did the oceans come from? If, as seems increasingly likely, Earth formed by the impact and accretion of cold planetesimal materials, whose impact produced heat, it seems probable that the oceans formed by outgassing from the planet's interior. Water is part of the chemical structure of many minerals, and, as heating and melting of the protoplanet occurred, it must have been released in huge quantities as steam from volcanic vents and fissures. It is also probable that water was contributed by incoming asteroids and other water-rich extraterrestrial bodies impacting the surface of the young Earth.

From these early days of Earth, and throughout all its subsequent history, the oceans have played a decisive role in every aspect of the planet's development and, not least, in the development of life and in the course of human history. The oceans have served as the pathway of exploration, discovery, settlement, trade and transport, and the search for wealth and empire.

The oceans have provided a major source of food and have served as the great regulator of planetary climate, keeping Earth's surface habitable by acting as a giant reservoir of heat and energy from the Sun that drives the circulation of Earth's atmosphere.

For most of human history, all but the shallowest parts of the oceans were inaccessible. The systematic scientific study of the oceans is of comparatively recent origin, having been greatly aided during the last fifty years by the development of new techniques for deep underwater survey and analysis. Efforts to harness the vast energy of the oceans and harvest its substantial mineral wealth are now in their infancy.

The oceans are younger than the Earth, though not, presumably, by much. We have, as yet, no clear indication of just when the earliest oceans formed. The oldest marine sedimentary rocks are some 3.8 billion years old, but that figure provides only a minimum estimate for the age of the oceans. As Earth's molten crust cooled, one can imagine a prolonged period of dense cloud cover and torrential rains.

In spite of our lack of detailed knowledge about this earlier history of the oceans, we know a great deal about their subsequent history, especially during the last 600 million or so years and about the distribution of ancient lands and seas. We know, for example, there have been huge changes in the distribution of land and ocean, with all the present continents having once been joined together in a single land mass—Pangaea—and all the oceans into a single, worldwide ocean—Panthalassa. We know, too, that much of the present surface of the continents has been covered at one time or another in the past by widespread seas, whose fossiliferous deposits still blanket the land in many places. These ancient events reflect changes in the form and structure of the ocean basins, and the extent of glaciers and ice caps, as well as major changes in the elevation of the land and in the relative level of the sea. For example, more than half of the three hundred largest port cities of the ancient world, built from 3000 BC to the fall of the Roman Empire in 55 BC, are now submerged.

The Face of the Deep

What would the ocean basins look like if we drained out all the seawater to expose the ocean floors? Although not all of the ocean floors have been mapped

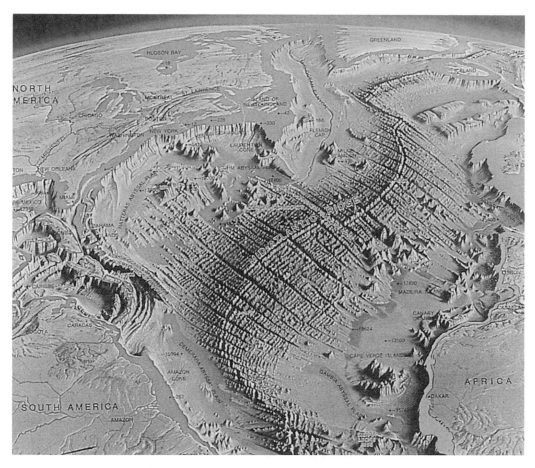

4.1 General view of Atlantic Ocean basin without water. The major feature is the Mid-Atlantic Ridge. Reproduced courtesy of ALCOA.

in detail, we have a reasonably good overall picture of their major features. The oceans have an average depth of about 2.35 miles, and though most of the surface is flat or gently undulating—forming the flattest places on Earth— the oceans are also marked by features of far greater size and extent than any found on the continents. The Grand Canyon, for example, is 5,580 feet deep, whereas the deepest ocean trench—the Mindanao, east of the Philippines—is 37,800 feet deep. The mid-ocean ridge system circles the globe, forming a mountain chain nearly 97,000 miles in length. Mauna Kea, a Hawaiian volcano, is the largest mountain on the planet, measuring over 32,808 feet from its base on the ocean floor to its peak, and having a diameter of over 62.14 miles. For all this extreme topography, however, the major features of the ocean basins are relatively few. (Figure 4.1.)

The Continental Shelf

The continents are surrounded by extensive *continental shelves*, which slope gently seaward at an angle of one to three degrees. These shelves are covered by shallow seas and are generally formed of young sediments derived from the land. The continental shelves, which cover about 7 percent of the total seafloor, may be wide, extending to as much as 620 miles in width at passive continental margins, which are marked by no tectonic activity or plate boundary. Or they may be narrow (less than a mile) at active continental margins, as along the western coast of South America, where the Nazca plate is being subducted into the Peru-Chile trench. The continental shelf is a zone of abundant marine life and active sedimentation. (Figure 4.2.)

The outer limit of the continental shelf is marked by a sudden change in slope from the 3° of the continental shelf to about 30°, where the *continental slope* plunges down toward the *abyssal plain*, which marks the deepest oceans. The continental slope, covering some 9 percent of the ocean floor, is characterized by darkness and high pressure, and is marked in many places by sediment slumps, "landslides," and deep underwater *canyons*—sometimes marking the extension of river systems—excavated by sediment-laden *turbidity currents*.

The *continental rise* follows the base of the continental slope and generally represents a wedge, formed by the accumulation of sediments at the edge of the abyssal plain.

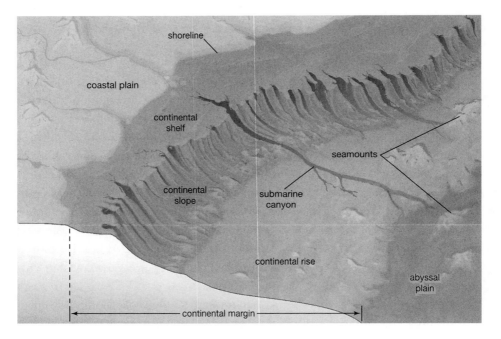

4.2 Typical profile of continental margin, showing continental shelf and slope. Reprinted with permission from the *Encyclopaedia Britannica,* copyright 2010 by Encyclopaedia Britannica, Inc.

The *abyssal plains* occupy most of the ocean floors. These are areas of total darkness, high pressure, and minimal sedimentation, consisting chiefly of fine particles of clay and the remains of microscopic organisms.

Three other features interrupt this simple picture of submarine bathymetry. The *mid-ocean ridges*, as we have already seen (chapter 3), are long, linear mountain chains, which are the sites of intense geological activity, produced by the moving apart (divergence) of two plates along their length. *Trenches*, in contrast, are deep, linear or arcuate troughs, generally developing along the margins of the continents, at which a crustal plate or slab is pulled downward (subducted) into Earth's interior.

Ocean rises are formed in places by sea mounts, guyots, and other submarine volcanic structures, as well as barrier reefs, that fringe the coasts of tropical land masses.

Above this varied topography of the seafloor, the waters of the oceans are in constant motion. At their interface with the land and the atmosphere they form the great reservoir on which the interchange of the *water cycle* depends. Solar heating evaporates seawater, as well as water from the land, to form clouds, from which rain and snow are ultimately precipitated, to be recycled again as these fall on the oceans and the land. (Figure 4.3.)

Within the oceans themselves the water is in constant movement. The upper waters of the ocean are marked by currents, some of which are local and short-lived, reflecting responses to local and seasonal conditions. Others are essentially permanent and cover the whole ocean. Surface waters, in response to the drag of prevailing winds and, to a lesser degree, the rotation of the Earth and the distribution of the continents and islands, move horizontally in huge circulating systems—"gyres."

These, together with smaller component currents, move across the face of the oceans, moving in a clockwise direction in the Northern Hemisphere and a counter-clockwise direction in the Southern Hemisphere (figure 4.4). In the polar regions, these directions tend to be reversed. This directorial difference reflects the same Coriolis effect of the Earth's rotation to the east that deflects the wind directions.

Where the two major currents converge near the equator, both flow westward (the north and south equatorial currents) and then are separated by a reverse flow, the equatorial countercurrent. Smaller parts of these major systems are identified by names associated with adjacent landmasses, such as the Labrador current or the Peru current. One of the best-known currents is the Gulf Stream—named by Benjamin Franklin—which carries warm ocean waters into the North Atlantic, bringing warm airflow to Britain, France, and Norway. Surface currents, such as the Gulf Stream, move at speeds from about 0.5 knots to 2.5 knots. Two centuries ago Alexander von Humboldt calculated—correctly—that it would take some fifty-eight months for the Atlantic gyre to make a complete circuit.

The ocean's surface waters affected by these currents make up only about 10 percent of the ocean's water, and occupy only the upper 660–1,290 feet of the

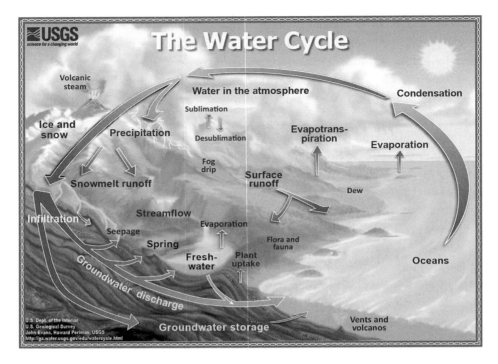

4.3 The water cycle. Courtesy of USGS.

ocean. They rest on deeper, denser water at a boundary marked by rapid change in the temperature and salinity of the underlying water.

The *deep water currents* move chiefly under the influence of gravity acting on differences in density of seawater. These density differences reflect differences in temperature and salinity and are typically marked by sinking of denser water in high latitudes. Upwelling of colder, nutrient-rich waters occurs in some areas, especially along western coastlines, where offshore winds push surface water aside. These relatively small areas of upwelling cold water support about half the world's fishing activity.

Vast as the oceans are—their volume is calculated as 325 million cubic miles—their volume is still insignificant in comparison with that of the Earth: only about 0.0008 of the volume, in fact. The ocean's average depth—some 2.35 miles (about 12,400 feet)—makes them a relatively thin film of salt water covering the surface of the Earth. So these features of the ocean basins—great as some are—are barely perceptible as surface irregularities, given the much greater diameter of the Earth.

Yet for all their subdued scale, these features of submarine topography are of immense significance, for their form reflects the direct linkage between the effects of continuous movement deep within the planet and crustal development on the one hand, and land surface erosion and marine deposition on the other. This ceaseless interaction, which constantly reshapes and re-creates the surface of the planet, is complemented by the continuous interchange between the oceans and the atmosphere, as well as by the interaction of both air and water with the surface of the

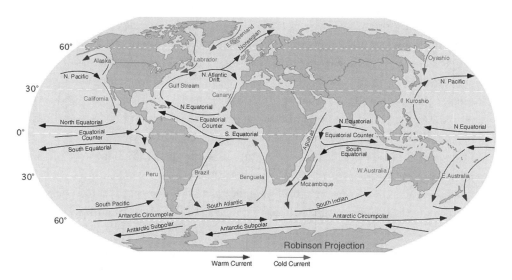

4.4 Ocean currents. Courtesy of Michael Pidwirny for U.S. govt., www.physicalgeography.net.

continents, and with all living things. From Earth's deep interior to the higher atmosphere, these endless cycles of interchange—of heat, water, nutrients, and material—with their continuous pattern of intermixing and interaction, driven by Earth's parent, the Sun, support and sustain us every moment of our existence. As a result of their long interaction, we, the children of Earth, emerged. As a result of their continuing efficacy, we subsist. Only in harmony with them can we hope to survive.

The Matrix of Life

We dwell on a blue planet: a watery planet. And that water, which irrigates the land, sculpts the crust, forms the oceans, and permeates the atmosphere, is also the primary and essential component of life. It is essential because every living thing not only requires water but also is made of water: our cells, our organs, our bodies consist largely of water.

It was from the water that our remote ancestors, the amphibians, emerged to take their first tentative steps on land. And it was from the water that their earliest ancestors—the lowly forms that marked the emergence of life itself—arose. Though we have no direct knowledge of just how this happened, several lines of evidence suggest a possible sequence of events. But how did the oceans and atmosphere develop?

Now, if Earth's atmosphere and oceans were derived largely from volcanic emissions of molten material and gas by early volcanic "outgassing" from the interior of the planet, one might expect some resemblance between the composition of fumes from existing volcanoes and the composition of the atmosphere. So how do they compare in composition? A typical Hawaiian volcano emits gas that contains about 70 percent water, 4.6 percent carbon dioxide, less than 1 percent each of hydrogen and carbon monoxide and very small amounts of other gases (nitrogen, chlorine, sulfurous gases, and argon). That could certainly produce the water required to fill the oceans, supplemented perhaps by impacting ice-bearing comets. But how could such emissions, however widespread or prolonged, ever produce Earth's present atmosphere, which consists largely of nitrogen and oxygen? Clearly they could not, but Earth's oldest atmosphere was quite different in composition from its present. I discuss this in some detail in chapter 5, but we need a thumbnail sketch here, as we consider the oceans.

As Earth accreted from the cloud of stellar gas and dust that formed the solar system, its earliest atmosphere probably consisted of the typical components of interstellar gas: hydrogen, helium, argon, neon, and perhaps silicate vapor and other gases. Only after this light, early atmosphere escaped and was lost is it likely that a new, secondary atmosphere was formed by the violent spasms of volcanic activity that marked a great period of outgassing of the planet. Earth also received material from impacting comets. This new atmosphere—Earth's second—presumably consisted of water vapor, nitrogen, carbon dioxide, and other gases that are typical of volcanic emissions. Much of this later became part of the oceans.

But if seawater comes from the steam of volcanoes, why is the sea salty? Ferren MacIntyre has answered that "according to an old Norse folktale the sea is salty because somewhere at the bottom of the ocean a magic salt mill is steadily grinding away. The tale is perfectly true. Only the details need to be worked out."[1] The "mill" as MacIntyre points out, is the Mid-Ocean Rift, stretching for 40,000 miles across the ocean basins. Accompanying the eruptions of volcanic rock that flow from the rift is juvenile water, erupted from deep within the Earth. This water contains elements that are precisely those that are "missing"—bromine, chlorine, and iodine—if we were to assume that all the solids dissolved in seawater were solely derived from the weathering of Earth's surface rocks.

Of the ninety-two naturally occurring elements, over seventy are present in seawater, though most are in minute concentrations. Eleven constituent elements make up 99.9 percent of the total dissolved material in ocean water, and its composition is remarkably uniform across the globe, except for areas immediately above and affected by spreading ocean ridges.

It has been estimated that if all the salt in the oceans could be removed and spread out over the Earth's land surfaces, it would cover them to a depth of more than five hundred feet, which is about the height of a forty-story skyscraper.

Vast as that volume of salt is, it seems probable that the ocean's "saltiness" comes chiefly from erosion and solution of salt contained in the rocks of the

---

1. Ferren MacIntyre, "Why the Sea Is Salt," *Scientific American* (November 1970), 104.

continental crust, as well as from volcanoes and undersea vents. It is estimated that the oceans receive huge quantities of solids—both dissolved and suspended—from the rivers that flow into them. One estimate is that rivers in the United States alone discharge 225 million tons of dissolved solids and more than twice that in suspended form into the oceans each year. The evidence of fossils suggests that the salinity of the oceans reached a "steady state" over 600 million years ago, and that since then, there has been a broad balance between inflow, deposition, and precipitation of dissolved salts.

In spite of these slow changes in their chemistry, the oceans still provide a more broadly protective and stable environment for life than land does. And that is one of the reasons many observers conclude that the oceans are the most likely place for the initial emergence of living things.

Just where and when the great transition from nonliving to living matter took place is a topic of continuing inquiry and debate. Though some have suggested thermal springs or geysers on the continents as suitable sites, many others prefer an oceanic site. Some suggest the nutrient waters emerging from "smokers" of the deep ocean, while others have suggested a more temperate, perhaps more widespread shallow marine environment, where abundant clay minerals may have served as templates or catalysts for the emergence and assembly of living things. We should not exaggerate the complexity of these early forms. These were, one supposes, distinguished from the dozens of other different organic molecules round about them chiefly by their enclosing membrane and their ability to benefit and develop from the organic "broth" around them.

In time, we shall no doubt have a clearer picture of this remarkable development, but one thing at least is already clear: it was only in water that this great transformation could have taken place. For all our earthliness, we are creatures of the water. And that water is, in turn, a product of the younger days of the "earthy" planet on which we dwell.

The Regulator of the Planet

The oceans and the atmosphere act as Earth's great regulator, or thermostat, moderating its overall long-term climatic patterns, and shaping, modifying, and sometimes disrupting, its local weather. The oceans provide two great conveyor belts that moderate Earth's climatic patterns. First, they carry heat from the tropics toward the polar regions, as warm equatorial waters are carried toward higher latitudes by the prevailing winds. Approaching these high latitudes, the warmer water gradually loses heat, becomes denser, and slowly sinks to the ocean bottom, contributing to the great, cold-water bottom currents flowing toward the equator. This endless cycle moderates the broad weather patterns of the planet. (Fig. 4.4.)

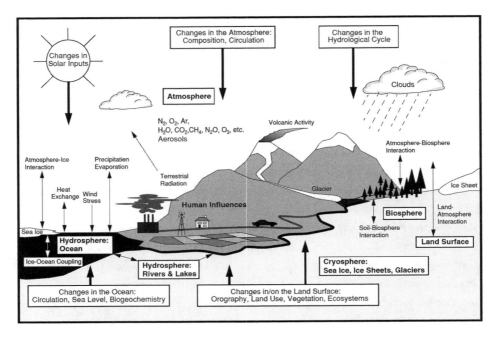

4.5 Action of oceans and atmosphere in regulating the Earth's climate, with some examples (boxes) of physical and biological processes related to climate and climate change. Reprinted with permission from Xavier Rodó and F. A. Comin, *Global Climate: Current Research and Uncertainties in the Climate System* (New York: Springer, 2002).

The second great cycle in which the oceans are involved is the removal and storage of carbon dioxide from the atmosphere. The endless flux in which Earth's oceans, atmosphere, organisms, and landmasses interchange energy and recycle materials underlies every aspect of the sustenance of life. Profound disturbance of those cycles of flux—by such things as increasing CO_2 emissions, widespread destruction of tropical forests, damage to the ozone layer, or other human activity disturbing the natural long-term cycles of the system—could produce severe consequences in some animal and plant populations. It is some aspects of such changes that we explore in subsequent chapters.

The Pantry of the Planet

Although so much of Earth's surface is covered by oceans, the variety of ocean life is conspicuously less than that on land. Plants, for example, abundant, varied, and exuberant on land, are represented in the oceans only by lowly seaweeds and microscopic plankton. Of the three-quarter million or so insect species, fewer than ten are truly marine. Few birds, mammals, and reptiles are marine. It is the fish and the great army of invertebrates that are the most distinctive creatures of

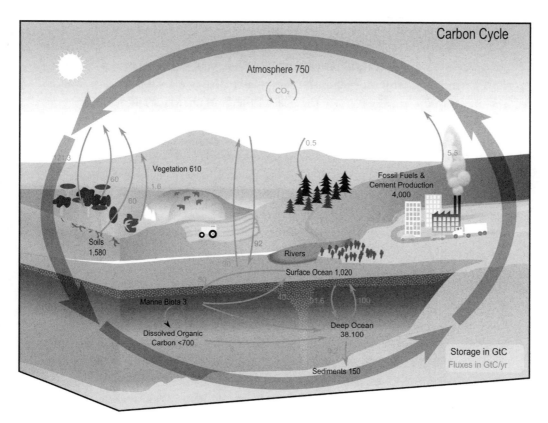

4.6 Carbon cycle diagram showing the storage and annual exchange of carbon between the atmosphere, hydrosphere, and geosphere in gigatons—or billions of tons—of carbon (GtC). Courtesy of NASA.

the seas. And though the number of species is fewer than their counterparts on the land, they exist in numbers of individuals that dwarf their terrestrial relatives.

Some major groups of invertebrates—sponges, sea urchins, starfish, jellyfish, and corals, for example—are confined to the sea. Other groups—mollusks and worms, for example—though they have land-living representatives, are dominantly marine creatures. These invertebrates lack the highly developed organs of their terrestrial neighbors. Sea anemones, corals, and jellyfish, for example, are saclike organisms, consisting largely of a digestive sac, surrounded by a ring of tentacles armed with stinging cells, by means of which they capture their prey. Though they have a network of nerve cells, these creatures lack any central nervous or circulatory systems, and they lack sight and locomotive organs. Sponges are even simpler than these coelenterates. Other marine animals—mollusks, worms, starfish, and crustaceans—are more advanced, with varying structures of digestion, reproduction, locomotion, and sight.

Most of these invertebrates are benthic, living fixed to or crawling over the ocean floor, though many pass through a free-floating larval stage. The fishes of the oceans, in contrast, are chiefly nektonic, free swimming in the open oceans. They exist in great variety and vast numbers. Many marine fish have complex migration patterns, partly related to temperature, spawning, and seasonal changes.

Plankton—microscopic animals and plants that live in the surface waters, drifting with the currents—form the base of the food chain for all the other ocean dwellers. They exist in such vast numbers that the microscopic tests ("shells") of some create a "snowfall" that forms the dominant sediment over great areas of the ocean floor.

The relative size and depth of the oceans with respect to the continents means that they provide the planet's biggest habitat. The interconnectedness of this habitat allows much more interchange between marine organisms than those living on the land, which is why the number of marine species of animals and plants is smaller that those of the land, even though the number of oceanic individuals is far greater. But conditions within the oceans vary greatly, both with latitude and with depth, and communities also show great variation, ranging from tropical reefs to temperate waters, to the open ocean, the deep sea, and the cold waters of the polar regions. Each habitat has its own distinctive fauna and flora, and though most organisms live in the uppermost five hundred feet of the oceans, even the deep, cold, dark regions of the ocean depths teem with life, some of it remarkably adapted to the particular environment in which it is found.

The oceans have long been a source of human food, and today some 5 percent of total human protein consumption comes from the seas. Much of this still comes from the ocean fisheries, but increasingly aquaculture is being seen as an important food source. The rate of fish farming is growing by 11 percent a year and is now set to overtake cattle ranching. Growth in beef production, which in the past increased greatly to provide the chief source of animal protein for the rapidly growing human population of the last century, has now leveled off. Ocean fishing has also leveled off, with some major fishing grounds seriously depleted. Lester R. Brown has commented that future growth "now depends on placing more cattle in feedlots or more fish in ponds."[2] Because cattle require about seven pounds of grain to produce one pound of live weight, whereas some fish can add a pound of weight with less than two pounds of grain, fish-based diets are seen as increasingly attractive. Grain itself, of course, also poses production costs because it too requires water and nutrients; in fact, it takes at least one thousand tons of water to produce every ton of grain. (Figure 4.7.)

Over two hundred species of fish and shellfish are raised commercially, and with overfishing of many of the world's most productive areas, aquaculture is likely to increase, although it cannot by itself supply the growing food needs for a burgeoning global population, and is not without its own environmental impact.

---

2. Lester R. Brown, "Fish Farming May Soon Overtake Cattle Ranching as a Food Source," Earth Policy Institute Plan B Updates, October 3, 2000, http://www.earth-policy.org/plan_b_updates/2000/alert9.

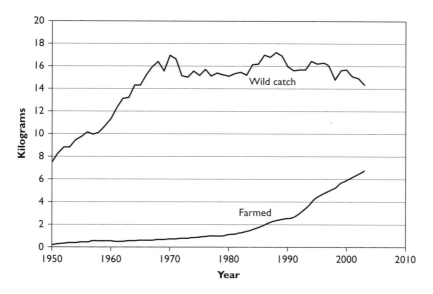

4.7 World fish production per person, 1950–2003. Courtesy of FAO.

Salmon, for example, a commonly raised fish in many parts of the world, is a carnivorous species, requiring a diet of five pounds of fish-based food for every pound of meat produced, as well as producing significant waste products. Fish farming is now practiced chiefly in the developing countries, where herbivorous species are favored, enabling more efficiently integrated aquaculture and farming.

Beyond current fishing and farming techniques, however, the needs of a growing human population will require additional new food sources in quantities far exceeding our present production capacity. It is here that the oceans, with their abundance and variety, surely hold promise for new thinking and new resources. We need to be as creative in harvesting the bounty of the oceans as we have become in harvesting the bounty of the land.

But there is a limit to this bounty, just as there is a limit to the feeding capacity of the land. It has been estimated that, using present methods, the eventual global production of fish from the oceans might be increased to 160 million metric tons annually, containing about 20 percent of usable proteins. So, short of new methods and techniques, the oceans are likely to be an important supplementary source for protein rather than the major supplier. And because the oceans are also a poor source of calories, they are unlikely to provide more than a small fraction of the total energy requirements for a growing global population.

Acidifying the Oceans

The biosphere, lithosphere, atmosphere, and oceans, as we have seen, form one great interacting system. Modify the atmosphere and you will modify the oceans, and that is what some fear may be happening. As carbon dioxide levels increase, the oceans become more acidic. One index or measure of acidity is the pH, or the hydrogen ion concentration, which decreases as acidity increases. Measurements suggest that ocean acidity has increased by almost 30 percent above preindustrial levels, and projections of future emissions point to substantially higher levels.

The effect of this acidification is likely to be felt by the many organisms that secrete shells or skeletons of calcium carbonate (calcite or aragonite). These include not only many edible shellfish and other mollusks but also corals, sea urchins, lobsters, and many microscopic surface organisms, such as for aminifera and coccoliths, which play a vital role in the food chain and in photosynthesis.

There is little agreement among specialists as to just how serious a threat ocean acidification poses. The effects are likely to be far from uniform, not only between organisms—some of which may even prosper—but also between different parts of the ocean. It is, however, clearly a significant factor, among others, that must be weighed within the larger debate on climate change.

The Benevolent Oceans

The ocean has nourished us. Within its ancient waters, the first living things developed. To it, life was confined for the earlier years of its existence. From it, some 400 million years ago, our remote ancestors, the lobe-finned fishes, made their first tentative ways into the freshwaters of estuaries and rivers from which their offspring, the amphibians, later clambered onto the land. Now, established there, we—their descendants—survive on land because it is the ocean that has continued to moderate the climate of our planetary home. Unlike the searing oven of Venus or the chill desert of Mars, Earth's hospitable climate exists by courtesy of the oceans. On a daily basis, Earth's oceans power the great climatic water cycle that produces our weather. The oceans provide a major source of food and materials, and even a place of disposal and ultimate burial for much of the effluent of our rapidly expanding industrialization.

But the ocean does much more: it is beneath the oceans that the whole planet is recycled. It is at the mid-ocean ridges that new land is created. And it is in the deep ocean trenches that older land is subducted and consumed, to be melted and ultimately reformed and recycled in the endless cycle of convection that underlies and powers the slow reformation of the Earth that is plate tectonics. This is the grand global process, the most basic of all Earth's cycles, that we have reviewed in chapter 3.

So our very existence and every aspect of our tenancy of this, our home planet, reflect the legacy and influence, not only of our landlord Earth, but of our sealord Ocean.

CHAPTER 5

The Veiled Planet

Seen from space, Earth is a marvelous azure-blue sphere, illuminated with jewel-like clarity by the light of the sun. But it is a sphere that is always partly veiled, with changing, diaphanous filaments of cloud, some small, and some so vast that they seem to encircle the planet. These clouds are the visible part of the atmosphere. Mars has an atmosphere so thin that the red planet is never cloud-covered, Venus an atmosphere so chokingly dense that its rocky surface is never visible. But Earth's remarkable envelope of air makes life possible by enfolding the planet in a protective embrace that shields and sustains life. We are as much creatures of the atmosphere as are fish creatures of water. It provides the air we breathe, it shields us from the Sun's lethal ultraviolet radiation, it regulates and moderates Earth's temperature, it provides rain, it drives the endless cycle of erosion that produces the soil that supports our food supply, and it provides the carbon dioxide on which plant growth depends. It is the atmosphere that paints the sky and oceans with their changing shades of blue, unlike the sky above the moon, which, lacking an atmosphere, is always black.

The Layers of the Atmosphere

The atmosphere is a layered blanket of air that gradually thins out until it disappears at about 350 miles up, high above the Earth. Flying at 1,800 miles an hour,

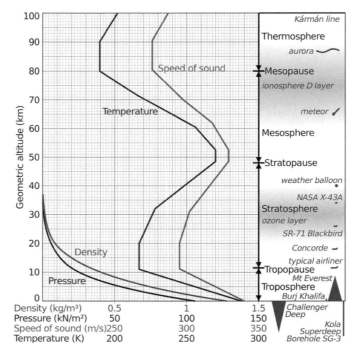

5.1 Layers of the atmosphere: comparison of the 1962 U.S. standard atmosphere graph of geometric altitude against density, pressure, the speed of sound, and temperature, with approximate altitudes of various objects. Created by Cmglee for Wikimedia based on NASA graph.

the space shuttle takes ten minutes to leave Earth's atmosphere, whose layering is clearly visible from space. But for all its limited extent, each layer of the atmosphere contributes to the ceaseless interchange that shapes the face of the planet and shields and supports its myriad forms of living things. The lowest level of the atmosphere, the *troposphere*, has an average thickness of about seven miles, ranging from five to nine miles, with its greatest thickness at the equator, and decreasing toward the poles. This lower level is the site of Earth's weather and it contains some 80 percent of all the atmosphere's mass. Air temperature generally decreases toward the top of the troposphere, with an average of about 50°F near Earth's surface and −71°F at the upper surface of the troposphere, with air density and moisture content also decreasing upward.

Above the troposphere lies the *stratosphere*, extending to a height of about thirty-one miles above Earth's surface and containing the thin ozone layer, which, by absorbing ultraviolet light, shields the Earth and its creatures from the harmful effects of ultraviolet radiation. Above this layer, extending from about thirty to fifty miles out lies the *mesosphere*. It is this layer that protects Earth from most incoming meteorites, which burn up in its cold, thin air. Above the mesosphere lies the *thermosphere*, which, though extending some three hundred miles above Earth and being by far the thickest layer of the atmosphere, contains only 0.00001 percent of its mass. The lower part of the thermosphere, the *ionosphere*, contains electrically conducting layers that reflect radio waves, making radio communication possible over great distances on Earth. It is in the upper portion of the thermosphere that the northern lights are formed, and it is also in the upper thermosphere that the International Space Station orbit lies. The atmosphere has no precise upper limit: it simply thins away into space.

Earth is not alone in supporting an atmosphere; all the planets have atmospheres, though they vary greatly in composition. We have seen that Venus, for example, has a dense, stifling atmosphere of carbon dioxide, with an atmospheric pressure about ninety times that of Earth. Mars, too, has an atmosphere of carbon dioxide, but in contrast to Earth, its surface is a freezing desert and the pressure of its thin atmosphere is only about 1 percent that of Earth. The atmospheres of each of the planets reflect not only their location, composition, and gravity but also their history. The atmospheres have evolved, and are still evolving, along with their parent planets. That, as we will later see, is a fact of considerable significance because of its influence on climate.

Origin of the Atmosphere

We have already seen that Earth's atmosphere has changed drastically over time. In its earliest form it probably consisted of clouds of silicate vapor as earth-forming materials accumulated, developing later into an atmosphere of hydrogen, helium, and traces of other elements. These gases, though abundant throughout the universe, are rare on Earth today, probably because they were lost early in Earth's history, driven off by solar radiation and intense heating.

The composition of Earth's present atmosphere is distinctive: it contains about 78 percent nitrogen and about 21 percent oxygen, with traces of other gases, including argon, carbon dioxide, and water vapor. How did this atmosphere originate? Two distinct possibilities exist. One explanation is that gases from prolonged volcanic activity on the young planet may well have enveloped the globe. Existing volcanoes provide a clue to the emissions that may have arisen from this outgassing. They produce a mixture of gases that includes 50–60 percent water vapor, as well as other gases. But as we have seen, even with the intensive volcanic outpourings of the young Earth, it seems questionable whether this outgassing could have provided enough water to account for the present oceans and atmosphere. A second possible source may have been provided by impacts from incoming comets—made of water ice, carbon dioxide, carbon monoxide, ammonia, and other materials—which would have been intense in the earlier history of the

5.2 Chief constituent gases of the atmosphere (dry air, by volume, in percent)

Element	Percentage
Nitrogen	78.09
Oxygen	20.95
Argon	0.93
Carbon dioxide	0.03
Neon	0.0018
Helium	0.000524
Krypton	0.0001
Hydrogen	0.00005
Xenon	0.000009
Ozone	0.000005

Source: NASA, *U.S. Standard Atmosphere, 1962* (Washington, DC, 1962), 9.

solar system. These two potential sources of Earth's atmosphere are not mutually exclusive: both could well have contributed to it.

But, as we saw in chapter 4, the missing component from the terrestrial atmosphere formed in this way is oxygen. The presence in ancient (Archean) rocks of minerals such as pyrite (FeS_2) and uraninite (UO_2), which are unstable under present atmospheric oxygen conditions, is clear evidence that the early atmosphere lacked oxygen. So where did the present atmospheric oxygen come from? A small part—1–2 percent, perhaps—probably came from the breakdown of water molecules under the intense ultraviolet radiation of the early preatmospheric ozone Earth. Some of this oxygen probably subsequently formed the beginnings of the ozone (O_3) layer, which, when it later developed, then shielded Earth from this ultraviolet radiation. This protection in turn allowed the first development of photosynthesis, in which the earliest organisms—cyanobacteria and later life forms—converted CO_2 and water in the presence of sunlight into organic components and oxygen. This process, photosynthesis, released oxygen, which slowly changed the composition of the early atmosphere.

So, ironically, Earth's life-supporting atmosphere was itself created by the emergence of simple, living organisms. This seems the most likely source of free oxygen, and there is some evidence from the rock record of this great atmospheric transition. Iron is highly reactive with oxygen, which is why steel rusts so readily. Banded iron formations (BIFs), consisting of alternating layers of iron-rich (Fe_3O_4) and iron-poor silica-rich strata, are known only from rock sequences deposited between 2.8 and 2.0 billion years ago, and most were formed about 2.3 billion years ago, when Earth's atmospheric oxygen probably rose to about one-third of present levels. These rocks could have formed only

GLOBAL RADIATION

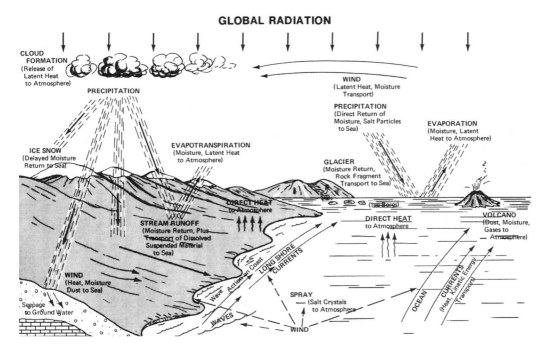

5.3 The processes of the geosystem by which energy, matter, and momentum are continuously exchanged among land, sea, and air. From George R. Rumney, *The Geosystem: Dynamic Integration of Land, Sea, and Air* (Dubuque, IA: W. C. Brown, 1970).

in an ocean rich in dissolved ferrous iron, and such an ocean could exist only in the absence of atmospheric oxygen. On the other hand, immediately after that period, bedded, red, sedimentary rocks formed under conditions of continental weathering and consisting of highly oxidized hematite (Fe_2O_3) are found in rocks younger than those of the banded iron formations, that is, formed during the last 2 billion years or so. This indicates that the oxygen content of the atmosphere has increased with time, probably reaching its present level only in the last 400–500 million years.

If the earliest organisms were heterotrophs—obtaining nourishment from the "broth" of synthetic molecules and later from the reaction of hydrogen and methane—the remaining early atmospheric hydrogen would slowly have been depleted, to be subsequently replaced by oxygen, as development of photosynthesis took place. As living things exploited the rich possibilities of oxygen-based respiration, gradual release of oxygen into the atmosphere led to the further development of the ozone layer. This layer, which lies about fifteen miles above Earth's surface, absorbs lethal ultraviolet solar radiation, and by doing that, it provided a shield for the emergence of the later metazoans. This development would have been accompanied by the gradual accumulation of sufficient free oxygen to support the metabolism of later multicellular organisms, as well as probably adding to the concentration of the ozone layer.

Thus, slowly, a habitable Earth developed, with the emergence of life protected, and the later development of life sustained, by the growing presence of oxygen. There is, however, a paradox in this. The earliest organisms could not have formed in the presence of free atmospheric oxygen. Oxygen prevents the growth of the most primitive living bacteria, for example, and inhibits the basic reactions that produce amino acids. Thus it seems probable that the earliest living things developed in anoxic environments, just as today's simplest forms do.

One other atmospheric puzzle needs to be addressed: Earth's atmosphere is distinctive, not only in the presence of oxygen, but also in the relative scarcity of carbon dioxide. Where has all the carbon dioxide gone? That may seem a strange question, given our concern about greenhouse gases. But after all, both Earth's neighbors—Venus and Mars—have atmospheres made up almost entirely of CO_2. Surely Earth must once have had an atmosphere much richer in CO_2 than its present 0.036 percent concentration. That seems likely, and it seems equally likely that this "excess" CO_2 is now locked up in the vast sinks represented by Earth's sedimentary rocks, such as limestone and dolomite, and its extensive forests.

So the great Earth cycles of interchange between the atmosphere, lithosphere, and hydrosphere interacted to nurture the emergence of the "fourth estate," the biosphere. This zone of living things interacts with all the rest, and includes within its countless species one, which by its very success and explosive growth in numbers, places stresses on many of the others, modifying the atmospheric matrix that gave it life and whose soft embrace still suffuses its every breath.

Atmospheric Movements

The land, the oceans, and all living things are part of a continuing cycle of interaction with the atmosphere. Heat and moisture are exchanged by convection in a ceaseless flux, with warmer air from the equator rising upward and spreading outward toward the poles. If the Earth were a uniform sphere with no rotation, this movement of the cooler, denser polar air spreading and sinking toward the equator would produce surface winds that would then move toward the equator at a uniform speed. In practice, the Earth's rotation produces a deflection of the wind toward the right in the Northern Hemisphere—as seen by an observer on the Earth—and the left in the Southern; a feature known as the Coriolis effect.

This simple pattern of airflow is complicated by the general disposition of land and oceans and the marked differences in topography, vegetation, and other features of the land. Local winds are caused by convection, where warm and cool air masses move in response to different absorption of solar heat by land surfaces.

The atmosphere behaves as a great thermostat for Earth, acting as an insulating blanket, protecting it from extreme daylight heating and extreme nighttime heat loss. Earth's temperature range is thus far narrower than that of, say, the Moon, which, having no atmosphere, has a surface temperature range of some 450°F. Because sunlight falls vertically on equatorial areas and at a steeply inclined angle at the poles, the air at the equator is always warmer than that at the poles.

Two movements of the Earth also influence the weather. The annual movement of the Earth around the sun produces the seasons and their typical weather. Because Earth's axis is tilted at 23.5° to the plane of its orbit, each hemisphere is successively tipped toward the sun, and so receives more concentrated heat and longer days. When the North Pole is tipped toward the sun for part of the year it has summer, while the South Pole, tipped away from the sun, has winter. Six months later, the positions of the hemispheres are reversed, as are the seasons. Earth's daily rotation on its axis gives us day and night and so also influences the major wind patterns.

The general pattern of winds, though modified by many of the features we have described, is broadly consistent (figure 5.4), being represented in both

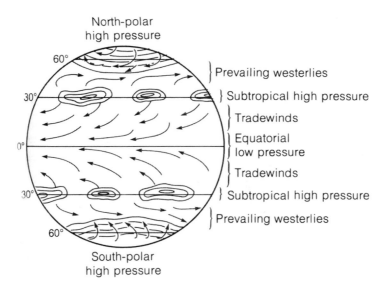

5.4 The global pattern of belts of surface winds and barometric-pressure cells. The subtropical high-pressure cells move somewhat and change size from summer to winter. Arrows show generalized wind directions. From Frank Press and Raymond Siever, *Earth*, 4th ed. (New York: W. H. Freeman and Co., 1986).

hemispheres and defined by three broad zones separated by two distinct belts of calm air.

For convenience we will describe the winds of the Northern Hemisphere. The *polar easterlies* consist of cooler, polar air moving southward until it meets the *prevailing westerly winds*, carrying warm, moist air from the south. The "front" between these two air masses moves back and forth, often producing severe weather.

The *prevailing westerlies* represent a steady airflow from a broadly westerly direction.

The *trade winds* (blowing from the northeast in the Northern Hemisphere and the southeast in the Southern Hemisphere) lie nearest to the equator and provide the steady sailing winds that were so important for earlier mariners and traders.

The *equatorial doldrums* represent a zone of calm, generally windless areas where the trade winds, moving toward the equator, meet increasing temperatures that cause them to rise and so produce lengthy periods of calm. In summer, rising hot air in these regions creates hurricanes and typhoons.

The *horse latitudes* represent belts of light winds and hot, dry, calm conditions, located at about 30° latitude in each hemisphere, between the equatorially directed trade winds and the opposing prevailing westerlies. The name reflects the fact that Spanish ships sailing for the West Indies would be becalmed for so long they were compelled to throw their horses overboard.

When two air masses collide, they rarely mix with one another, unless they are very similar in temperature and moisture content. Instead they interact, and this interaction has a major effect on weather.

Weather

"Everyone complains about the weather," Mark Twain and Charles Dudley Warner are both reported to have said, "but nobody does anything about it." And that's just as true now as it was in Twain and Warner's day. Weather is partly predictable, but it's subject to influences far beyond our control. Edward Lorenz, one of the pioneers of chaos theory, for example, described what he calls the butterfly effect—the notion that a butterfly stirring the air today in Brazil can set off a tornado in Texas.

Weather arises from the incoming energy Earth receives from the Sun, seeking some condition of equilibrium. Weather reflects the combined effects of heat, temperature, wind, and moisture in the atmosphere. Of the total radiation Earth receives, about 30 percent is reflected back into space by clouds, snow, and other surfaces, about 20 percent is absorbed by the atmosphere, and about 50 percent is absorbed by Earth's surface. Regional differences in cloud cover, and the variety and irregularity of Earth's surface, affect the degree of reflection and absorption and lead to instability and air movement on a global scale, which, in turn, affect local climate. It is this incoming solar heat that drives the weather.

Weather differs from climate: weather is local and immediate, while climate is the weather at a given place over an extended period, a general statement of a region's weather, averaged and generalized over time. Climate varies from place to place, even within a given region, reflecting differences in temperature, precipitation, and terrain. Microclimates of very limited extent arise because of the local influence of mountains, local vegetation, development, and other features of temperature and rainfall.

We have already seen something of the wind and temperature interactions that produce "typical" patterns of climate and weather. There are, however, extreme climatic events that are far less predictable and sometimes far more dangerous than these more typical patterns. These extreme climatic events can produce severe effects on human populations. Floods, prolonged heat waves, and drought, for example, can have devastating effects on communities over large areas, destroying crops and livestock, threatening livelihoods and property, displacing populations, and spreading famine and disease. Against these there is only limited protection. Good engineering, flood protection, prudent building codes, and farsighted planning all help, but in many poorer areas of the world, the use of such measures is limited. Even in the developed world, they may be of limited value, as Hurricane Katrina has painfully reminded us.

Other atmospheric events are more limited in their effects, and in some cases are more manageable. Lightning, caused by the attraction of unlike electrical charges between thunderclouds and the Earth, produces short but powerful discharges in which the total energy of one thunderstorm can exceed that of an atomic bomb. The grounding of buildings and sensible precautions by individuals have been shown to prevent both destruction and loss of life. But lightning still kills over two hundred people every year in the United States.

Some weather events are frequent and more or less predictable, but can still sometimes produce severe effects and create hardship, damage, and casualties.

Monsoons are continental winds that represent air movement produced by differences in temperature over land and water. They reverse direction with the seasons. In winter, the air over the oceans, being warmer than that over the continents, rises, and winds flow toward the ocean from the colder land. Summer reverses the process, with rising, warmer continental air being replaced by landward winds of moisture-laden air from the oceans. In such places as India and the east coast of Asia, these winds can produce summer rainfalls of over one hundred inches. *Chinook winds*, a distinctive feature of the eastern Rockies, are caused by the downslope movement of cool air from mountaintops to lower levels. *The mistral winds* of Europe represent the flow of cold air from the Alps down the Rhone Valley to the warmer areas of the Mediterranean. The fabled *Santa Anna winds* of Southern California, formed by rising air pressure in the high elevations of the Great Basin, sweep through the canyons toward the coastlines.

Other wind and weather patterns, though seasonal and more or less predictable, are far more extreme and can cause widespread devastation. *Hurricanes* and *typhoons* are tropical *cyclones*, formed by the collision of warm and cold air masses. Denser, cold air moving under warm air produces a low pressure area, with more or less circular air movement, as the cold air, moving more rapidly than the warm, lifts the warm off the ground, producing a low pressure cell of spiraling air. Hurricanes differ from cyclones in occurring over the warm oceans, and in having far more intense winds of seventy-four miles per hour or more. Although hurricanes generally break up soon after reaching land, their effects can be devastating. Hurricanes over the southwestern part of the North Pacific (China, Japan, Korea, and adjacent areas) are known as typhoons. In addition to the Pacific, the other areas of most frequent hurricane activity are the Indian Ocean, especially around the north Australian and east African coasts, and the western Atlantic, which includes the Caribbean and southeastern United States. Elaborate tracking systems are used to follow the course of these hurricanes and thus allow prediction and warning of their impending impact.

The destructiveness of hurricanes is caused by their winds, which sometimes reach two hundred miles per hour, together with storm waves, tides, torrential rains, and floods. Hurricane Katrina provides a tragic example of the scale of devastation that hurricanes can produce. Early warning systems allow preparation and evacuation in the face of approaching hurricanes, but improved coastal

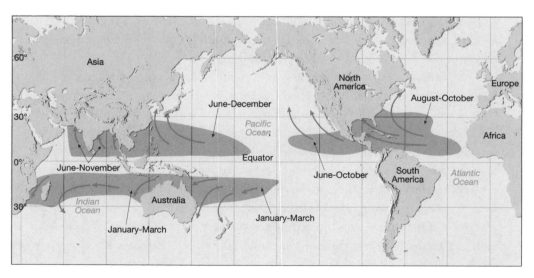

5.5 Hurricanes. Adapted from Frederick K. Lutgens and Edward J. Tarbuck, *The Atmosphere*, 7th ed. (Upper Saddle River, NJ: Prentice-Hall, 1998).

defenses, shoreline preservation, stringent building codes, and emergency planning are also required to prevent human misery and suffering.

Tornadoes are much more limited in their destructive impact than hurricanes or typhoons, but they still cause substantial damage. The United States alone typically experiences over 120 each year. They are formed by moving, twisting, swirling funnel-shaped masses of air, accompanied by lightning and torrential rains that mark unstable air masses. Winds within the vortex can exceed three hundred miles per hour, while low pressure in this area, combined with a powerful updraft, can cause great destruction. The strength of this updraft is illustrated by tornadoes that occur over water and produce *waterspouts*, towering columns of water that rise high into the air. Waterspouts are most common over the tropical oceans but are known in other areas as well. Though spectacular, they are so widely dispersed as to rarely be destructive. We examine weather hazards in more detail in chapter 6.

In all these phenomena, the atmosphere acts as a gigantic heat engine, exchanging heat and water vapor between land and oceans everywhere on Earth. In an endless cycle of heating, evaporation, cooling, and precipitation, water is recycled, heat is absorbed and redistributed, Earth's surface is sculpted and shaped, and life is sustained. It is the atmosphere that has sheltered life, and it is

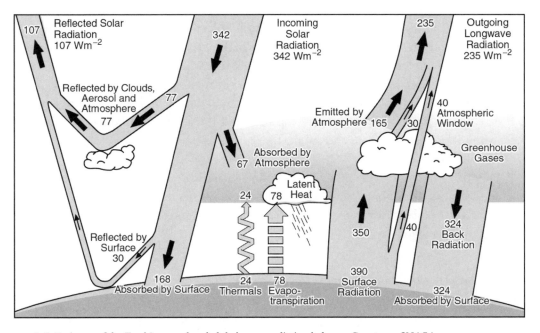

5.6 Estimate of the Earth's annual and global mean radiation balance. Courtesy of NASA.

the atmosphere's turbulence and interchange with land and oceans that influence the distribution and day-to-day existence of all Earth's creatures.

Atmospheric Changes

The atmosphere is in constant motion and so undergoes constant change. Over one hundred thousand cubic miles of water are evaporated every year from Earth's surface into the atmosphere, for example. There are, however, three longer-term changes representing worldwide trends that have become causes for concern. Though each involves environmental factors, they reflect rather different causes and raise somewhat different concerns. We will consider them in turn.

Acid Rain

All rain is acidic, because naturally occurring carbon dioxide interacts with pure water to produce weak carbonic acid.

$$H_2O + CO_2 \rightarrow H_2CO_3$$

Although the term "acid rain" was coined in 1872, it is only since the 1970s that it has become a matter of public concern. The acidity of rain has been increasing in several heavily industrialized areas of the globe, including the eastern United States. This has been the result of the prevailing easterly winds carrying airborne contaminants from the manufacturing areas of the Midwest. Two industrial gases seem to be the chief culprits: SO_2, sulfur dioxide, which interacts with water to produce a weak solution of sulfuric acid, and nitrogen oxides (NO_2 and N_2O_3), which produce weak nitric acid. The effect of acid rainfall is widely seen in damage to tree growth and forest development and the acidification of lake waters, with toxic effects on plant and animal life, as well as in corrosion of buildings and metallic surfaces.

Stringent controls of energy production—including removing sulfur from coal and oil, and "scrubbing" gas emissions from power plants—have led to substantial reduction in acid rain over the last decade or two and some reversal of earlier environmental damage.

Damage to the Ozone Layer

In 1981 the British scientific station at Halley, in West Antarctica, made a startling discovery: the continuous recordings they had been making over a period of some twenty-four years showed a remarkable reduction in the ozone concentration in the upper atmosphere. This observation was subsequently confirmed by comparisons of readings at other scientific stations. It was wholly unanticipated, not immediately explicable, significant in scale, and—though it did not mean the sky was falling down—it was potentially serious, because the "ozone layer" in the stratosphere was known to shield the Earth from harmful ultraviolet radiation. What did this discovery mean?

Ozone is a form of oxygen in which the molecule is made up of three atoms (O_3) rather than two, as in the "normal" oxygen (O_2) molecule. It is toxic to living things but it exists only in very low concentrations in the atmosphere—about 0.000005 percent by volume—so that generally it has little effect. So why worry then? Wouldn't a reduction in ozone concentration—even a significant reduction—be a good thing? But it was not ozone concentration in the lower atmosphere that the readings reflected: it was ozone concentration in the upper atmosphere. That was the concern.

Now ozone is an unstable gas—it can be produced synthetically just as it is produced in nature by an electrical discharge. It is also a useful gas, and it is manufactured and used commercially as a decontaminant and disinfectant. In nature it is formed in the stratosphere, not only by lightning, but also the action of ultraviolet sunlight on oxygen (O_2), which combines with atomic oxygen (O) to form O_3. It is the absorption by the ozone layer of ultraviolet radiation that protects the Earth from its harmful effects, particularly as a cause of skin cancer and cataract formation.

Long-term studies at the South Pole showed that by the mid-1980s the ozone layer was beginning to thin, with seasonal holes appearing over the polar regions. It was concluded that the cause of this damage in the ozone layer was the increased worldwide use of CFCs (chlorofluorocarbons), which were widely employed as aerosol propellants, in refrigeration, and in insulating foam. Though the CFC molecules are heavier than air, they are eventually carried into the stratosphere, where they are broken apart by solar radiation, releasing chlorine atoms, which react with ozone. One chlorine atom can disintegrate more than one hundred thousand ozone molecules. A few other molecules used in industrial products also damage the ozone layer.

Although chlorine is widely present in the oceans and in volcanic gases, long-term measurements make it clear that it is "human" chlorine, especially from CFCs, that has created recent damage to the ozone layer.

By 1978 the United States had already banned the use of CFCs in spray cans, and in 1987 the Montreal Protocol was developed, by which its signatory nations—of which there are now over 180—agreed to reduce, and later, in the light of subsequent scientific findings, to ban, the production and use of CFCs and other ozone-damaging substances. Alternatives for all these materials are now widely available.

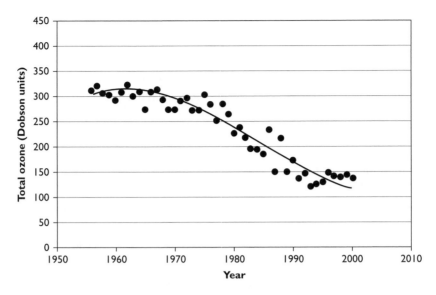

5.7 Ozone depletion. Courtesy of EPA.

Such international agreement and response to environmental hazards can serve as a model for dealing with other concerns. The benefits, however, will not be immediate. Although measurements show that stratospheric chlorine levels have peaked and are no longer increasing, CFCs are stable and thus long-lasting compounds, and it is estimated the ozone layer will not fully "recover" until about 2050.

The Greenhouse Effect

The atmospheric blanket of air that encircles the Earth traps energy, and thus the atmosphere slowly warms, much as a greenhouse retains heat. This is a natural and a beneficial process. Without it, Earth's temperatures would average −18°C (−0.4°F) rather than their actual 15°C (59°F). This greenhouse effect is created by a group of "greenhouse gases," consisting especially of water vapor, carbon dioxide, nitrous oxide, and methane. It is now clear that the concentrations of these gases and Earth's surface temperature have increased in the last 150–200 years, and human activity seems to be a significant part of the cause.

Deforestation and fossil fuel combustion seem to be the chief sources of these increased greenhouse gases, the first because of the decline in CO_2 absorption by trees, the second because the increase in CO_2 production from temperature increase is around 0.6°C (0.9°F) out of a total of some 0.8°C (1.2°F) in the last 150 years, though there is still some spirited debate on this point.

Continuing increases in global temperatures are likely to produce an increase in extreme weather events, including droughts, glacier melting, sea level rise, flooding, and changing seasons, together with significant impact on food and water supplies and the spread of disease.

There is broad, though not unanimous, agreement that we need to stabilize this temperature increase by limiting and reducing greenhouse gas emissions (GGE). That is not a difficult goal to establish, but it is, however, a very difficult goal to achieve, because the appropriate response to global climate change has profound scientific, technical, political, social, and economic aspects. We shall discuss it in more detail in chapter 13.

The atmosphere is Earth's least conspicuous component: less durable than the rocky crust, less tangible than the streams and oceans. But it is the atmosphere that is our common habitat: not the rocks, not the waters, but the air. It is the air that surrounds, supports, shields, and sustains us. If we are to be good tenants of Earth, we need to become responsible custodians of this thin, enveloping mantle of air.

CHAPTER 6

The Hazardous Planet

Born in a Hazardous Place

Life is a hazardous business, but our species has elevated survival to a high art by making individual health care and public health national priorities. We live sensibly, we eat prudently, we exercise doggedly—at least some of us do. We legislate elaborate safety standards and costly safeguards, and we devote sustained attention both to protecting and prolonging life. We make a huge investment—more than 10 percent of the gross national budget—in health care. And so we should. Our views of human worth and individual dignity demand it.

And the results are impressive; the benefits real. Human longevity has increased dramatically in the last two hundred years and especially in the developed nations. In the United States, for example, the average life span increased from forty-nine years in 1901 to seventy-seven years at the end of the last century. In China and India, it is reported to have risen from about forty years in 1950 to about sixty-three years at the end of the last century. Factors that have led to this huge improvement include the steep reduction in infant mortality, improved sanitation, public health and nutrition, and especially the control of infectious diseases. In most countries, women have a greater life expectancy than men. But for most of the species with whom we share the planet, life is far more hazardous than it is for us. For most species, life is short and hard, and survival prospects are bleak, and, in some cases, becoming bleaker. Longevity records are sparse and information is often contradictory. In birds, for example, the adult life expectancy of small temperate-zone songbirds is about ten months. The European robin has an average life span of 1.1 years, the European blue tit 1.5 years, the starling 2.5 years. The mortality rates for the young of many bird species are staggering and may be as high as 75 percent in many species after they have fledged.

Larger domesticated animals tend to have longer life spans. Cats have a life span of 15–20 years, dogs 8–20 years, horses, camels, and cattle 30–50 years. A few animals, of course, live much longer. Elephants are well-known for their longevity. The African subspecies has an average life span of 50–80 years. Some turtles live 150 years, and some whales (bowhead) 200 years. But many familiar species—even long-lived ones—that were once widespread and abundant are now in decline. There are now only 35,000 Indian elephants left in the world, for example, and fewer than 1,500 tigers in India. A century ago, it is estimated that there were over 100,000 tigers in the world. Today the total may be closer to 5,000. Three of the eight subspecies of tiger have become extinct in the last sixty years. Some 20 percent of the 564 species of sharks are threatened with extinction.

In many cases human influence, direct or indirect, is a major feature in species decline and, as human population grows and settlement expands, these pressures on other species will increase.

But back to humans, and back to this hazardous planet we share. Some hazards are of our own making: an automobile accident, a house fire, or a factory explosion, for example. And although they have potentially tragic consequences, they are not natural hazards. They are hazards we, and our patterns of life, create. And some global hazards—things like wars and even hunger and famine—are calamities for which we ourselves are also largely responsible, directly or indirectly. We have the means to prevent or alleviate them, however complex the issues involved, and however difficult the means of alleviation may be.

But many other hazards are outside our control: these are natural hazards—the things insurance companies describe as "acts of God." Of course, we can work to understand them and to minimize their destructive impact: we can build levees and lightning conductors, as every community does. We can adapt to them—by building on high ground, for example—and communities worldwide take these precautions, but we can rarely prevent or eliminate all risk.

These natural hazards range all the way from the inconvenient—a summer downpour that washes out a family picnic—to the catastrophic—the 7.6 intensity earthquake of 2005 in Pakistan that killed 80,000; the 7.9 earthquake in Sichuan, China, in 2008 that killed more than 69,000 people; or the January 12, 2010, earthquake in Haiti that left an estimated 230,000 dead, 300,000 injured, and

1,000,000 homeless. From the local—a landslide that blocks a highway—to the global—the tsunami of December 26, 2004, that killed 240,000 people, left hundreds of thousands homeless, and affected eleven countries. It is these natural hazards I want to discuss.

But even to discuss them is to recognize a paradox. These hazards all arise from the same Earth processes to which we owe our own existence and on which our daily survival depends. Consider volcanoes, for example. We tend to think of them as hazardous, and so they are. The explosion of Mount Saint Helens in 1980, after 123 years of inactivity, produced widespread local devastation and killed more than sixty people; the explosion of Mount Pelée in Martinique in 1902 killed 28,000. But volcanoes are our ancestors. We exist by courtesy of ancient volcanoes, which produced the atmosphere and oceans during Earth's early history. Some ore bodies and industrial minerals on which we depend result from the same deep-seated igneous activity that gives rise to volcanoes. Geothermal heat in some areas, such as Iceland, New Zealand, Japan, and the western United States, provides energy. The typhoon that ravages coastal Asia is part of a larger weather system that sustains the agriculture of the whole continent. The global warming that we battle and deplore is, in part, a continuation of a climatic trend that allowed our ancestors to emerge from the frozen tundra wastes of the last Ice Age. However threatening these natural disasters, they are part and parcel of the property we call Earth, created by the very forces that make Earth unique and by the ongoing processes that make Earth habitable. The action of plate tectonics, for example, gives rise to mountain ranges, which trigger rainfall, which results in crustal erosion, which in turn provides nutrients to farmland, forests, and oceans. And although we ourselves have contributed to the impact of some hazards, as in the case of global warming, we can neither prevent them nor even reduce them in most cases. What we can do is understand the hazards with which we have to live, adapt as best we can to them, and so minimize their destructive effects.

To some extent, we choose our own hazards. The particular threat with which we live depends on our zip code. There are no blizzards in Florida, for example, though there are hurricanes; there are no volcanoes in Kansas, though there are tornadoes. Hazards are not universal threats: they are local and regional, and, though we cannot yet predict their occurrence, we can identify the areas that are prone to various hazards based on the climatic and geologic characteristics of particular locations. And there are no hazard-free areas, no risk-less states. A few years ago *USA Today* published a risk map of the United States, with states marked one to eight on the basis of increasing risk of natural disasters. Nevada presented the lowest risks, the only state to receive a ranking of one. At the high end of the risk scale, Texas was ranked an eight; several midwestern states were ranked in category seven. And to emphasize the margin of error of such generalized rankings, Louisiana, the site of the subsequent Katrina catastrophe, was awarded a two. Hazard categories cannot meaningfully be lumped together: we have to study them one by one, weighing the causes and nature of the processes, assessing their impact, and developing ways to reduce their harmful effects.

Climate change will also change the nature of hazards. Global warming will bring benefits to some areas—longer growing seasons, milder winters, and so on—even as it brings hardships and hazards to others. New York City's Holland Tunnel entrance, for example, is only 9.5 feet above present sea level; La Guardia Airport is 6.8 feet above sea level.

There are four broad categories of natural hazard that threaten our species. They vary from place to place, they change over time—some are sudden, some prolonged, some are seasonal, others are not—and they vary with local climate, geology, and topographic conditions. But no spot on Earth is completely immune or hazard free. A meteorite is just as likely to land on the beach at Nantucket as on the top of Mount Rushmore.

Weather and Water Hazards

The first category of hazard is weather related. The atmosphere that sustains us is turbulent, and encompassing. Atmospheric disturbances are the most familiar and most widespread group of hazards. Weather is everywhere. Every place on Earth has weather of some sort, and it never quite suits us. The English have made a social art of discussing the imperfections of the weather. But there's weather and *weather*. A touch of frost is one thing; a blizzard is another. Temperature extremes—heat waves or ice storms—can cause severe hardship and have serious secondary effects. High winds and record snowfalls killed 154 people in New Jersey, Ohio, Pennsylvania, and West Virginia in a blizzard in January of 1996. A year earlier, Chicago recorded 465 deaths from a two-week heat wave, with daily temperatures of 100°F plus. Cities are more vulnerable to the effects of heat waves because they contain and reflect back heat.

High winds can be every bit as damaging as high temperatures. Tornadoes ("twisters") are funnel-shaped clouds that develop during spring and summer thunderstorms and move across the land with devastating effects. They are generally confined within a narrow north–south belt, stretching from Texas through the Great Plains to the Midwest.

Hurricanes and cyclones, as we have seen, are swirling, spinning storms that form over the tropical areas of the oceans—though not between 4° north and south of the equator—having wind speeds exceeding seventy-five miles per hour and bringing torrential rains. They consist of a spiraling vortex of moving air, which may have a diameter of fifty to five hundred miles, and have an "eye"—a quiet central area up to twenty miles in diameter.

Hurricane Katrina in August 2005 was one of the most serious of all natural disasters ever to strike the United States, leaving at least 1,836 people dead and thousands homeless, while causing estimated damage of $81 billion. (Figure 6.1.)

Hurricanes and cyclones are often associated with torrential rains, and these can lead to widespread flooding and water damage. Low pressure in the eye of a hurricane can also create a local rise or bulge in sea level, leading to a tsunami-like surge or wave of seawater. Because both winds and water can be so destructive, there is now a mountain of legal cases following Hurricane Katrina. Traditional home insurance covers wind damage but not flood damage. Untangling which is which, especially when property has been demolished, is a difficult task. But insurance is one positive force for responsible planning and zoning, by refusing coverage for property in flood-prone areas and so discouraging construction in areas of potential flooding.

Even in arid areas, heavy rainfall and rapid runoff can produce flash floods of devastating destructive power. River valley flooding arising from these storms can be immensely destructive, especially in the Upper Mississippi Valley.

Thunderstorms, ice storms, and even droughts each contribute their own distinctive threats to particular regions. Drought, for example, leads to crop loss, soil erosion, wildfires, and wildlife destruction. And storms at sea can have devastating consequences on land. The "perfect storm" of 1991 created waves that brought widespread destruction to many coastal areas in the northeastern United States.

Floods

Floods, though less spectacular and less newsworthy than earthquakes and volcanoes, can create widespread damage. Even in "typical" years, floods kill an average of 140 people in the United States and cause over $6 billion in losses. In the United States, floods are the most destructive of all natural disasters. Although improved warning systems and protective structures have greatly reduced the loss of life since such disasters as the Johnstown, Pennsylvania, flood that killed an estimated 2,200 people in 1889, population growth has created widespread river and coastal development and urbanization, which have tended to increase property losses.

Floods are created by rapid water flow that overspills established drainage channels. They can be produced by a variety of natural causes—torrential rainfall, coastal storm surges, and dam failure, for example—and, in unpopulated or sparsely populated areas, floods are often welcomed as bringing new sediment and new fertility to agricultural land. The earliest known agricultural settlements—the Tigris and Euphrates rivers of Mesopotamia and the Nile Delta—depended on annual floods to replenish the soil. But today, continuing population growth, new settlement, and new construction disrupt natural drainage patterns and add to the destructive power of floods. Floods involve the trade-off between the benefits of building new communities near the coast or rivers and the hazards of inundation. The pressure on coastal development and river sites comes chiefly from population growth, especially in densely populated countries of both the developed and the developing world. The danger from flooding near the coast is just as real as flooding from rivers, as the North Sea disasters of January 31, 1953, demonstrate. At that time, a combination of high winds and high spring tides produced a tidal surge, over ten feet high, which overwhelmed coastal areas in Holland and southern England. Some 2,142 people were reportedly killed on land, and more than two hundred others were lost at sea. A subsequent North Sea flood in 1963 involved winds of over seventy-eight miles per hour, creating a storm surge that overwhelmed coastal areas of Germany and surged up the River Elbe for sixty-two miles, as far as the city of Hamburg. In all, 315 people lost their lives and some sixty thousand homes were damaged.

The floods described here are no different in their effects from those created by hurricanes, monsoon rains, or tsunamis. All can carry destructive force, against which there can be only limited defense.

One traditional method of flood protection is by the construction of dikes and levees, but this frequently offers protection against the dangers of flooding in upstream communities at the expense of those further downstream. Early warning systems, disaster planning, and improved coastal defenses all offer increased protection against these hazards. The developed countries have designed expensive schemes—such as the Thames Barrier and the Dutch Delta Works, built after the North Sea floods of 1953 to protect major cities. Developing countries remain especially vulnerable to future flooding and damage. But no countries—whether developed or developing—are immune to the hazards of storms or flooding, and continuing population growth, increasing urbanization of coastal areas, and global temperature change all increase the risks of flood damage.

Water is humanity's silent benefactor but it can also be a significant threat. Many of the world's greatest cities are built near water—either a coastline, estuary, or major river. The benefits are obvious, transport and commerce among them. But the hazards are equally real. It is estimated that the Yellow River of China has been responsible for 6 million deaths in the past century and a half, either directly or indirectly, from flooding of farmlands and subsequent famine. It is often referred to as the "Mother River of China," but its lower course has also been called "China's Sorrow." Cyclone Nargis in 2008 brought heavy rains and a powerful storm surge to many low-lying areas of the Irrawaddy Delta of Myanmar (Burma), with the loss of at least 130,000 lives, as well as many more thousands missing. Water hazards come not only from the location of thousands of communities

6.1 Hurricane Katrina close to maximum intensity. Courtesy of NOAA.

on river flood plains but also from the potential failure of reservoirs and lakes in earthquake-prone areas.

While the location of existing cities obviously cannot be changed, responsible future development and the construction of protective devices are of great importance. So also are the siting and prudent construction of future reservoirs.

Other Water Hazards

Still other water and weather hazards are less dramatic and less perceptible, but no less threatening. Take climate change, for example. We have already mentioned the impact of short-term climate change over which we exert significant impact. Long-term change in Earth's temperature—over which we have little control—could produce either global warming or another ice age. The last Ice Age, for comparison, blanketed much of the Northern Hemisphere under ice sheets, thousands of feet thick. In contrast, continued global warming would lead to a significant rise in sea level, flooding huge coastal areas along the edges of all the continents, producing extreme weather patterns and modifying crop production. Sea levels rose by 450 feet (150 meters) from about 18000 BC to 5000 BC, for example. Further damage to the ozone layer by rising greenhouse gas emissions could lead to reduced protection from lethal ultraviolet radiation. These trends result from both natural cycles, which we cannot control, and from human behavior, which we can influence.

Some hazards may arise from isolated accidents, though their effects may be widespread. Accidental release of toxic materials, for example, or the failure of a nuclear power plant, or the use of a nuclear weapon may create devastating and widespread effects. Other hazards may result from longer-term practices and ways of life: deforestation, overfishing, industrial pollution, groundwater loss, and so on. Some of the negative effects of these may be worldwide, as we have seen with the depletion of the stratospheric ozone layer, for example. Others may be local. And some economic groups may suffer far more than others.

Just as the nature of hazards varies, so must the response. It is now possible, for example, to design structures to control flooding and construct earthquake-proof buildings. And the most vulnerable flood and earthquake regions of the world are well-known. So building codes can be devised to meet the problem, and in vulnerable areas of Europe, California, Japan, New Zealand, and elsewhere, these codes work well. But many of the world's most hazardous flood-prone areas and earthquake zones are inhabited by the world's poorest people. Iran, Afghanistan, Turkey, and Pakistan, for example, include both earthquake hazards and areas of crushing poverty; in those places the implementation of such codes may be unrealistic.

Surface Hazards

A second category of hazards results from minor changes and adjustments in the Earth's surface rocks. The adjustments may be responses to the effects of surface events, such as weather or wave erosion, or they may be triggered by sudden or gradual subsurface events, such as movement on a fault or earth tremor or underground collapse into a cave or old mine workings. Landslides and mudflows, for example, develop from the downslope movement of rocks and soil, often on steep hillsides or cliffs, after heavy rainfall that lubricates their slippage. Many are very local: "washouts" of gravel roads are familiar to all farmers. Others can be regional and can have devastating effects. The landslide above the Vajont Dam in Italy in 1963 following heavy rains led to the deaths of over 2,500 people.

Landslides may also be regional in extent. A few prehistoric slides are known that cover areas of scores of square miles and involve rock debris more than seven hundred feet deep. Not all landslides move rapidly. Some creep so slowly that, while damaging buildings, they represent no hazard to life. Joint faces parallel to hillsides, or steeply inclined, soft yielding clays and shales, as well as triggering events—such as earthquakes or heavy rains or rock excavation, for example—may lead to sudden slippage. Large landslides and slumps on the ocean floor can trigger tsunamis.

Some landslide and flood hazards arise, not from nature, but from the pattern of human use of the Earth. The Vajont disaster is one of the most terrible of these.

6.2 The Vajont Dam, site of the 1963 disaster. Photo by Emanuele Paolini. Reprinted with permission.

Longarone is one of a cluster of peaceful villages, high in the Italian Dolomites, about sixty-two miles north of Venice. In the late 1950s a great dam, 950 feet high, was constructed in the narrow V-shaped gorge of the Vajont Valley above the village, in order to provide electricity to the growing cities of northern Italy, including Milan, Turin, and Modena. It was celebrated as one of the highest dams in the world. (Figure 6.2.)

As filling of the reservoir took place, small, "creeping" displacement of the nearby rock walls occurred, as well as minor landslides and cracks. The reservoir was lowered and slowly filled under carefully controlled and monitored conditions to determine the safe water level at which this sliding could be prevented. This was repeated three times, but on the night of October 9, 1963, disaster struck. In spite of all the precautions, a huge landslide of rock debris, representing almost a third of the southern wall of the reservoir, sped down the steep valley wall at over sixty miles per hour and crashed into the water below. The effect was to produce a terrestrial tsunami. The displaced water, estimated to have a volume of about a quarter of the capacity of the reservoir, formed a giant wave, two hundred feet high, which spilled over the top of the dam and crashed on the villages below. More than 2,500 people perished on that terrible night.

How could this have possibly happened? Analysis, debate, and inquiries followed. The dam itself remained largely intact, but about one-quarter of the water it supported had splashed over it, displaced by the landslide, and had engulfed the villages of Longarone, Pigaro, Villanova, and Rivalba below. There were suspicions that the company owning the dam, anxious to sell it to the newly nationalized state utility, had ignored or underestimated the significance of the early warning signs. Or perhaps their geological survey or engineering designs were inadequate. Or perhaps, in spite of conscientious plans and rigorous analysis, the terrible failure was not predictable. These were the questions that confronted the official inquiries. Those inquiries ultimately proved that site surveys had failed to detect thin layers of clay within the limestone of the valley walls. It was along these surface layers of clay that the catastrophic landslide had taken place. Before the filling of the reservoir, the rocks remained stable, as groundwater moved naturally within them. After the reservoir was filled, the weight of the water and the changed drainage pattern and flow of groundwater combined to create the landslide.

There is still some dispute as to exactly how that occurred. Although it is clear that slipping took place in very thin clay layers interbedded within the limestone, it is not clear whether the landslide was a reactivation of an older one. It seems probable, however, that the rising water level of the reservoir increased pore pressure in the clays, reducing their strength. Water, flowing in joints in the limestone, lubricated the sheer surfaces along which catastrophic failure took place.

There is, perhaps, in the tragedy of Vajont, a lesson applicable to all such giant engineering undertakings. The geological setting of the valley was carefully mapped and well understood. There was evidence of an ancient landslide on the northern wall of the valley, and it was recognized that a downfolded synclinal structure of Jurassic–Cretaceous limestone underlay the reservoir. This synclinal structure was seen as a potential weakness, but early test borings and seismic studies suggested that a major landslide was unlikely. It was only later that a large area of the left (southern) bank was found to be unstable. The designers concluded that they could not prevent a landslide of this vast area by traditional means—sealing the surface or injecting the rock with cement, for example—nor could they afford to trigger the slide in an attempt to control it. They concluded that they would attempt instead to control the slide movement by controlling the water level of the reservoir and by constructing drainage tunnels to relieve water pressure in the rock walls. At first this seemed to limit the rate of downhill movement, but by the time of the third lowering of the reservoir in late September and early October, slippage rates of more than seven inches a day were recorded. At 22:35 GMT on October 9, 1963, the landslide moved with catastrophic results, plunging into the reservoir and surging 460 feet up the opposite wall of the valley, producing a wave that destroyed the village of Casso.

It is easy later to look back at this terrible event and, with 20/20 hindsight, to allocate blame. But even with responsible precautions and effective controls our knowledge is never complete, our science never exact, our engineering designs never perfect. The geologists and engineers at Vajont understood foundations, analyzed the risks, evaluated the problems, revised the design, and sought to minimize the hazards. But they were wrong. The knowledge and analysis were inadequate.

And any enterprise involves some assessment of the degree of hazard. To commit an unlimited budget to risk assessment, to delay until every conceivable test has been completed may mean that a project is never undertaken. In some cases, that may be a proper outcome. But all major projects of this kind are presumably undertaken because they promise some public benefit. To cancel projects because of lack of certainty, or to delay them to the point that they become unproductive or unprofitable, also carries some significant cost in loss of public benefit.

Life itself is not a risk-free activity, and neither our understanding of the details of the context nor our prediction of the impact of our work will ever remove all risk. The responsible and prudent consideration of risk assessment, with all its imperfections, is the most that we can ask.

Other forms of ground movement reflect the subsidence of surface rocks over underground caves, formed by groundwater in limestones or by the collapse of surface rocks over old mine workings or areas of brine pumping. These are rarely destructive to human life, though they can create extensive structural damage. Some of these "sinkholes" that result from the collapse of surface rocks are filled with water; others are dry. They are often associated with caves and may range from a few feet to a few hundred feet in diameter. They can appear suddenly, causing damage to highways and buildings, or gradually, as in downhill movement of soil and rock that takes place almost imperceptibly over time. On steep slopes in areas of heavy rainfall, soil creep and small slumps may cause structural damage.

Although some subsidence is natural in origin, much is accelerated by engineering construction or mining activity. Venice is subsiding, for example, largely because the city is built on a foundation of soft strata that are slowly compacting, and the subsidence is aggravated by groundwater pumping. Restrictions on pumping have reduced the rate of subsidence. The central valleys of California and Mexico City are also subsiding because of excessive extraction of groundwater. Mexico City has sunk some twenty-five feet.

All these movements represent detailed aspects of the more general weathering and erosion of crustal rocks. Beach destruction and coastal erosion, for example, exist the world over, though they are rarely threatening to life. Erosion of hillsides goes on everywhere, but the most destructive surface movements represent the results of ground failure under man-made structures built under conditions of geological instability. The 1966 Aberfan disaster in Wales, for example, involved the collapse of a waste tip, or pile of rock waste removed from an adjacent coal mine. This "slag heap"—a mountain of shale, sodden and lubricated after a prolonged period of heavy rains—suffered a minor rotational slippage, collapsed, and slid down as a gigantic slurry of rock, dust and water into the village below, engulfing an elementary school and killing 144 people—116 of them children, most between the ages of 7 and 10. A subsequent inquiry found that the huge slag heap had been built over a known stream, down which the "liquefied" debris flowed into the valley. The owners of the mine, the National Coal Board, were found responsible for "bungling ineptitude" and a failure of communication. After the disaster, legislation established tough new requirements to ensure the safety of mine tips. Other surface movements, though less devastating, can still be serious. Houses and other structures, built on unstable hillsides, can develop foundation cracks and even collapse after heavy rainfall or floods.

Prevention of these disasters, and others like them, involves building location and design codes to recognize the potential hazards of certain sites and conditions. In the case of major structures—dams, bridges, mine workings, and the like—detailed site mapping and hazard analysis are critical to avoiding disasters.

Earthquakes, Tsunamis, and Volcanoes

The third category of hazard is created not by the surface weather and water or movements of superficial rocks and sediments but by deep-seated movements of Earth's crust. Earthquakes, volcanoes, and tsunamis, though immensely destructive on the surface of the Earth, are not triggered at the surface, but by forces acting miles below it. These are widespread in their effects, devastating in their destructive power, and significantly less predictable than any of the hazards associated with weather. They can occur at any season of the year and, though they develop in relatively well-defined zones, their effects can be widespread. Earthquakes may produce destructive tsunamis. The December 26, 2004, tsunami, originating in a localized 9.3 magnitude earthquake off the coast of Sumatra, killed an estimated 240,000 people and created havoc over such a large area that its waves caused loss of life on the coast of Africa. The eruption of Krakatau (Krakatoa) in 1883 in Indonesia killed 36,000 people.

Most earthquakes arise from the release of stress that builds up along the margins of the great crustal plates that pave the surface of the Earth (chap. 3). These plates are in slow but constant motion against each other at a rate of an inch or two a year. They may move apart—as at the mid-ocean ridges, where lavas spill out from fissures, such as those in Iceland—or they may converge, sliding over and under one another, producing volcanic islands, mountains, deep sea trenches, and earthquakes, as in Japan and Indonesia. Or they may move laterally against each other, as in the San Andreas Fault, "sideswiping" and "grinding" and sometimes locking as they go. Earthquakes are brief, sudden events, lasting only a minute or two, but the build up of stress from this "locking," whose sudden release creates earthquakes, may represent decades or centuries of surface inactivity.

Some earthquakes have been so destructive as to become etched in the broader pattern of history. Let me give two examples. Lisbon, Portugal, in the mid-eighteenth century was one of Europe's most admired capital cities, prosperous from its favored trading position at the mouth of the River Tagus, cosmopolitan from its lingering Moorish influence and bustling commercial activities, beautiful with its splendid palaces and the elegant homes of its wealthy merchants, and devout, with innumerable cathedrals, churches, monasteries, and convents. And then, at 9:40 a.m. on All Saints Day, Saturday, November 1, 1755, disaster struck. An earthquake of estimated magnitude of 8.7 to 9.0 devastated the ancient city. The first shock was followed by a powerful second shock, lasting two minutes, creating widespread destruction, to be followed by a third damaging shock. Even as the dust began to settle, fires broke out across the city, followed in turn by a tsunami, creating three successive towering waves that engulfed those fleeing toward the docks and waterfront to escape the horrors of the earthquake on land.

Early reports estimated that 10,000 to 15,000 people died, though later studies put the deaths at up to 90,000 out of a population of 275,000. There was also further loss of life from the tsunami, perhaps as many as 10,000 in Morocco. Fire destroyed far more than the buildings: manuscripts, libraries, countless works of art, jewelry, huge stocks of merchandise, and many maps, charts, and records of medieval Portuguese voyages were consumed.

But the loss of life and property—great as they were—were not the only effects of the Lisbon earthquake. Lisbon in the mid-eighteenth century was the embodiment of a proud and confident, outward-looking, intellectually engaged European civilization. The Lisbon earthquake shattered that confidence and destroyed the stability of an age described by the historian Basil Willey as "the nearest approach to earthly felicity ever known to man."

It was an age of exploration, scientific discovery, new learning, and growing prosperity, but, in a few brief minutes, it came to an end. The great Lisbon earthquake, preachers declared, was a judgment on the vanity of man, an act of divine retribution for human arrogance and pride. Perhaps, some argued, the world itself was coming to an end.

In the midst of the catastrophe, Lisbon was fortunate in one thing: it was led by an able prime minister—Sebastião de Melo, the Marquis of Pombal—who, with the support of an enthusiastic monarch, Joseph I, took charge of the task of reconstruction, caring for the survivors, dealing with public health, maintaining public order, burying the dead, clearing the debris, planning for the rebuilding, supervising new construction, and even directing a study of the underlying earthquake itself. He proved an exemplary leader.

Voltaire wrote a poem on the disaster ("Poéme sur le désastre de Lisbonne") and the satire *Candide,* in which he ridiculed the philosophy of "the best of all possible worlds." Immanuel Kant published three commentaries on the Lisbon earthquake, one of which is widely regarded as the pioneer writing on seismology.

If the Lisbon earthquake marked the end of an era of confidence and contentment, it also ushered in a new one: of instability, uncertainty, and of the quest for natural explanations.

The geological mechanism that produced the Lisbon earthquake is still debated. Most estimates suggest that the earthquake had a magnitude of perhaps 8.75, that it was centered offshore, some 120 miles south-southwest of Lisbon, and that it may represent the development of an Atlantic subduction zone.

San Francisco in 1906 was the largest city on the West Coast, the financial, commercial and cultural hub of the region, the "Paris of the west," the "gateway to the Pacific." But the shock that struck San Francisco at 5:12 a.m. on April 18, 1906, a century and a half after Lisbon's terrible earthquake, was every bit as destructive. It was one of the worst natural disasters in U.S. history. An earthquake of magnitude 7.8 triggered movement along a northern segment of the San Andreas Fault, a zone of weakness that extends eight hundred miles from the Salton Sea to Lake Mendocino. More than three thousand—perhaps as many as six thousand—died in the disaster, trapped under burning and collapsing buildings. More than half the city's population was left homeless. Fires raged unchecked for three days, as broken water mains left firefighters helpless, able only to dynamite neighboring buildings as a method of containing the blaze. Twenty-eight thousand buildings, over more than five hundred city blocks, were destroyed, covering an area of some 4.74 square miles, more than six times the area devastated by the Great Fire of London.

There were many heroes in San Francisco on that terrible day and in the days of turmoil that followed. But one hero was Andrew Lawson, a geologist from the University of California, Berkeley, who assembled a team to study the cause and effects of the earthquake, and who concluded that it represented the sudden release of pent-up forces along some three hundred miles of the eight-hundred-mile-long fault that he named the San Andreas Fault, and a group of related faults running parallel to the California coastline. The meticulous studies by Lawson and his colleagues laid the basis for the modern science of seismology. We now know that these forces are built up by small but continuous movements of two of the Earth's plates, sliding past one another at the rate of a couple of inches a year.

The stresses created by this movement gradually accumulate, and then snap every couple hundred years or so, to produce a major earthquake.

It is that finding that makes this area of California one of the most studied and monitored areas of the world. For other major earthquakes will surely come: the only question is "When?" For the 7 million residents of the Bay Area, that is a question of surpassing importance.

But to most people, the tragedies of Lisbon and San Francisco, terrible as they were, seem scarcely threatening. To those living in most parts of the world, earthquakes seem a remote hazard: something that hit San Francisco a century ago and that may create occasional problems in another part of the world—in Indonesia, say, or perhaps Japan—but events of only marginal personal concern. Meanwhile, earthquakes take their toll. The U.S. Geological Survey reports that since 1900 there have been 116 major earthquakes in the world, each resulting in more than one thousand deaths. In the year 2008, in the United States alone, there were 3,618 earthquakes recorded, of which 2,011 had a magnitude of 3.0 or higher. It is estimated that several million earthquakes occur throughout the world in any given year, many of them small and undetected in remote areas. About 20,000–30,000 earthquakes worldwide are recorded each year by the USGS National Earthquake Information Center.

These earthquakes are not randomly distributed but are concentrated at the junctions of the great crustal plates that form Earth's surface. Deeper earthquakes, with a focus at 62–435 miles' depth, are found along a belt that encircles the Pacific, running along the Andes of South America, the Antilles of the Caribbean, the Aleutian Arc, the Sea of Japan, the Indonesian islands, and the Vanuatu (formerly New Hebrides) Chain. These are areas also marked by deep ocean trenches, where crustal material is subducted, swept downward into Earth's interior. Shallow earthquakes are more widely distributed in these areas and in south Asia, southeastern Europe, and along the length of the mid-ocean ridges. (Fig. 6.3.)

Although earthquake magnitude is typically measured on the Richter scale, the degree of damage caused by earthquakes has also been calibrated and expressed as the modified Mercalli intensity scale (table 6.4). The effects of an earthquake decrease from its epicenter, so the Mercalli intensity varies from place to place.

Design of earthquake-"proof" buildings and choice of appropriate foundations and substrates are crucial to limiting earthquake damage. Buildings constructed of strong, reinforced floor slabs and weak supporting walls are especially damage prone. Wooden-frame buildings resist earthquake shocks, but adobe and brick buildings are vulnerable to damage. Bedrock is the best foundation for buildings in earthquake zones. Clay, silt, thick soil, gravel, and fill tend to amplify shock waves and, in some cases, can even liquefy.

The key to earthquake protection is strict building codes and regional planning that prevents building in areas where the underlying geology and surface

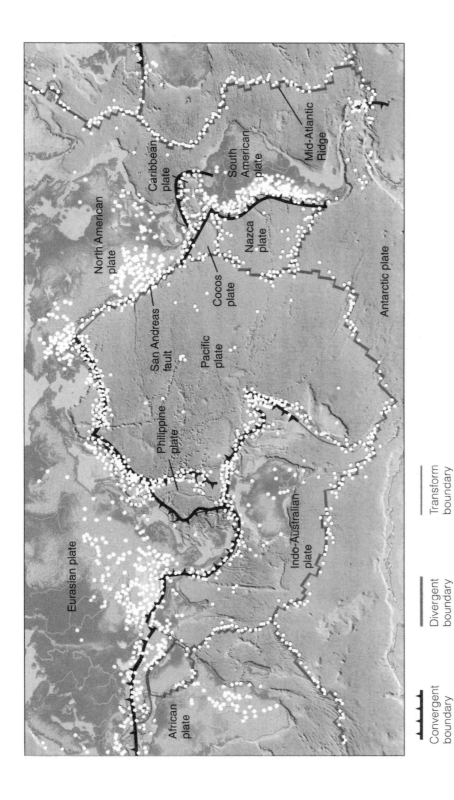

Eurasian plate

African plate

North American plate

Caribbean plate

San Andreas fault

Philippine plate

Pacific plate

Cocos plate

Nazca plate

South American plate

Mid-Atlantic Ridge

Indo-Australian plate

Antarctic plate

Convergent boundary

Divergent boundary

Transform boundary

6.3 Earthquake distribution: the relationship between earthquake epicenters and plate boundaries. From Reed Wicander and James S. Monroe, *Essentials of Geology (with GeologyNOW)*, 4th ed., copyright 2006 Brooks/Cole, a part of Cengage Learning, Inc. Reproduced with permission, www.cengage.com/permissions. Data from NOAA.

6.4 Modified Mercalli intensity scale and Richter scale

	Modified Mercalli Scale	Richter Scale
I.	Felt by almost no one.	2.5 Generally not felt but recorded on seismometers.
II.	Felt by very few people.	
III.	Tremor noticed by many, but they often do not realize it is an earthquake.	3.5 Felt by many people.
IV.	Felt indoors by many. Feels like a truck has struck the building.	
V.	Felt by nearly everyone; many people awakened. Swaying trees and poles may be observed.	
VI.	Felt by all; many people run outdoors. Furniture moved; slight damage occurs.	4.5 Some local damage may occur.
VII.	Everyone runs outdoors. Poorly built structures considerably damaged; slight damage elsewhere.	
VIII.	Specially designed structures damaged slightly; others collapse.	6.0 A destructive earthquake.
IX.	All buildings considerably damaged, many shift off foundations. Noticeable cracks in ground.	
X.	Many structures destroyed. Ground is badly cracked.	7.0 A major earthquake.
XI.	Almost all structures fall. Very wide cracks in the ground.	8.0 and up: Great earthquake.
XII.	Total destruction. Waves seen on ground surfaces; objects are tumbled and tossed.	

Source: North Carolina Geological Survey.

topography present hazards. California and Japan, for example, have just such codes.

Can earthquakes be predicted? Earthquake prediction is still in its infancy, though there have been encouraging experiments in fluid injection along fault zones to ease the "locking" of plates by allowing small accommodating movements. Although hazardous earthquake areas have been carefully mapped, the best predictions involve centuries rather than months or days. The best-known example of successful prediction involved the 1975 7.3 magnitude earthquake in Haicheng, China. During the months before the earthquake a series of minor earthquakes ("foreshocks"), changes in land elevation and standing water levels, and the unusual behavior of animals led to a general alert, which was upgraded to an evacuation order when the frequency of foreshocks showed a marked increase. Such careful documentation and prediction have not proved possible for most earthquakes. In spite of the successful Haicheng prediction, an earthquake of magnitude 7.6 killed an estimated 250,000 people in Tangshan only a year later.

It may never be possible to forecast the location, magnitude, and timing of large earthquakes. None of the traditional methods of prediction—changes in groundwater levels, animal behavior, foreshocks, changes in ground level, radon levels, weather conditions, changes in electromagnetic fields—have provided consistently reliable predictors, though satellite observations showed a sharp increase in ionospheric density and temperature shortly before a Japanese earthquake in 2004. The most promising efforts at present involve close monitoring of known earthquake areas and the development of maps showing earthquake possibilities in given regions. These, combined with stringent new building codes, reinforcement of existing buildings, and regional zoning restrictions of new construction offer the best, if limited, protection against earthquake hazards.

With or without better warning systems, there remains a critical need for improved public information, disaster training, evacuation, rescue and relief planning, and the infrastructure to support all these operations.

Tsunamis, volcanoes, and earthquakes are intimately related. Tsunamis are all too familiar to us, following the disaster on December 26, 2004, when an earthquake of magnitude 9, located about 150 miles west of Sumatra created shock waves that roiled the waters of the Indian Ocean, killing an estimated 240,000 people in eleven different countries and rendering millions homeless. The loss of life extended to the coast of Africa. Tsunamis are giant ocean waves, up to 150 feet high, triggered by earthquakes, volcanic eruptions, or submarine slides that displace the seafloor. The waves travel at a speed of up to four hundred miles per hour in the open ocean and hit coastal areas with devastating force. They arise chiefly in the great earthquake and volcano areas around the Pacific. The eruption of Krakatau in 1883, for example, was associated with tsunamis that killed over 36,000 people, and was felt and heard as far away as Cape Horn. The explosion was 13,000 times more powerful than that of the atomic bomb that destroyed Hiroshima, and the gases and particles released affected weather patterns and atmospheric conditions around the world for several years. Edvard Munch's famous painting *The Scream* was described by him as depicting the "blood-red" sky over Norway in 1883.

Tsunamis can cause destruction far from the site of the earthquake or volcanic eruption that produces them. An earthquake in the Aleutians can bring (and has brought) devastation to Hawaii, two thousand miles away. Tsunamis tend to be less dangerous on steep shores than on flat, and they can be most destructive when they strike narrow inlets or harbors.

Tsunamis are sometimes called "tidal waves" but they have nothing to do with the tides. The name *tsunami* is Japanese, reflecting the widespread destruction these storms have created in Japan over the years. Although tsunami waves travel widely—over hundreds or thousands of miles in all directions—and rapidly (up to 400 mph), they are not destructive in open ocean waters, where they reach only a few feet in height. Once they approach the shore, however, shallowing water causes the base of the wave to drag on the sea bottom and its height may build up to thirty or forty feet before it crashes on the shore. There waves come in series, and often the third to the eighth are the most destructive. The first wave may often be gentle, followed by a "draining away" of the sea, as the trough of the wave approaches, giving a false sense of security to those in the area.

Tsunamis cannot be prevented, but they can be predicted. Earthquake recordings, with precise locations and measurements of direction and intensity, together with surface motion detectors, can be used to give limited warning for evacuation, and recording ocean buoys and markers have now been placed across the world's oceans.

Volcanoes are intimately linked to the same mechanisms deep within the Earth that cause earthquakes and tsunamis, but they bring both benefit and ruin to humankind. As we have seen, they gave our planet its atmosphere and oceans, and they produce the most fertile soil and plentiful rainfall. In places, they provide geothermal energy. The forces that produce them are also the sources of some of our most important mineral deposits. They form some of the world's most splendid scenery—from Mount Etna, to Mount Kilimanjaro, to Mount Fuji, to Mount Erebus—and some of its most magical tropical islands. But their benefits come at a price. They can bring destruction to those who dwell under their shadow. About 550 volcanoes erupted on land during historic times. About 1,500 have erupted in the last ten thousand years. In any year, sixty or so are active, as well as many others beneath the oceans. The United States is third after Indonesia and Japan in the number of volcanoes: fifty have erupted in historical times.

Not all volcanoes are equally hazardous. Volcanoes of the open oceans tend to emit basaltic lavas and have limited explosive activity. These submarine volcanoes exist in great numbers. One area of the east Pacific discovered in 1993 showed 153 volcanoes and almost a thousand seamounts in an area about the size of New York State. In contrast, those of the continental margins and shallow seas tend to have silica- and gas-rich lavas, and tend to be explosive. These are the volcanoes associated with converging, destructive plate margins.

Basaltic volcanoes—those of Hawaii and Iceland, for example—though spectacular tourist attractions, are rarely threatening. When eruptions do take place, they tend to be nonexplosive, with lava advancing at a steady but slow pace, channeled in part by the local topography. The flow has been controlled in some cases by spraying water on the front and by diverting it from vulnerable areas. Such eruptions, though quiet, can, however, produce extensive amounts of volcanic ash, which can be carried far from its source. Disruption of transatlantic air travel in 2010 represented just some eruptions.

Volcanoes of the plate margins are another matter. Their lava is acidic, viscous, gas-filled, mixed with pyroclastic bombs and other glowing fragments, and their activity is explosive and unpredictable. They frequently produce clouds of incandescent gas (*nuées ardentes*) that speed downhill from their craters, and they often produce explosions, tsunamis, ashfalls, and lahars (huge mudflows produced by torrential rain, melting ice, or lake waters, mixed with soft unconsolidated volcanic ash and breccia). Each of these brings its own distinctive threat. It was clouds of incandescent gas, filled with red-hot ash, moving at perhaps sixty miles an hour, that destroyed Pompeii in 79 BC and that overwhelmed Saint-Pierre with the loss of 29,000 lives in 1902. It was a volcanic explosion in 1883 that destroyed much of the island of Krakatau.

Prediction of volcanic eruptions has shown some success, in part because they can be monitored more readily than extended earthquake zones. Shield volcanoes, such as those of Hawaii, are monitored for lava activity, ground tilt, seismic movements, gas analysis, and changes in gravity and magnetic fields. These methods have been particularly effective in monitoring the activity of Kilauea, perhaps the most closely studied volcano in the world.

Predictions of volcanoes at plate margins are more difficult. Their eruptions, though generally infrequent, are not yet predictable. All the monitoring efforts described above are used, together with detailed records of each volcano's "character" and history. Satellite imagery of thermal increases in volcanic vents and records of ground swelling provide what may be useful predictive data, as do changes in low-energy seismic activity in the area of the volcano.

In the world's wealthier volcanic areas—North America, Japan, and New Zealand, for example—such monitoring will continue to improve and will ultimately provide some degree of early warning of impending volcanic activity. But most of the Earth's active volcanoes lie in the developing world, where comparative poverty encourages people to settle on and around them because of the rich soils and plentiful rainfall they provide. In these areas, volcano monitoring is sparse; and even when it exists, communications are poor. Nyiragongo, for example, is one of several volcanoes in the African Great Rift Valley of eastern Congo that frequently erupt lava, flowing relatively quietly over long distances. In 2002 monitoring devices on the mountain showed sharply increased seismic activity for five days before an eruption, but no warning was issued and forty-five people perished, attempting to rescue their belongings from the encroaching lava flow.

Most of the millions of people living in hazardous volcanic areas are equally vulnerable.

Extraterrestrial Hazards

The fourth category of hazards cannot be blamed on our planet: some threats are extraterrestrial. Meteorites are fragments of asteroids and comets, dating from the origin of the solar system, that intersect Earth's orbit and—surviving their fiery passage through the atmosphere—fall on the Earth's surface. They range from fine dust to more than a mile in diameter. Most meteors are small, rare, and harmless. But it is widely believed that a giant meteorite or asteroid that fell 65 million years ago was a major factor in the extinction of the dinosaurs and other groups, perhaps by generating a vast dust cloud that filtered out sunlight for an extended period.

It has been estimated that Earth receives an average of one "Hiroshima-sized" impact every year, but we are usually unaware of these because the explosions generally occur at high altitude or the bodies fall in the oceans or uninhabited lands. The Tunguska event in 1908 in Siberia, which produced widespread devastation over an area of some 830 square miles, is thought to have been caused by the airburst of a comet or asteroid three to six miles above Earth's surface.

Living with Hazards

All the hazards we have reviewed—climatic, surface, deep-seated tectonic, and extraterrestrial—will continue to impact our planet. We cannot plug a volcano, freeze a tsunami, or deflect an asteroid. At least not yet. What we have to do is adapt, and that means understanding the hazards, predicting them as early as we can, and then modifying and adjusting our structures, settlements, and behavior to minimize their destructive impact. Some of those adjustments are relatively easy. Others—at least those associated with global climate change—are far more difficult.

CHAPTER 7

The Ancient Planet

Perhaps our forebears always had an intuitive sense that they inhabited a place of great antiquity. All civilizations seem to have had an interest in the beginnings of things, and each had its own account of origins. The ancient Brahmans believed the Earth to be eternal. So, it seems, did Aristotle, who thought it inappropriate to consider Earth to be just as much an aging creature as humans are. Others proposed a definite age, and as early as 450 BC Herodotus had suggested that the slow rate at which the waters of the Nile deposited the sediment that formed its delta, indicated that its building must have required many thousands of years. Babylonian astrologers suggested that humans appeared half a million years ago.

Not all lines of evidence suggested such great antiquity. Archbishop James Ussher of Armagh, primate of All Ireland, always gets bad press in this regard. Let me explain why. By the seventeenth century there was growing speculation concerning the age of the Earth and evidence was sought from whatever source might be fruitful. The great astronomer Johannes Kepler (1571–1630) combined the study of science—using cycles of solar eclipses to date the darkness of the crucifixion—and the Genesis narrative of creation to calculate a date of 3993 BC for the origin of the Earth. Sir Isaac Newton, perhaps the greatest scientist of all time, refined that to 3998 BC. It is here that Ussher took up his calculation. Ussher was not only primate of all Ireland, but also provost of Trinity College,

Dublin, and has been described as one of the greatest Hebrew scholars of his day. His extensive library formed the core of the great library of Trinity College. In his *Annals of the World*, published in 1650, he suggested a date for the origin of the Earth as "the entrance of the night preceding the twenty third day of October, in the year of the Julian Calendar, 710." This was, he said, the best date that could be deduced from all the substantial array of ancient documents he had collected. The year 710 of the Julian calendar is equivalent to 4004 BC. The indulgence with which Ussher's calculation is now treated overlooks the serious scholarship behind his work. Ussher and Kepler based their chronology on a combination of historical and scientific records—surely a not unreasonable approach—but Ussher also made one major assumption: that the age of the Earth was the same as the age of humanity. That assumption, we now know, was wrong, but the very concept of time and the experience of history are profoundly influenced by one's view of life.

Two experiences seem always to have been intertwined: first, the sense of life's cyclicity, of day and night, of sleeping and waking, of monthly cycles, of phases of the moon, of movements of the planets, of seedtime and harvest, of spring and fall and the regularity of the seasons. Perhaps, then, the Earth itself, perhaps life itself, was cyclical, seasonal, changing, yet unchanging. That was the Buddhist and the Hindu view. It was the view of Plato, who suggested a great cycle of 72,000 years, first of advance, then of decline. It was the view far more recently of a group of eminent astronomers—Fred Hoyle, Hermann Bondi, and Thomas Gold—who in 1948 and later suggested a cosmic pattern of "continuous creation."

But against that view was the sharp reality of birth and death: of childhood, youth, maturity, and age; of parents, children, family, and friends, whose lives showed, not cyclicity, but directionality, linearity, from a beginning to an end. It was this view that was embedded in the great Abrahamic faiths of Judaism, Christianity, and Islam, based as they are on the history of divine creation and providential purpose. And that's why, one supposes, the first serious attempts to construct a chronology of Earth came from those who were not only well skilled in science but also well versed in the scriptures. Johannes Kepler and Sir Isaac Newton were among the most respected biblical scholars of their day, and are also counted among the greatest scientists of all time.

But no calculation based only on human history could calibrate the span of prehuman history, though even the Genesis narrative bore witness to such a period. Increasingly, the more one looked at Earth itself and the processes that shaped its surface, the more those observations implied its antiquity.

For some, perhaps, it might have been enough to learn that we live on a very old piece of property: older than we are, older than our parents. But for most, that's not enough. The questions "How old?" or "Just how ancient?" soon emerge. After all, to live in a house built in the 1930s may involve a rather different pattern of life from living, say, in a Victorian home, or a colonial villa. In most things, age matters; whether in cheese or in wine, age makes a difference.

So how can one possibly hope to measure the age of the Earth? How is it possible even to begin to calculate the age of the rocks beneath our feet that make up this planetary home? Well, we might perhaps find a way to identify Earth's oldest rocks and somehow analyze them. But even here there's a snag. We don't have Earth's most ancient rocks: they are nowhere to be found. The oldest rocks available to us on the Earth's surface are not the same as Earth's oldest rocks. The oldest rocks we can find are themselves the metamorphosed remains or products of still older rocks and processes. Earth's earliest rocks have long ago been resorbed and recycled within the Earth. Earth must be older, appreciably older, than its oldest observable rocks.

Curiously enough, it was by looking at some of Earth's younger rocks that some clues began to emerge. Observers and collectors over the centuries had been attracted to the striking fossils that these rocks contained. Some saw these fossils as curiosities or "sports of nature": inorganic objects that just happened to resemble living things. A few saw them as creations of the devil, intended to mislead simple souls into supposing them to be evidence of former living creatures. And others saw them for what they ultimately turned out to be: remains of creatures, long dead, buried, and thus preserved in the sediment that now entombed them: "medals of creation" as some early observers called them. Long before Archbishop Ussher made his calculations, Herodotus (484–425 BC), for example, and Aristotle (384–322 BC) had recognized fossils as the remains of once living things. Herodotus even went on to demonstrate that the presence of fossil marine mollusks in the rocks of Egypt implied an earlier submergence of the land by the sea. So land and sea were less permanent than they seemed.

It took centuries for that view to gain acceptance. Not until the great revival of curiosity and learning of the fifteenth century did others, including Leonardo da Vinci (1452–1519), reach the same conclusion. Centuries later, using the layering and order of superposition of bedded rocks, an English land surveyor, William Smith, demonstrated that fossils could be useful in identifying and correlating stratified rocks. By the mid-nineteenth century the fossiliferous strata of Europe had been grouped into systems based on their order of succession and classified by their relative age. The publication of Charles Darwin's *On the Origin of Species* in 1859 rekindled serious interest in fossils as a record and calendar of ancient life. Meticulous studies of the succession of fossiliferous rocks, especially in western Europe, began to give a broad outline of the history of life. And

Darwin's theory of gradual evolution, if true, required time—great time—to account for life's history. (Figure 7.1.)

But to reconstruct life's history was one thing. To date it was another. Certainly, the sediments that made up these rocks must have accumulated relatively slowly. Darwin estimated the total thickness of fossiliferous rocks—the so-called Phanerozoic—in Britain to be some 13¾ miles. But two problems emerged in attempting to turn thickness into years. How rapidly had the sediments accumulated? And had some earlier ones not, perhaps, been lost by erosion, or even, perhaps, redeposited as new sediment? In spite of efforts to estimate an average rate of deposition, most concluded that sedimentary depositional rates were highly variable from place to place and that any estimates of age based on thickness of sedimentary rocks were likely to be both low and unreliable. Darwin's suggestion of 13¾ miles in thickness, for example, for the Phanerozoic rocks, has been increased by recent observers to about eighty-six miles. Even though this may be a more accurate estimate, an "average rate of sedimentation" still seems a meaningless figure when measured rates can vary a hundredfold from one environment to another.

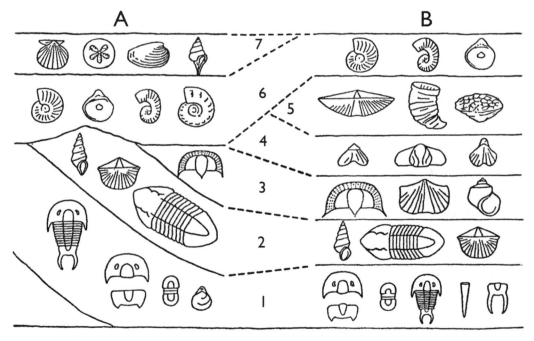

7.1 The use of fossils in rock correlation. The diagrammatic sections A and B show rock successions at two localities, 200 miles apart. From Frank Rhodes, *Evolution of Life* (New York: Penguin 1962). After R. C. Moore, C. G. Lalicker, and A. G. Fischer, *Invertebrate Fossils* (New York: McGraw Hill, 1952). Used with permission.

So, great antiquity could be established. Precise age could not.

Similar limitations arose from another method that was attempted. Sir Edmund Halley, the astronomer who discovered Halley's comet, among other things, argued that if the oceans were originally freshwater and the salt in the present oceans came from the erosion of salt in the rocks of the Earth's crust, the oceans could be neither very young nor infinitely old. If very young, they would be much less salty; if very old, they would resemble the Red Sea in its high salinity. John Joly, professor of geology at Trinity College Dublin, attempted in the 1890s to quantify this, and calculated an age of 80–90 million years for the age of the Earth. But the snag with Joly's calculations was the same as the snag with Darwin's: the recycling and redeposition of salt was just as much of a problem as that of erosion and redeposition of rocks. Any estimate reached by that method was likely to be far too low. Some other method must be sought. But what method could provide a more reliable figure?

Years before, in the 1700s, a remarkable Frenchman, Comte Georges-Louis Leclerc de Buffon, concluding that the Earth's dense interior must be made of iron, experimented with the rate of cooling of iron spheres, which, he argued, if one scaled up the dimensions accordingly, could provide a model for the rate of cooling of the young Earth. The figure he produced was 75,000 years: far too low a figure to satisfy the interpretations of nineteenth-century geologists. In the 1860s the great physicist Lord Kelvin revived the debate. Kelvin had little patience with nonquantitative reasoning and a disdain for the descriptive approach of some contemporary geologists. He used the same reasoning as Buffon, supplementing it with recent calculations by Hermann von Helmholtz to argue that the rate of cooling of both the Sun and the Earth indicated an age of not more than 20–400 million years, and more probably closer to 20 million. Over the next thirty years he revised these limits to 20–40 million years. Kelvin's estimate assumed that all Earth's heat came from its original molten condition and that no new source of heat existed. If those assumptions were granted, his estimates seemed impeccable, backed up by the authority of the greatest physicist of the age.

Here then were two contradictory scientific estimates. John Joly calculated an age for the Earth of not less than 80–90 million years, and new, growing evidence tended to increase it, while Kelvin suggested a figure of no more than 20–40 million years, and preferred to decrease it. The impasse was not resolved until the discovery of radioactivity in uranium salts by the French physicist Henri Becquerel in 1896. Radioactivity proved the existence of a completely unsuspected new source of energy, for which Kelvin's calculation had made no allowance. It was this discovery that not only proved the existence of a new source of internal heat and thus established the greater antiquity of the Earth, but also provided the basis for a new kind of Earth clock, which, for the first time, could provide a reliable timescale for the history of the planet. All our present accurate measurements of the age of the Earth and of its various rocks are the result of that discovery by Henri Becquerel. (Table 7.2.)

Most elements on Earth are stable, unless they are subjected to physical or chemical change. In contrast, a handful of elements—uranium, thorium, rubidium, and potassium among them—are unstable under *any* conditions, breaking down spontaneously into more stable daughter elements at a rate that is both measurable and constant, whatever the external conditions. Uranium, for example, breaks down to produce lead at a constant, measurable rate, and it can be demonstrated that, whatever the changes in temperature, pressure, or chemical conditions, the rate remains constant. Knowing that rate, which is known as the half-life period—the time it takes for 50 percent of the element to break down—it is possible to use the measurement of the ratio of "undisintegrated" uranium (^{235}U) to the amount of "disintegrated" lead (^{207}Pb) to calculate the age of the mineral containing the element in question. For some elements, these "half-life" figures are very large: rubidium (^{87}Rb) has a half-life of 47 billion years, uranium (^{238}U) 4.5 billion years. Others are much shorter: carbon (^{12}C), for example, which breaks

7.2 The most commonly used isotopes in radiometric age dating

Isotopes		Half-Life of Parent (years)	Effective Dating Range (years)	Minerals and Other Materials That Can Be Dated
Parent	*Daughter*			
Carbon-14	Nitrogen-14	5730	100–70,000	Anything that was once alive: wood, other plant matter, bone, flesh, or shells; also, carbon in carbon dioxide dissolved in ground water, deep layers of the ocean, or glacier ice
Potassium-40	Argon-40 Calcium-40	1.3 billion	50,000–4.6 billion	Muscovite Biotite Hornblende Whole volcanic rock
Uranium-238	Lead-206	4.5 billion	10 million–	Zircon
Uranium-235	Lead-207	710 million	4.6 billion	Uraninite and pitchblende
Thorium-232	Lead-208	14 billion		
Rubidium-87	Strontium-87	47 billion	10 million– 4.6 billion	Muscovite Biotite Potassium feldspar Whole metamorphic or igneous rock

Source: From Graham R. Thompson and Jonathan Turk, *Earth Science and the Environment (with EarthScienceNow and InfoTrac)*, 3rd ed., copyright 2005 Brooks/Cole, a part of Cengage Learning, Inc. Reproduced with permission, www.cengage.com/permissions.

down to ^{13}C, has a half-life of only 5,730 years, making it valuable in measuring the ages of younger rocks and fossils. In practice, some safeguards and assumptions need to be made, the most important being that the ages calculated for igneous rocks are those of the time of their crystallization, so that any melting and recrystalization of the rock after its initial formation will give too low an estimate of its original age.

The rocks of the Earth's crust vary greatly in age. Several thousand rocks from many parts of the Earth have now been measured, and the oldest—the so-called Acasta Gneiss near Great Slave Lake in northwest Canada—prove to be some 4.03 billion years old (Ga). Slightly younger rocks are known from west Greenland (3.7–3.8 Ga), northern Michigan (3.5–3.7 Ga), and western Australia (3.4–3.6 Ga). Zircon crystals found as detrital fragments in sedimentary rocks in western Australia have been dated at 4.04 billion years, implying the existence of an even older parent rock. Different methods of dating these ancient rocks, using several different radioactive mineral clocks, give remarkably consistent results. Only a few localities have proved to have such ancient rocks. Most of Earth's older rocks have been reworked and re-formed under the forces that continually shape the crust.

Is this figure—the age of the oldest discovered rocks—also the age of the Earth? That seems unlikely, for just the reason we have already given: these rocks are themselves metamorphic rocks, lavas, and sediments, the products of a fully functioning Earth, not its most primitive crust. They show Earth to be more than 4.03 billion years old, but the most ancient rocks must surely have been metamorphosed, resorbed, and reworked as Earth developed. We can, however, use the same radioactive clocks to explore the ages of two other groups of rocks: those of the Moon and the meteorites that have fallen on Earth's surface, which are generally regarded as part of the debris from which Earth itself formed.

That assumes, of course, that the Earth and all the other solid bodies in the solar system formed at the same time, which seems a reasonable proposition. If one accepts that, then lunar rocks might provide some confirmation of the age of the Earth because the surface of the Moon has not been modified by the action of plate tectonics. That's why its deeply scarred surface is so conspicuous. Few lunar rocks have yet been returned to Earth and those that have differ in age. The oldest have ages between 3.1 and 4.6 billion years, however, and thus seem to confirm the minimum age calculated for Earth.

Following the same assumption, the ages of meteorites should give us a comparable age to that of Earth and the Moon. More than seventy different meteorites, representing various types, have now been dated by radiometric analyses and all give ages between 4.53 and 4.58 billion years. These ages are remarkably consistent with those of the oldest rocks of the Earth and Moon. By analyzing Earth's

7.3 Geologic time scale (absolute ages shown are in millions of years) in relation to evolution of organisms and major physical events

Eon	Era	Period	Epoch		Life Forms	Major Events
Phanerozoic (phaneros = "evident"; zoic = "life")	Cenozoic	Quaternary	Recent, or Holocene	Age of Mammals	Spread of modern humans Extinction of many large mammals and birds	Eruption of volcanoes in the Cascades
			Pleistocene		*Homo Erectus* Large carnivores	Worldwide glaciation Fluctuating cold to mild in the "Ice Age" Uplift of the Sierra Nevada Linking of North and South America Beginning of the Cascade volcanic arc
		Neogene (Tertiary)	Pliocene 1.6		Earliest hominid fossils (34 – 3.8 mya)	
			Miocene 5.3		Whales and apes	Beginning of Antarctic ice caps Opening of Red Sea
		Paleogene (Tertiary)	Oligocene 23.7		Large browsing mammals; monkey-like primates; flowering plants begin	Rise of the Alps; Himalaya Mountains begin to form Volcanic activity in Yellowstone region and Rockies Ice begins to form at the poles
			Eocene 36.6		Primitive horse and camel; giant birds; formation of grasslands Early primates	
			Paleocene 57.8		Extinction of dinosaurs and many other species (65 mya)	Collision of India with Eurasia begins Eruption of Deccan basalts
	Mesozoic	Cretaceous 66.4		Age of Reptiles	Placental mammals appear (90 mya) Early flowering plants	Formation of Rocky Mountains
		Jurassic 144			Flying reptiles Early birds and mammals	
		Triassic 208			First dinosaurs	Breakup of Pangaea begins Opening of Atlantic Ocean
		245				

Era / Eon	Period	Age (mya)	"Age of"	Biological events	Geological events
Paleozoic	Permian		Age of Amphibians	Coal-forming forests diminish	Supercontinent Pangaea intact
		286		Coal-forming swamps abundant	Culmination of mountain building in eastern North America (Appalachian Mountains); extensive glaciation of southern continents
	Pennsylvanian (Carboniferous)			Sharks abundant	Warm conditions, little seasonal variations; most of North America under inland seas
		320		Variety of insects	
	Mississippian (Carboniferous)			First amphibians	
		360		First reptiles	
	Devonian		Age of Fishes	First forests (evergreens)	Mountain building in Europe (Urals, Carpathians)
		408		Early land plants	
	Silurian			Invertebrates dominant	Beginning of mountain building in eastern North America (rest of North America low and flat)
		438		First primitive fishes	
	Ordovician		Age of Marine Invertebrates	Muticelled organisms diversify	Extensive oceans cover most of North America
		505		Early shelled organisms	
	Cambrian				
		570			
Precambrian — Proterozoic ("Early Life")				First multicelled organisms	Formation of early supercontinent (~1.5 billion years ago)
				Jellyfish fossil (~670 mya)	Abundant carbonate rocks being deposited; first iron ore deposits
		2500			Oldest known sedimentary rocks
Precambrian — Archean ("Ancient")				Early bacteria and algae	Primitive atmosphere begins to form (accumulation of free oxygen)
					Earth begins to cool
		~3800			
Precambrian — Hadean ("Beneath the Earth")				Origin of life?	Oldest known rocks on Earth (~3.96 billion years ago)
					Oldest moon rocks (~4 billion years ago)
				Formation of the Earth	Earth's crust being formed
		~4600			

Source: Stanley Chernicoff and Ramesh Venkatakrishnan, *Geology: An Introduction to Physical Geology* (Richmond, UK: Worth, 1995).

oldest lead ores, and comparing these with the lead 207–lead 206 (^{207}Pb–^{206}Pb) ratio in uranium-free phases of meteorites, we can also calculate the time of their origin. This gives an age of 4.54 billion years for the time at which these elements, uranium and lead, were swept up into meteorites and the Earth. This age of 4.54 billion years is consistent with recent calculations for the much older age of our Milky Way galaxy (11–13 billion years) and the larger universe (13–15 billion years), based on the recession of remote galaxies.

Perhaps it's only by analogy that we can get any sense of how vast these numbers are. Suppose that a cosmic historian began to write at the rate of one line every century when Earth was first created. Suppose this same historian completed a set of volumes, each of five hundred pages and each with fifty lines to a page. That would see the completion of a book every 2.5 million years. There would now be a series of more than 1,800 such volumes, and they would stretch along a library shelf more than half the length of a football field. A volume representing a history that began when the oldest known fossils appeared 3.5 billion years ago would now be 1,400 volumes long.

Or suppose this: suppose a tree was planted that grew only at the rate of one-tenth of an inch every thousand years. At that rate, planted when the oldest fossil traces of living things appeared in the rock record, 3.5 billion years ago, it would be about 29,167 feet high: higher than Mount Everest. Planted when the first member of our species, *Homo sapiens*, appeared 195,000 years ago, it would be 19½ inches high. The same tree, planted at the start of the Christian era, would now be two-tenths of an inch high.

Or suppose this. Suppose we think of a cosmic calendar in which January 1, New Year's Day, represents the origin of the Earth, 4.6 billion years ago, and midnight on the last day of the year, December 31, represents the present moment. Each second of our cosmic calendar would then represent 146 years of Earth history, each hour 525,000 years, each day some 12,600,000 years.

The oldest indirect evidence of life would appear on March 4, and the oldest known forms of living things—bacteria and algae—would appear about March 27. The oldest abundant hard-bodied animals—mollusks, arthropods, and so on (500 million years ago)—would appear on November 21, the oldest reptiles (340 million years ago) on December 4. The oldest mammals (195 million years ago) would appear on December 15, and the oldest members of our own species on December 31 at 11:50 p.m. The whole of recorded human history would then take place in the last forty seconds before midnight on December 31.

So ancient is the Earth on which we dwell: ancient beyond our imagining. So long ago were the resources formed that we now consume at an ever-growing rate. So remote are the events that first made this planet habitable, and which set in sequence the processes that undergird our existence and well-being today.

So we are tenants of a property that is ancient; that was occupied before us by parents, grandparents, great-grandparents, and all their distant forebears, stretching back to the earliest manlike creatures (hominids) 20 million years ago and the countless other species that have formed a part of the great branching lineage of creatures that preceded us: anthropoids, early primates, ancestral placentals, mammal-like reptiles, reptiles, amphibians, crossopterygian fish, protochordates, mollusks, arthropods, and all the rest. There are those who seek to deny this long ancestral cavalcade, seeing it as something of an embarrassment, reducing our stature and diminishing our place in the great terrestrial scheme of things. But others see it as a lineage of wonder and nobility, as an indication of the incorporation of all living things within a larger harmony, dependent alike on the benevolence of the planet we share and on the resources of its changing crust. "There is," wrote Charles Darwin, "grandeur in this view of life, with its several powers, having been originally breathed by the Creator into a few forms or into one; and that whilst this planet has gone cycling on according to the fixed law of gravity, from so simple a beginning endless forms most beautiful and most wonderful have been, and are being evolved."

And the grandeur we glimpse is reflected, not only in the sharing of genetic materials, but also in the sharing of endlessly recycled materials of our planetary home. Amphibians about 375 million years ago gulped in the same atoms of the air we breathe today. Harlow Shapley once wrote as he captured this great continuity:

> Since about 1 per cent of your breath is argon we can determine approximately the number of atoms in your next argonic intake. In your next determined effort to get oxygen to your lungs and tissues you are taking in, besides nitrogen and oxygen, 30,000,000,000,000,000,000 atoms of argon; in briefer statement 3×10^{19}. Every saint and every sinner of earlier days, and every common beast, have put argon atoms into the general atmospheric treasury. Your next breath will contain more than 400,000 of the argon atoms that Gandhi breathed in his long life. Argon atoms are here from the conversations of the Last Supper, from the arguments of diplomats at Yalta, and from the recitations of the classic poets. We have argon from the sighs and pledges of ancient lovers, from the battle cries at Waterloo. Our next breaths, yours and mine, will sample the snorts, sighs, bellows, shrieks, cheers and spoken prayers of the prehistoric and historic past.[1]

---

1. Harlow Shapley, *Beyond the Observatory* (New York: Scribner, 1967).

Nor is this all. It is not just that we share a common homestead with all our neighbors—human and nonhuman—or inherit our genetic material from a legion of ancestral forms; it is also that our day-to-day existence is possible only by courtesy of the countless other organisms that have been contributors to, as well as consumers of, Earth's bounty over the aeons of geologic time. The oxygen that sustains us was first created by simple microorganisms some 2.5 billion years ago. The oil and natural gas that power our industrial societies were formed by the gentle accumulation of the remains of the myriad of microscopic organisms, accumulated generation upon generation over the floors of ancient seas. The coal-fired generating power plants that supply most of the world's electricity burn the remains of ancient tropical forests that flourished across vast deltas of Eurasia and North America 300 million years ago. Our fuels, our food, our materials, our bodies: all these are derived from Earth's crust and the contributions of its earlier inhabitants, as well as from the air and oceans that encircle it. In an endless cycle of resorption and renewal, Earth and its inhabitants interact and interchange with one another, linked in mutual dependence, which, one-sided though it is, has been as vital to our past as it is fundamental to our future.

CHAPTER 8

The Bountiful Planet

We are a young species, a recent arrival among the creatures with whom we share the planet we call home. But our lineage is venerable: we are youthful members of an old and successful family, long established on an ancient homestead. Like most youngsters, we still owe much to our parents; we are reliant to some degree on the creativity and industry of our forebears, and wholly dependent on the rich variety of products and bountiful fruits of the expansive estate and fertile homestead we have inherited from them.

For all our casual sophistication, declared independence, and supposed self-sufficiency, we still are dependents, living off our inheritance, existing moment by moment, surviving day by day only on the materials and provisions supplied by this benevolent and bountiful planet. In its youth, our species made only limited—though critical—demands on the planet. Small in numbers, limited in distribution, our early ancestors breathed the air, drank the waters, scavenged the plants and hunted the game of their ancient surroundings. Only later, as they multiplied and spread, did they begin to burn its timber for fuel, chip its flints and fashion its stones for weapons and tools, till its soils and fabricate its native metals. Later still, they learned to search out and smelt its ores; they cleared its forests to make charcoal; and so multiplied and grew in numbers that, in places, they depleted its vegetation and wildlife, polluted its lakes and streams, and exhausted its fragile soils. So, over the centuries, as human development "progressed," humanity's demands on the family homestead steadily increased. And so, too, did the growth and spread of human population and the size of humanity's footprint on the lands it occupied. In our own times, exploding human numbers, rising levels of consumption, and growing demands for materials have combined to increase the rate of depletion of some of Earth's nonrenewable resources, as well as increase pollution of the environment. We shall need creative planning to ensure that prosperity and progress do not come at the price of killing the goose that laid the golden egg.

We, like most of our ancestors, are inheritors of a vast storehouse of human ingenuity and knowledge, heirs to a vast estate of once fertile lands, pristine waters, and bountiful resources. Yet in a few brief generations, we—our species—have so proliferated in numbers, have so indulged our appetites, and have so squandered our riches that we now threaten the inheritance of our children and grandchildren. Lord Ritchie-Calder has described our dilemma as "mortgaging the old homestead."[1] Will and Ariel Durant put it plainly: "History is subject to geology."[2] We desperately need to ponder the price of prosperity: to weigh the real benefits of development against the cost of depletion. We need to consider the costs of our global appetite against the balance of our global checkbook.

This will not be easy, for we sometimes ignore how closely our human history is tied to the use of Earth's resources. Walter Youngquist has written of our "geodestiny: the ongoing flow of events which continually relate nations and individuals to the Earth's resources on which they depend for survival."[3]

But unless we consider the costs of the convergence between burgeoning population, rising levels of consumption, growing resource depletion, and accelerating climate change, we have little hope of developing effective guidelines to assure the survival of our species.

Our history has been one of living off the land, the bounty of the planet. The odds against the survival of early humans must have been formidable. Life's most basic requirements—food, water, shelter, warmth—must have involved our forebears in a ceaseless quest. Reflect on how any one of us would cope if deposited on the African savanna, with no food, no water, no clothes, no tools, no weapons, no shelter. It is probable our earlier ancestors lived in just such conditions, surviving as nomadic plant eaters, carving out an existence from fruits, leaves, roots, and tubers. Such an existence is known to provide little energy: perhaps as little as 2–5 kilocalories a day. In contrast, an individual in present industrial societies derives about thirty times that amount from his or her food.

Meat-eating habits probably arose from scavenging, with dead or wounded animals providing food. Human teeth are poorly equipped for killing or eating meat, and the earliest tools were probably naturally shaped stones used to fend off

1. Lord Ritchie-Calder, "Mortgaging the Old Homestead," *Foreign Affairs* 48, no. 2 (1970): 207–20.

2. Will Durant and Ariel Durant, *The Lessons of History* (Roseburg, OR: Simon and Schuster, 2010), 14.

3. Walter Youngquist, *GeoDestinies: The Inevitable Control of Earth Resources over Nations and Individuals* (Portland, OR: National Book Company, 1997), x.

carnivores or to kill them for food. Such "handy" weapons—"eoliths" or pebble tools, conveniently shaped or chipped by nature—have been found associated with 2.6-million-year-old prehuman *(Homo habilis)* fossils and animal remains—zebra, antelope, pig, and other species–in East Africa. But tool users were subsequently replaced by toolmakers, and the oldest known man-made tools are chipped and flaked hand axes—some 1.8 million years old. These hand axes (figure 8.1) were frequently made of flint or flintlike rocks. Flint is a homogeneous rock formed of the mineral silica (SiO_2), which fractures with conchoidal (shell-like) edges, rather

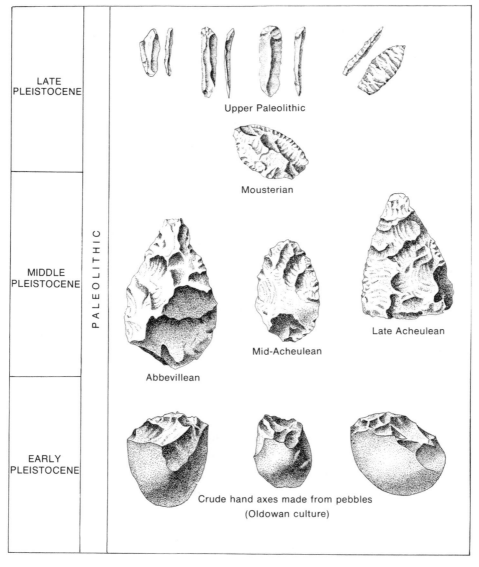

8.1 Progressive improvement in making tools from stone during the Pleistocene. From Harold L. Levin, *The Earth through Time,* 9th ed. (Hoboken, NJ: Wiley, 2009), fig. 17–25, p. 554. Used with permission of the publisher.

like glass. These axes are remarkably similar in appearance over a wide geographic range, suggesting perhaps a long period of development and little experiment in design once a basic form had been developed. The flakes from this flint chipping were sharp edged and were almost certainly used for skinning and flensing game.

Humans are not the only tool users. Darwin discovered finches in the Galápagos Islands that use a cactus spine to pick out insects. Other birds, the Egyptian vulture, for example, use stones to break open the eggs of other species; sea otters use stone anvils to open shellfish; ant lions create and "use" craterlike sand traps to catch their prey. Some chimpanzees are said to be toolmakers in shaping twigs to catch termites.

But toolmaking as the conceptual design of tools—from axes to computers—has characterized human history. From hand tools that served as extensions of the forelimbs, which themselves took on functions performed by the teeth in our ancestors and in other animals, human history has shown a striking pattern of tool design, manufacture, and use. Crudely chipped stone implements persisted unchanged for a very long time—perhaps because food was abundant—but later gave way to carefully shaped and finely chipped stone tools and weapons, which were later replaced by ground and polished implements. Pointed spearheads, harpoonlike bone points, fish hooks—all these are found successively later in a sequence of deposits, ranging from 1.5 million years to 18,000 years in age.

Not all these developments took place at the same time in all areas. The oldest pottery is found in Japan in deposits some twelve thousand years old. The oldest pottery in North Africa and the Middle East, by contrast, appeared 8,000–10,000 years ago, and later still in Europe.

Other distinctive aspects of human culture can also be traced in their development. Agriculture, for example, developed about ten thousand years ago and was followed by the growth of settlements and constructed dwellings. Needles, implying the use of clothing, appeared about two thousand years ago, although they may have existed before that. This development of physical materials and tools went hand in hand with cultural changes. The use of fire, for example, can be detected in ancient hearths, and was followed by many other cultural developments that can be traced in the fossil record. Linking all these cultural changes was the development of speech, which had incalculable implications for social life and development. When this developed is not clear. Neanderthal people, the contemporaries of early humans, seem to have had only limited voice capacity, though they may have had a repertoire of communication sounds and signs.

Our own species, *Homo sapiens sapiens,* appearing almost two hundred thousand years ago in Ethiopia, can produce a vast range of sounds and articulate thousands of words. With the arrival of our own species, we see a dramatic growth of societies and ceremonies. Necklaces, bracelets, carefully carved bone, wall paintings, and sculpture all appear. Weapons, rings, buttons, ceremonial burials, pigments, fire, flutes, and whistles give us glimpses of ancient lives and early cultures. Language, tradition, creativity, memory, social groupings, ceremony, necessity: all these and more seem to have played a part in humanity's earliest development. But linking all these changes was the continuing interaction between our species and its habitat: a vulnerable creature in a harsh environment. Early humans reached out to grasp the nearest thing at hand—stones—as a means of defense and attack in the hazardous world in which they lived. For most of human existence, the most important commodity in the world was stone. Pits and mines were developed to excavate flint nodules. Obsidian, a volcanic glass, was traded over considerable distances.

The Stone Age began over 2 million years ago. It lasted until recent times in some remote isolated communities, but about 5,500 years ago copper and, later, bronze implements were developed in what are now Iran and Iraq. It is possible that bronze was developed independently in other areas. With the use of bronze, a new era began, not only in toolmaking but also in broader human capacity to modify, adapt, and create. Stone, for all its qualities, is limited in the ways in which it can be crafted or shaped. Metals, in contrast, can be fashioned into endlessly varied forms to serve particular uses. But unlike stone, which occurs almost everywhere, lying on the surface of the ground, metals are not easily found. Even the most common are comparatively rare in Earth's surface rocks. And, with few exceptions, metals are not found in their pure native form, uncombined with other elements, but in mineral ores, consisting of chemical compounds. In fact, before bronze was used, native gold had been used in places for ornament, and

8.2 Abundance of metals in Earth's crust

Element	Crustal Abundance (% by weight)	Concentration Factor
Aluminum	8.00	3–4
Iron	5.8	5–10
Copper	0.0058	80–100
Nickel	0.0072	150
Zinc	0.0082	300
Uranium	0.00016	1200
Lead	0.00010	2000
Gold	0.0000002	4000
Mercury	0.000002	100,000

Source: From Frank Press and Raymond Siever, *Earth*, 4th ed. (New York: W. H. Freeman and Co., 1986).

native copper, which is found in the Middle East, Cyprus (from which the name "copper" comes), and North America, was mined and used for trinkets, tools, and weapons. The oldest copper tools are about nine thousand years old. (Table 8.2.)

The use of native copper required new techniques for metalworking. Unlike stone, which can only be chipped, ground, or polished, copper can be beaten or melted into new shapes and forms. The discovery of this new technology, and the later ability to smelt copper ores to produce native metal, fostered the adoption of copper tools and weapons. The malleability of copper, along with its strength, durability, and ease of working, accounts for its widespread use. It was copper that was used to craft the first metal weapons, as well as tools. Copper tools were used to shape the stone blocks of the pyramids of Egypt. It was copper, because of its high conductivity, that was used much later to carry electricity and telephone messages around the world. When combined with tin to produce bronze, copper has additional useful properties. Bronze is easily cast, melts at a low temperature, is resistant to corrosion, and is harder than pure copper. Bronze is an alloy, a mixture of elements, as are brass—made of copper and zinc—pewter—made of lead and zinc—and steel—made of iron and various other elements, such as nickel, vanadium, or carbon. These alloys required new techniques of smelting and casting.

The Early Bronze Age also saw the development of urban communities and the growth of larger social groups, including nation-states—the Assyrians, Babylonians, and Hittites. It also saw migration and widespread trading. Ancient Bronze Age weapons and ornaments were crafted with great skill and are often of elegant design. Bronze Age boats are known from a few localities.

The gradual replacement of bronze by iron took place in about the twelfth century BC in Greece, India, and the Near East, in the eighth century BC in central Europe, and in the sixth century BC in northern Europe. In a few places, such as Anatolia, the earliest iron used was from meteorites (both the Sumerian and Egyptian words for "iron" describe its heavenly origin), and for some time its use continued side by side with that of bronze. Iron implements offered several benefits over bronze: they were harder, could take a sharp edge, had a higher melting point, and iron was far more widely distributed than tin and copper. But there was no abrupt transition from Bronze Age to Iron Age across the inhabited world. The development of smelting and fabrication techniques in Anatolia and neighboring areas led to the growth of iron use in the region, encouraged perhaps by a growing shortage of tin. The use of iron swords is frequently credited with the rapid rise of the Hittite Empire, and especially their victories over the Assyrians and Egyptians.

Alongside the development of the use of metals and high technologies, the shaping and firing of clay, dating from at least twelve thousand years ago in Japan and eight thousand years ago in Anatolia, had two transformative influences on human development. First, unlike the nomadic hunter-gatherers, who would eat on the spot what they had hunted and gathered, the growth of agriculture required safe storage and transportation of cereals. The crafting of pots as vessels for food and water allowed storage, and with it, settled existence, the growth of communities, and the development of trading and exchange. Second, clay tablets became the basis for the earliest writing—markings incised and then baked on the clay—and written records of trade, history, and instruction in turn allowed a revolutionary change in the pace of human development. No longer would skills die with the craftsman or knowledge with the knower: from now on uncertain oral tradition and fallible human memory were reinforced by written records. With the creation of writing, humanity moved—both literally and figuratively—from the prehistoric to the historic. The visible recording of language allowed the circulation and preservation, not only of facts, but of ideas, of impressions, of feelings—in fact, of all the finest sensitivities and creative efforts that define our humanity. Some eight thousand years ago, recorded human history began.

But the firing of clay was symptomatic of a wider change in the manipulation and use of Earth materials. The earliest use of stone involved chipping and flaking to change its shape and size—sometimes facilitated later by heating to assist the process. But heating clay changed not its shape but its properties. It created a new material, with new properties: rigid and brittle, when once it had been soft and yielding. Not only could clay be shaped into useful forms: it could also be transformed by firing into a new material—a ceramic, whose rigid shape and general properties differed markedly from those of the soft, earthy material from which it was derived.

The relationship between human history and Earth resources has continued to be as profound and intricate in historical times as in prehistoric. Greek civilization triumphed, in part, because the Greek armies defeated the Persian forces at the Battle of Marathon by the use of a new weapon—bronze swords and shields, against the leather protection of their enemies. It was gold from Mount Pangaeus that financed the conquests of Alexander the Great, and it was silver from the deposits of Laurium (Laurion), near Athens, that allowed the Greeks to build a navy of triremes that overcame the Persian king Xerxes at the Battle of Salamis in 480 BC and presaged the decline of the great Persian Empire. But the Greek success in smelting the lead-silver ores of Laurium was only partial. Roman smelters reworked the waste tips (piles of rock waste removed from a mine), in the third century AD, and they were reworked again fifteen hundred years later. Human history is marked, not only by the introduction of new techniques, but also by the refinement and improvement of existing ones. The treatment and use of metals illustrates this trend.

Copper was found either as native metal or in ores (malachite—the rich green carbonate and the dazzling blue azurite, another carbonate) from which it could easily be extracted or smelted. Copper was first smelted about six thousand years ago in what is now Iran. Because a wood fire is not hot enough to reduce copper ore to metallic copper, the earliest smelting was probably done in pottery kilns, helped by using charcoal—which is richer in carbon than wood is—as a fuel.

In later times, about 1300 BC in Asia Minor, the replacement of copper tools by those of iron allowed the development of stronger, more durable, and sharper tools of every kind, so that agriculture, forestry, manufacturing, and every aspect of daily life were slowly transformed. Though iron was harder and more abundant than copper—it is, after aluminum, the second-most-common metal, making up about 5.8 percent by weight of the Earth's crust, whereas copper makes up only about a thousandth part of that—there was one snag to its widespread use: it does not occur as a native metal but always in compounds, and it requires higher temperatures to extract it from its ores. Smelting iron requires a temperature of 1,538°C (2,800°F) that was not easily obtainable in wood or charcoal fires, even with the help of bellows to blast air. At the lower temperatures obtainable in early furnaces, it softened, rather than melted, but it was found that the softened mass—the bloom—could then be forged or beaten to shape and harden it, thus producing wrought iron. This skill gradually spread from Iran and Turkey, where it seems to have originated, to other areas, so that by Roman times wrought iron was widely used to create small objects and weapons.

As early as 300 BC, Chinese workers were forcing air into furnaces to increase their working temperature. These "blast furnaces" were not introduced into Europe until the fourteenth century, but their adoption provided a sufficiently high temperature to melt iron and thus allow its casting into a new range of cast iron objects.

So successful did the new technology become that the growing demand for charcoal created massive deforestation across Europe and a growing crisis in fuel supplies. The demands for charcoal on the one hand, and timber for shipbuilding on the other, became so great that in the sixteenth century, Queen Elizabeth of England issued royal decrees limiting the use of woodland for charcoal production. It was this shortage that indirectly encouraged the greater use of an alternative energy source: coal. But coal itself created some problems, for it gave off impurities, such as sulfur, which combined with—and weakened—the iron that was produced. In the early eighteenth century, Abraham Darby, a Shropshire industrialist, made the discovery that coke, which contains some 80 percent carbon, burns slowly with high temperature and little flame, and can be manufactured by the destructive distillation of coal in high temperature ovens. Coke thus provided a far more efficient fuel than coal for iron smelting. This discovery led to the rapid growth of underground coal mining in Britain, and ironware became widely and cheaply available. The subsequent development of steel, an alloy of iron, with added carbon and small portions of other chosen elements, produced a harder, stronger, more flexible, and more easily worked metal than iron, and gave iron an even greater versatility in its usage. The combination of coal and iron, coke and steel, formed the basis of the Industrial Revolution, which, originating in Britain in the mid-eighteenth century, rapidly spread to other areas. That "Revolution" marked the rapid transition from hand tools to growingly elaborate machinery, transforming every industry, from agriculture to manufacturing, and remaking and relocating social life, from rural to urban communities.

In approximately 1700 the population of the world was growing at the rate of 21,000 new people per year, and until the mid-eighteenth century, wood had been the only fuel, while wind, water, human, or animal power were the only other sources of energy. But the improved mining techniques, growing use of coal, its proximity to abundant iron ore deposits, and the invention of the first commercial steam engine by Thomas Savery in 1698, its adaptation in 1712 by Thomas Newcomen to drive a pump to remove water in mines, and its refinement by James Watt and John Wilkinson in 1769, all contributed to a revolution in industrial development in Britain. With it came a rise in entrepreneurism, the rapid growth of factories, cities, roads, and canals, and the development of railroads and steamships. A huge global network of manufacturing and urbanization, construction, commerce, and capitalism changed the face of the land, first in Britain but later elsewhere, and transformed the character of society.

It is worth noting that none of the four architects of this technical revolution was a "professional" scientist or engineer. Savery was an army officer from Cornwall, Newcomen a blacksmith from Devon, James Watt a Scottish instrument maker, and Wilkinson a Staffordshire iron maker. It is also worth noting how private funding from Matthew Boulton, a Birmingham industrialist, enabled the development and manufacture of the steam engine and its widespread application in manufacturing, agriculture, and transportation. The engine, by employing coal to produce steam, converted heat energy into mechanical energy by using the huge expansion of steam to drive pistons that in turn drove crankshafts to provide rotary motion. In a few decades, stagecoaches and horse-drawn wagons were replaced by railroads; sailing ships by steamships; and hand-operated, horse-, water-, or wind-powered cottage industries by steam-powered factories. No one person can be credited with starting the Industrial Revolution, but this small group who developed the steam engine were central to its creation, and to the still-continuing social and economic transformations it has produced.

With the blossoming of the Industrial Revolution in Britain came also a stark reminder, though few noticed it at the time. Blessed though Britain was in having coal and iron supplies that were both adjacent to one another and plentiful, as well as deposits of lead, tin, and copper, such deposits were not inexhaustible. In the nineteenth century, this small island was successively the world's largest producer of coal, iron, lead, tin, and copper, supplying much of the global demand for these materials. But increased mining costs and dwindling mineral deposits

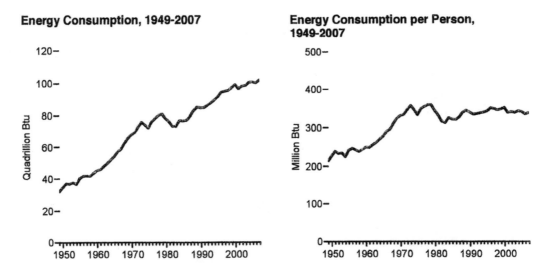

8.3 Energy consumption per capita in the United States from Energy Information Administration *Annual Energy Review 2007.* Courtesy of EIA and DOE.

led to rapid depletion: peak lead production was reached in 1856, peak copper in 1863, peak tin in 1871, peak iron ore in 1882, and peak coal in 1913. Britain now produces no significant amounts of any of those vital commodities.[4] There is in this a reminder that all resources are finite: very few of the resources we use are renewable. This distinction between renewable and nonrenewable resources is, however, complex. Even renewable resources, such as soil, can be lost by exhaustion or contamination (salting, for example). On the other hand, some nonrenewable resources can be replaced by substitute materials, such as bronze by iron. The explosion of human population, our growing appetite for material goods, and our burgeoning energy demands already place a significant strain on Earth's limited resources. (Figure 8.3.) This is a theme to which we return in chapter 9.

The Industrial Revolution is still with us. In fact, we still live in the Iron Age. Think of the endless array of tools, appliances, vehicles, buildings, bridges, tunnels, and all of the complex machinery used to produce them, on which our day-to-day existence depends. The difference is that, in our times, the use of iron has been supplemented by the use of a great range of other substances, from metals—aluminum, nickel, manganese, tungsten, chromium, mercury, titanium, cadmium, vanadium, and many other metals—to lime and gypsum, and from petroleum, asbestos, and phosphates to synthetics, plastics, and composites. The addition of these materials to the repertoire of useful resources has involved

---

4. Ibid., 19.

continuing experiment, invention, and refinement in discovering, extracting, fabricating, and using them; it has also required new energy sources and made new demands on water supplies. Though many of the new materials we now use are synthetic (manufactured), they still depend on Earth resources for their creation: plastics—nylon, Teflon, Kevlar, and all the rest—are derived from petroleum; pharmaceuticals from a host of other compounds and materials; and composites from a wide range of glass or other fibers and other materials.

But undergirding our dependence on these and other materials on which our existence relies, there are energy and water. Without them, no exploration for new resources, no extraction of raw materials, no manufacturing, no transportation, no food—indeed, no life—could exist. For all our creative invention of new materials and new techniques, it is energy and water that continue to sustain us and support all the multitudinous activities of our 7.0 billion fellow humans. And these, too—energy and water—are Earth's products, provided ceaselessly by the materials and stability of this rare and benevolent planet.

Society has changed much since the Industrial Revolution. New energy sources—petroleum, natural gas, nuclear, and solar among them—new materials, new processes, and new products have revolutionized our day-to-day living. But in one fundamental sense, nothing has changed. We are still Earth's dependents, existing because of the long ancestral sequence of Earth's creatures that have given us birth, surviving moment by moment only by our use of the source materials and the air and water of this bountiful planet that is our home.

CHAPTER 9

The Finite Planet

Earth's Dimensions: A Study in Constancy

We dwell on a changing planet whose surface is in constant flux: atmosphere and oceans, land and sea, air and water, rain and soil, all interact in ceaseless cycles, some very long and some very short. Cliffs crumble, coastlines erode, new volcanic islands appear, ancient cities are buried under blowing sand, estuaries and harbors become silted up. But these surface exchanges, and others like them, though locally conspicuous, are scarcely significant on a global scale, or even in a human time frame. Certainly, over geologic time, over millions of years, mountains come and go, and vast landscapes are submerged by spreading seas; but in human terms, the great shapes of the continents, the forms of islands, the floor of the oceans, the height of mountain chains, undergo changes that, though real, are almost imperceptible. In terms of human history, the Earth's radius does not change, its total land area is essentially constant, and the geography of land and ocean—though slowly changing—is broadly fixed. We live on a finite planet, whose various features and dimensions we have come to comprehend.

That finitude is both a blessing and a bane: a blessing because, in spite of earthquakes, tornadoes, floods, and droughts, it is Earth's broad stability on which we depend, its constancy of features and regularity of seasons on which we rely. But it is a bane because that's all there is. Earth's finitude allows no expansion of available land, only replacement of use; no increase in arable acreage, except by clearing forests; no space for new or expanding settlement, except by encroaching on farmland, irrigating deserts, or "reclaiming" fragile coastal areas. It's no good talking of living on the Moon or colonizing Mars: at least not yet. For now, Earth's all we have.

Earth's Resources: A Study in Constraints

Because we live on a finite planet, Earth's resources are as finite as its form. That's true of renewable resources, such as the atmosphere and hydrosphere, Earth's air and waters, even though they are recycled over time. If we contaminate them or pollute them, we have nowhere else to turn, and their recovery can take a long time. Make a hole in the ozone layer—as we have—and recovery time is forty to fifty years after we remove the cause of damage. We've no spare atmosphere available. Overfish the oceans—as we have—and recovery of some areas may take a generation or more. Pollute trout streams or salmon rivers—as we have—and some may never come back. And these are *renewable* resources.

With *nonrenewable* resources such as oil, gas, coal, and metals—most of our present energy sources and raw materials—there is no recycling, no natural replacement, except over cycles of tens of millions of years. Mine the copper deposits of Chile, and they're gone. Extract the cobalt of the Congo or pump the natural gas of the North Sea, and they're exhausted. Yes, there are other occurrences, other deposits in other places, but often less plentiful, less extensive, less concentrated, less readily extractable, less easily purified, more remote, and more costly to exploit. For a while, at a price, these will be developed. After all, there's estimated to be 6.6×10^7 tons of cobalt dissolved in seawater, but the energy required for the separation of these low concentrations is significant, and energy itself is in short supply. There is plenty of energy at the mid-ocean ridges, but we have yet to harness it to extract metals from seawater.

Now, some nonrenewable resources are far more abundant than others. Iron and aluminum, for example, have reserves that will last several hundred years at present rates of extraction. Natural gas exists in sufficient quantity to give us another hundred years or so of economic production, oil perhaps fifty years, coal more than one hundred. But all are finite, *nonrenewable* resources. Renewable energy sources—solar, wind, biomass, and hydropower—make up only about 7 percent of all world energy use at present.

This would be challenge enough if the size of our human population remained steady, but it does not. We continue to add to the world's population. In 1960, there were 3 billion people. In 2000, there were 6 billion. There are now more than 7.0 billion. By 2050 there are likely to be some 9.4 billion. We shall not be able to continue that rate of growth indefinitely. If we did, "there would not be

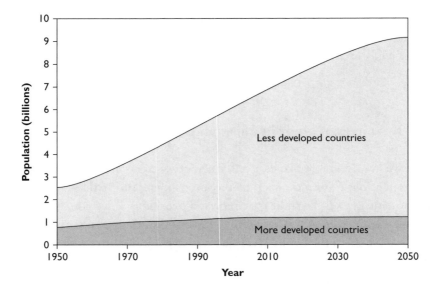

9.1 World population information, 1950–2050. UN Population Division, *World Population Prospects: The 2008 Revision*, medium variant (2009). Courtesy of UN.

standing room for [our] progeny." That's not a new conclusion: it's an old one. That phrase was used by Charles Darwin in 1859 to describe the effects of natural selection. (Figure 9.1.)

In practice, of course, population is likely to peak at something like 9.4 billion and then level off to some 8.5 billion. That does not seem an unsupportable number for the planet, but it will require continuing improvements in food production, conservation, alternative energy supplies, and free trading of goods and services.

So our human future will involve a study in constraints: we are constrained in available land area, in land use, in available water, in producible food, in sources of energy, in essential raw materials. But we are, above all, a creative and inventive species, and our creativity must not be constrained if Earth is to support its growing numbers.

Earth's Inhabitants: A Study in Contrasts

As Earth itself changes so slowly as to appear unchanging, so too do its inhabitants. We mourn the loss of the dodo, the passenger pigeon, the ivory-billed woodpecker, perhaps, but there are, after all, a multitude of other species of animals, plants, and microorganisms with which we share the planet. Populations fluctuate, distributions change, but most of the species we know seem to get along reasonably well. But even that may be changing. The once ubiquitous house sparrows and starlings of Britain have now been officially added to the list of declining or threatened species, their numbers half what they were twenty-five years ago. This list of threatened species, incidentally, is twice as long as the comparable list ten years ago. E. O. Wilson has noted that over one hundred rare bird species from the United States have become extinct since 1973, when Congress passed the Endangered Species Act.[1] But there is one species—our own—whose activities now have the capacity to influence the fate of every other. The direct stresses that uninhibited growth in human populations imposes on the habitats and food supplies of other creatures are magnified by our larger indirect impact on water quality, air purity, and land use. We have evolved with our present neighbors—animals and plants—on a moving stage of Earth, air, and water, which has allowed the development of a broad equilibrium between us. Once perturbed, that earlier equilibrium is not likely to be easily restored. The doubling or quadrupling of numbers in any one species imposes requirements and pressures, both physical and biological, on the environment and on other species that may represent profound stresses to the system as a whole. We have yet to learn, for example, what may be the long-term impact of the hole we created in the ozone layer or of the fallout from Chernobyl.

But what about other species? To what extent does our activity affect them? Although there are only some 1.9 million or so species yet recognized, described, and named, many biologists estimate that the "real" total may be at least 10 million.[2] One high estimate puts the possible total at up to 100 million.[3] And one recent approximation by E. O. Wilson suggests that the annual rate of extinction may be as high as 33,000: that's about one species lost—one extinction—every fifteen minutes of the day and night. Specialists will, no doubt, debate the accuracy of this projection, but even if it is too high by a factor of ten, it is still a sobering figure.

This is neither a doomsday scenario nor a call for return to the pastoral life. But it is a reminder that we live on a finite planet, whose carrying capacity is also finite. Doubling our numbers in forty years—as we have recently done—or trebling our energy use—as we have also done—are trends that for a while we can sustain. But only for a while. If we continue that growth unchecked, every species is endangered, including our own. Natural selection will take its toll. That is the reality of life within limits.

---

1. Edward O. Wilson, *The Creation* (New York: W.W. Norton, 2006), 74.

2. Ibid., 148.

3. See http://www.livescience.com/4593-greatest-mysteries-species-exist-earth.html and http://www.nytimes.com/2011/08/30/science/30species.html.

Yet it need not be so. If resources are limited—as they are—human creativity is not. If land area is finite—as it is—human resourcefulness is not. What is required is the human will to exercise restraint: the willingness to assume the role, not of thoughtless plunderers, but of responsible tenants. We need a tenant's agreement, a lease. We need to comprehend the nature of the property and understand its limits; we need a document that outlines the terms and conditions of the lease. And then we need to sign our names, provide a witness to our signature, and live up to the terms of agreement. We need to move from opportunistic exploitation of Earth to responsible tenancy.

Earth's Prospects: Study in Resourcefulness

Responsible tenancy is a daunting challenge, but it is possible. In this dark prospect there is one bright ray of hope, and it comes from the very species that is the source of most of our problems. Humankind, whose rising numbers and rapacious consumption have created most of these problems, is also the source of the one commodity that is both renewable and potentially unlimited: creativity. That creativity produces knowledge, and knowledge allows enlightened action. We are both our own worst enemies and our own best hope. When truly aware, we have the capacity to understand. When perplexed, we have the capacity to analyze and evaluate. When endangered, we have the capacity to correct, to change course. When mobilized, we have the capacity to cooperate. When moved, we have the capacity to be our neighbor's keeper.

Knowledge is the one unlimited resource: boundless, capable of application to endless uses and of serving multiple purposes. It is undepleted by its use. It expands, even as it is shared. It is refined, even as it is questioned. Knowledge is the resource that can enlarge every other resource and harness every other force.

But we are more than knowledge creators. Though, at our worst, we are selfish exploiters, at our best, we can be enlightened and caring custodians, behaving as responsible stewards, and accountable tenants of a planet whose benevolence sustains us and whose bounty allows us to choose our own path.

PART II

EARTH PAST

The Changing Planet

CHAPTER 10

The Singular Planet

Within the vastness of our solar system, Earth is, as we have seen, a singular planet: singular in its temperature, singular in its atmosphere, singular in its oceans, singular in its composition, singular in its crust and its core and their distinctive interaction, singular in the activity we call life. For all we know, it may well be singular in its degree of singularity. Though we know of other planets beyond our solar system, we know almost nothing, as yet, about their characteristics.

But what we do know about Earth is that its singularity owes everything to its location: third planet out from the Sun, situated between the choking, boiling surface of Venus and the cold, red, desert landscape of Mars. Atmosphere, oceans, crust, core—even life itself—all derive, as we have seen, from Earth's orbital location. And we know that each distinctive characteristic interacts with and influences all the others. In a great terrestrial dance, impinging cycles of all its constituent parts interact in an endless dynamic equilibrium, as Earth itself spins hurtling through space.

But our present Earth is a planet with a past. Earth is a product of an eventful history, extending back to its origin, 4.6 billion years ago. Our survival depends on our effective adaptation to the constraints and conditions of the planet that is our home. But if we are to be able to live in harmony with the planet of the present, we need to understand enough of its past to appreciate the events that have shaped its present character.

In this search for Earth's history, it is generally maintained that an understanding of Earth's present character is the key to our understanding of its past. And that assumption has proved time and time again to be true. By understanding the dynamics of Earth's present rivers, for example, we can recognize the erosional effects and depositional products of rivers of the distant past. By analyzing the activities of existing volcanoes, we can detect their former existence and unravel their effects in the past.

In a more general sense, conversely, the study of Earth's past can elucidate Earth's present. For example, the imprint left on ancient volcanic rocks of changing directions in Earth's magnetic field helps explain not only some features of Earth's present magnetic field, but also the past and present distribution and relationships of the continents. And by understanding the scale and frequency of past changes, we can understand the context of present trends or changes.

Can we also use the past to predict the future? Though we can, and do, use the history of Earth's past to understand the workings of Earth present, we have only limited ability to predict the details of Earth's future.

So Earth's long history is more than a prelude to the present. It is a key to our understanding of the present. It is to that history that we now turn.

CHAPTER 11

The Uninhabitable Planet

The universe, as we have seen, is some 13.7 billion years old; the solar system is much younger, having been formed, together with the Earth and other planets, about 4.6 billion years ago. This date is based on the oldest ages of meteorites, which are thought to represent the oldest solid material in the solar system. The oldest known Earth rocks, however, were formed "only" some 3.7–3.8 billion years ago. The time interval between these two ages—3.7–4.6 billion years ago— is known as the Hadean Era, from the Greek word *Hades*. This era is well named, for the early Earth must have been an inferno: very hot, blistered with volcanoes, buffeted by impacting meteorites, subject to mighty upheaval from internal forces of change.

The early Earth was lifeless, devoid of even the simplest of living forms. It was, as we have seen, not only uninhabited; it was uninhabitable, having none of the distinctive and benevolent qualities that support life today. All the evidence indicates that Earth came into existence by the collision and steady accretion of very cold stellar material that formed the Sun, which probably emerged as a slowly rotating, immense but diffuse cloud of dust and gas. The continuing bombardment and shattering impact and interaction of many colliding bodies gradually raised the temperature of both the early Sun and the emerging protoplanets circling around it, as did the growing compression of the innermost material and the breakdown of radioactive elements within the emerging planetesimal bodies.

Over a period of perhaps 100 million years, the young, newly formed Earth began to melt, with the denser material slowly sinking toward the center and ultimately forming its iron-rich core. Over time, with this density settling, the lighter material gradually rose to the surface of the youthful planet, forming a thin surface crust, separated from the denser core by an intermediate zone of semimolten material: the mantle. It was this slow process of gravity settling and differentiation, produced by heating, that created our present layered Earth, in contrast to what must have been the more or less homogeneous body of the earliest Earth, from which it developed. This process—distant though it now is—had enormous significance for every subsequent part of the planet's history. Without it, none of the processes that sustain a "living" Earth could have come into being.

Although Earth's present character makes it clear that this early differentiation of the Hadean Era must have taken place, just how it took place is still a matter of conjecture. But by at least 3.7–3.8 billion years ago the physical state of the Earth was much like that of the present, with continents, oceans, the hydrologic cycle, and so on.

Over time the early Earth's high temperatures were reduced by the slow loss of heat, both by conduction and later by convection within the planet, as soft or semimolten material below the crust began slowly to overturn in great convective cells within the mantle. The slow sinking and concentration of denser materials (especially iron and nickel) into the core produced, not only density layering, but also a chemical zoning within the Earth. Thus iron, which makes up over 30 percent by weight of the elements in Earth as a whole, represents only 6 percent of the crust, while such elements as oxygen, silicon, and aluminum are conspicuously more abundant in the crust than they are in the "whole Earth." This chemical zonation reflects both the density of the elements themselves and the densities and other properties of the compounds into which they are typically combined. This is the reason some heavy elements, such as uranium and thorium, which typically form light compounds, are now concentrated in the crust, in spite of their density.

But a solid crust is one thing: continents and oceans are another. How did Earth's present surface features come into being? We lack direct evidence, but it seems likely that continuing volcanism, with repeated fracture, down-warping, burial, heating, and gradual surface weathering and erosion of Earth's early crust led, not only to its melting and reworking, but also to the steady aggregation of blocks of lighter material as "islands." The constant accretion and growth of this lighter material over time produced the continents, while the continuing volcanic outgassing of the planet provided the water that ultimately formed the oceans.

If somehow we could transport ourselves back in time, if we were able to look prospectively before Earth came into being, all this history would look more and more improbable. The history we've already discussed is remarkable enough: the distinctive composition of the planet, its favorable position in space, its antiquity and the elaborate history of its formation and development, its prebiotic melting and differentiation that slowly "cooked" it toward a configuration where sunlight and rainfall, oceans and continents, volcanoes and mountain chains act and interact endlessly one on another. But that these same interactions could give rise to molecules that in time became self-replicating and, in turn, led to the rise of plants, invertebrates, fish, amphibians, reptiles, mammals, and thus to sentient bipedal hominids, pushing the bounds of Earth itself—the prospect of all this would appear too much, even for the most extreme reach of science fiction. But that is the long path that brought us here. To glimpse it is to marvel. Some see it as the outcome of cosmic chance upon chance, others as the hand of divine Providence. We are not required to reach agreement on such profound matters in such a discussion as this.

But perhaps we can, at least, reach agreement on a more limited conclusion: this remarkable heritage has given us both opportunity and responsibility. This planet, which, once uninhabitable, became habitable and so provided nurture and habitat for all its teeming creatures, is not unlimited in its resources. We, among its newest dwellers, have so exploded in our numbers and increased in our demands that we threaten now to deplete some of the very resources—soil, fuel, and water among them—that have allowed our emergence and supported our sustenance. The planet itself is not endangered. It has endured far greater change; ice, fire, and bombardment have left but superficial scars. It is we, its creatures, who continue to be threatened by the consequences, not only of our own indulgence, but also of our behavior toward members of our own species. We bring on ourselves and our suffering fellows the consequences of our divisions, of war, famine, disease, poverty, and forced migration, almost all of which are preventable.

It's not difficult to nod in agreement with the needs for responsible tenancy. But we should recognize that, for all our apparent freedom, we face constraints. We exist as part of a web of living things, wholly dependent on a paper-thin envelope of soil, water, and air. Without those, we perish. It is to the history of those life-support systems that we now turn.

CHAPTER 12

The Living Planet

In the Beginning

"In the beginning . . ." For all its haunting familiarity, it's a phrase that touches life's most obscure mystery and raises its most profound questions: "Where did we come from?" "How did we come to be?" "How did life come into existence?" "Are we alone in this vast, cold universe?"

We know nothing, as yet, of life elsewhere in the universe. Does life—whatever its form, whatever its precise definition—exist elsewhere in the far reaches of the cosmos? Has it ever existed on one of our neighboring planets? May it be, perhaps, a frequent feature of planets like ours? Or are we alone? Are we truly unique? Does "the simplest creature on Earth," as John Sepkoski asks, "embody a chain of causes each so unlikely that it takes a cosmos, with its billions of planets, to permit it even once?"[1] We do not know. We cannot, as yet, even speculate.

But on our own planet, there is life. *That* we can study. Paleontology—the study of prehistoric life—provides only some limited, though still significant, answers to the profound questions we all ask. It does not carry us back to "the beginning," but it does give us a record—imperfect as it is—of the development of

---

1. J. John Sepkoski Jr., "Foundations: Life in the Oceans," in *The Book of Life*, ed. Stephen Jay Gould (New York: W. W. Norton, 1993), 37.

life that went before us. Most of the evidence comes from fossils: the remains of, or direct indication of, prehistoric life. We've seen that fossils are formed by the burial of animals and plants in sediments that are themselves later buried, consolidated, hardened, and subsequently uplifted. Some of these sedimentary rocks, as these sediments become, have preserved impressions or traces of the structure of these organisms, though generally of only their hard parts. This preservation is *partial*—soft tissues and delicate structures are only rarely preserved, for example. It is *selective*—thick-shelled invertebrates living in the shallow seas are more likely to be preserved than, say, large terrestrial mammals. And it is *rare*—the vast majority of the countless organisms that once lived have left no trace of their former existence. Add to this the fact that, even when organisms are preserved as fossils, they face the hazards of subsequent metamorphism or erosion or nondiscovery, or inaccessibility because the rocks that entomb them lie under a cover of younger rocks or vegetation, and it is small wonder that—for all its riches—the fossil record gives us only glimpses of the long history of life. It provides snapshots of a long process, photographs in a family album, most of whose pages are missing, not a movie of life's past. But for all that, it is a remarkable history.

The Oldest Cells

The earliest glimpse we have of the presence of once-living things comes from 3.8-billion-year-old rocks of the Isua Group of Greenland and from rocks in South Africa that are reliably dated as about 3.35 billion years old. In these latter rocks, the so-called Onverwacht Group, the ratio of stable carbon isotopes ^{12}C to ^{13}C rises sharply. Now, of carbon's two isotopes, ^{12}C is slightly lighter and is more easily absorbed than ^{13}C by living cells in photosynthesis of plants, so this is highly suggestive of the presence of photosynthesis when these ancient rocks were formed. From the same part of the world, flintlike cherts—sedimentary rocks formed by the precipitation of silica (SiO_2)—in the so-called Fig Tree Cherts, deposited some 3.5 billion years ago, yield traces of individual cells of blue-green algae and bacteria. Slightly younger rocks from Zimbabwe—the 2.9-billion-year-old Bulawayan Group—contain the fossil remains of layered, calcareous, matlike deposits formed by cyanobacteria (often called blue-green "algae," though they are not true algae). These stromatolites are formed by sticky microbial sheets and mounds, whose living descendants are still found in saline lagoons along the coast of western Australia and other continents.

Simple as these ancient organisms are, they do not represent Earth's earliest living things: they are already too late in time and too complex in structure. What those most ancient organisms were, we can so far only conjecture. It seems probable, however, that the environment in which they developed was very different from that of the present, and that these earliest organisms must have been very simple, probably surviving by "feeding" off the chemical broth from which they formed. Charles Darwin wrote in 1871, in a letter to his friend and confidant Joseph Hooker, that it seemed likely that

> all conditions for the first production of a living organism are present, which could ever have been present. But if (and Oh! What a big if!) we could conceive in some warm little pond, with all sorts of ammonia and phosphoric salts, light, heat, electricity, etc. present that a protein compound was formed ready to undergo still more complex changes, at the present day such matter would be instantly devoured or absorbed, which would not have been the case before living creatures were formed.[2]

Darwin's statement identified three aspects of this emergent life that still promote debate: the ingredients of which it was formed, the environment in which it developed, and the energy source that triggered its formation. It is also probable that we need to add two more requirements: some catalytic help was probably required to sustain the chemical reactions involved, and some protective mechanism needed to preserve the newly formed complex organic molecules from destruction by harmful radiation.

There is continuing experiment, analysis, and spirited debate about each of these aspects of the emergence of the simplest living things, but the broad theoretical possibilities of their development from an inorganic broth are now reasonably well established. Though the details are far from settled, and debate continues to be intense, most students would accept that the earliest living things probably developed relatively soon after Earth's origin, in a chemical environment made up largely of water vapor, nitrogen, and carbon dioxide. In what particular physical environment this took place is unclear: it might have been near deep-sea smokers, or in tidal pools, or warm tropical seas, for example. Solar radiation, perhaps assisted by lightning or volcanic activity or other energy sources, probably promoted chemical reactions that, facilitated by catalysts of various kinds, produced amino acids—the basic components of living molecules—which later formed proteins. Proteins are large, complex molecules, found in all living cells, consisting of carbon, hydrogen, and oxygen and varying amounts of other elements. These large molecules, which "grew" from the chemical broth around them, must ultimately have exhausted the broth from which they emerged, but two things could have allowed their survival. First, at some stage, the development of photosynthesis introduced the ability of certain cells to use the energy of sunlight to produce

---

2. *The Life and Letters of Charles Darwin*, vol. 3, ed. Francis Darwin (London: John Murray, 1887), 18.

organic compounds from water and carbon dioxide. Second, once photosynthesis developed, free oxygen was produced. As this escaped into the atmosphere, Earth's earlier atmosphere was slowly transformed and the ozone layer came into existence, shielding organisms from the harmful effects of ultraviolet radiation. It was, ironically, this same type of radiation that may well have contributed to triggering the chemical reactions that brought life into existence.

We get only the rarest glimpses of the subsequent development of these simple organisms, though in the 900-million-year-old, flintlike Bitter Springs Chert from central Australia, William Schopf has found a rich biota of microfossils, representing thirty species, including a dozen forms of cyanobacteria, as well as algae, fungi, and bacteria. He has also discovered indications of eukaryotic cells (cells with a nucleus, as opposed to the simpler prokaryotic cells, which lack a nucleus) that seem to represent fungi (which are known to reproduce sexually), and some of these microfossils show distinctive sexual cell division. Sexual reproduction provided a new degree of variability in animal and plant structure, with the combination and shuffling of parental sets of genes allowing a degree of variation not found in asexually reproducing organisms. In a constantly changing world, this versatility is no small advantage.

The cyanobacteria of Bitter Springs include forms that resemble living representatives, and they are associated with calcareous stromatolites, showing that this association is both ancient and remarkably resilient. Other animal and plant dynasties have flourished and declined, but the lowly cyanobacteria endure.

This speculative account of life's origin remains long on hypothesis and short on facts. That does not make it implausible, but it does make it speculative and tentative. We are dealing, remember, with events that took place almost 4 billion years ago, occurring somewhere on the surface of a planet so unlike that of the present that we would scarcely recognize our Earth, and with molecules so simple that we should perhaps only provisionally classify them as living. Yet for all our speculative explanations, for all the ambiguity of our understanding, we cannot give up the quest for life's origin, the most perplexing and one of the most intriguing questions of our existence.

Soft Bodies and Hard Shells

Australia is a continent of surprises. From koalas to kangaroos and platypuses to cockatoos, its creatures fascinate and delight us. The creatures that occupied the shallow seas over 500 million years ago, above what are now the Ediacara Hills of south Australia, were equally remarkable. It is there we catch our first glimpse of metazoans (animals made up of more than one type of cell). Jellyfish floated across the surface waters, wormlike creatures burrowed or crawled across the sandy sediments of the seabed, and anemones and sea pens—not unlike those still living today—swayed in the currents, rooted to the seafloor. Other creatures in that ancient ocean were quite unlike any living today. Indeed, some paleontologists regard all the Ediacaran creatures as unrelated to living forms. All these animals—both those familiar and those unfamiliar—were soft-bodied invertebrates. Of the host of later shellfish, lobsters, crabs, sea urchins, starfish, fish, mammals, birds, and all the rest that now populate the oceans, there is in this sample of ancient life no trace. This same assortment of creatures—the so-called Ediacaran fauna—is known also from rocks of similar age in some thirty localities on five continents. Particularly significant collections have been described from England, Namibia, Iran, the Urals, the Chang (or Yangtze) River Valley, and a dozen sites in North America. These indicate a distinctive worldwide development of a soft-bodied invertebrate fauna, living some 560–580 million years ago. There are also other traces of animals: burrows and trails in the once soft but now hardened sediment and fecal pellets testify to the presence of a rich—though to our eyes "simple"—assortment of marine animals. In the still older, 800-million-year-old so-called Buckingham Sandstone of northern Australia we find similar traces of burrows and trails, suggesting a long but otherwise unknown history of these early burrowing and crawling creatures. In a few other scattered localities, similar trails and burrows have been found. Of other features of these forms of life we know nothing as yet, which is, perhaps, not wholly surprising when we consider the long odds against the preservation of any trace of these soft-bodied creatures.

But between the 900-million-year-old microscopic, single-celled organisms of Bitter Springs and these "simple" one- or two-inch-long multicellular creatures, there is a huge gap in time, in the fossil record, in our knowledge of how life developed, and also in the level of complexity represented by the younger animals. Though the affinities of some Ediacaran fossils are obscure, others (jellyfish and sea pens) are almost certainly coelenterates, related to, but distinct from, modern forms living today. These simple invertebrates, which either float passively or live rooted to the seafloor, have a two-layered body structure, in which tentacles waft food into a digestive body sac. The Ediacaran flatworms, in contrast, have a bilateral symmetry and were active, mobile scavengers, the ancestor, probably, of many later forms of life.

It is not until some 540 million years ago that we find any significant indications of the development of more complex prehistoric life. In many different rocks on several different continents an extraordinary blossoming of new kinds of animal appeared, quite unlike the Ediacaran fauna in variety of form, in complexity of structure, and in the possession of hard body parts. The Ediacaran fauna now is essentially gone. So great is this change from the generally barren rocks of the older Phanerozoic ("hidden life") to these fossiliferous "younger" rocks of 540 million years ago that they are classified as marking the dawn of a new era of geologic time: the Paleozoic, the time of ancient life. Small, shell-bearing creatures of great variety are found. Six wholly new, major groups of animals appear in rocks of so-called early Cambrian age, formed some 483–542 million years ago. Brachiopods (lampshells), sponges, mollusks (snails and clams), arthropods (trilobites and ostracods), echinoderms, coral-like archaeocyathid reef builders, edrioasteroids, and nine other groups all appear in a short span of geologic time in an exuberant profusion of new forms.

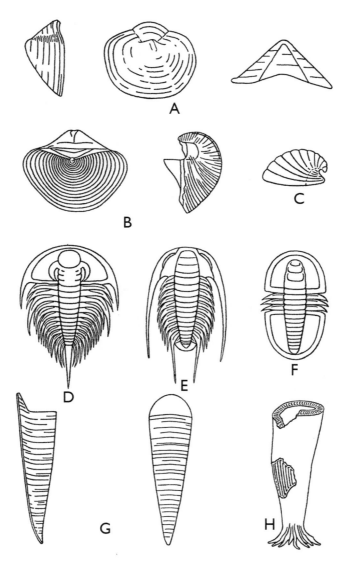

12.1 Typical early Cambrian fossils. A–B, Brachiopods; C: Gastropod; D–F, Trilobites; G: Pteropod; H: Archaeocyathid. From Frank Rhodes, *Evolution of Life* (New York: Penguin, 1974). Used with permission.

What is remarkable about this Cambrian assemblage is not only its variety of form, complexity of structure, and diversity of lifestyles, but also the variation in the chemical composition of the hard parts secreted by its members. Sponges secreted an internal skeleton of microscopic siliceous spicules, trilobites an external armor of chitin—a cellulose-like polymer—some brachiopods external shells of calcium carbonate and others of chitin, and archaeocyathids reefs of calcium carbonate, for example. Many small phosphatic fossils are known, some representing groups of obscure affinities. And to remind us that all this is but a limited glimpse of a vastly greater but otherwise unknown cavalcade of once-living things, high up on the flanks of Mount Wapta in British Columbia, a century ago, Charles Walcott of the Smithsonian Institution discovered a rich fauna of some 120 different species of soft-bodied animals, all with detailed structures exquisitely preserved. More recently other exquisitely preserved soft-bodied animals have been discovered in rocks of similar age in northern Greenland and southern China, but almost all of these soft-bodied forms are otherwise unrepresented and unknown in rocks of similar age in other places. Elsewhere in rocks of Cambrian age, some one hundred families of trilobites are known. The fossil record thus shows ancient life to have been extraordinarily prolific. But the fossil record is also incomplete and very selective. For all its richness, the fossil record has yielded evidence of only perhaps 0.001 percent of all creatures that ever lived.

The creatures of the ancient Paleozoic seas changed slowly over time. The bottom-feeders and browsers were joined by a host of new forms, many of them filter feeders or active predators, most that we know having hard parts of varied form, composition, and complexity. To the small invertebrates—brachiopods, trilobites, snails, clams, and corals—that thronged the ancient shores were added some of extraordinary size—nine-foot-long eurypterid arthropods, for example, and ten-foot-long, straight-shelled, nautilus-like cephalopods. Most organisms were solitary, but some, such as graptolites and corals, for example, were colonial. While many creatures, such as brachiopods, corals, and clams, were sessile, others, such as trilobites, browsed the seafloor; cephalopods were active swimmers, while others, such as jellyfish, for example, were passive floaters. For all their variety, however, the environment they shared was marine. Among these ancient creatures, we know of none that lived in freshwater or on the land.

Most of the earlier forms of marine invertebrates were bottom-feeders, but some early and many later forms fed on material suspended in the water: floating plankton and larvae, for example. To the earlier forms of ancient life we have described, new groups, such as starfish and sea urchins, were added over the 350 million or so years of Paleozoic time, even as some existing groups—such as trilobites—declined. There were episodes of mass extinction, perhaps as many as five of them in the 60 million years of Cambrian time. And all this organic change took place on and was profoundly influenced by the slowly changing Earth, as the great dance of ocean ridge spreading, continental movement, volcanism, and mountain building unfolded, changing patterns of land and sea, basin and shelf. This was the moving stage that both supported and influenced the unfolding development of invertebrate life in the oceans. It's a broad pattern of life that persists to our own day in the teeming hosts of invertebrate animals thronging the shallow seas that lap the shorelines of the world.

The Cambrian explosion saw the appearance of essentially all the major groups (phyla) of animals within a time span that represents less than 1 percent of geologic time. Eleven still-living phyla appeared, as well as another, now extinct. How did this Cambrian explosion—this vast blossoming, not only of new forms of life, but of new ways of life—take place? It's worth remembering, perhaps, that the Cambrian covers a very long geologic period—some 60 million years or so—and that only a few of these many forms appeared during its earlier days. This was not one big bang. But it's also significant that—simple microorganisms apart—the shallow seas of those distant days were essentially unoccupied by competitors. Once hard-shelled multicellular organisms appeared, competition increased and extinction rates became conspicuous. That seems, in fact, to be one of the few generalizations we can make from the history of life: open environments, whether arising from the extinction of earlier occupants or the invasion of new areas, lead to rapid colonization and explosive evolutionary change.

The broad patterns of marine invertebrate life persisted for some 300 million years, though in the Ordovician Period (443–490 million years ago) new groups of filter feeders—hinged brachiopods, bryozoans, as well as stalked echinoderms, graptolites, and, later, corals—increased in numbers. Jawed fish and coiled cephalopods appeared in later (Devonian) times and became major predators.

The invertebrate patterns of life have proved to be extraordinarily successful not only in the oceans but also above and beyond them, in estuaries, rivers, lakes, and, later, on the land and in the air. Mollusks—snails, clams, and similar forms—are present almost everywhere in freshwater streams, ponds, and lakes. But the most strikingly widespread invertebrate group is the arthropods, whose members, from crayfish, scorpions, spiders, and beetles, to the countless forms of insect, have invaded every niche of the water, land, and air.

The insects are the most varied and abundant of all animal groups, making up fully three-quarters of the million or so known animal species. They are known from rocks as old as those of the Rhynie Chert of Scotland, formed some 396–407 million years ago, and though their fragile form makes them poor candidates for fossilization, they are represented by a great variety of forms throughout the fossil record.

The Greening of the Land

The oldest fossil organisms known to us are plants: simple plants, to be sure, but plants nonetheless. Algae, bacteria, and fungi have, as we have seen, existed in the seas for over 3 billion years. Perhaps simple plantlike forms such as these were the first to gain a foothold on the land. We do not know, though it seems probable that no land organisms—whether animal or plant—could have long survived until the development of the ozone layer, which shielded them from destructive ultraviolet radiation.

The earliest known land plants were a far cry from those of the present day tropical forests or alpine meadows. Just as animal life on land required structures distinct from those required for living in the water, so plants required a means of utilizing food found only in the soil. The ultimate development of roots and stems provided a means of tapping and distributing nutrients, a strong woody supportive structure, and protection against desiccation. The oldest known fossil land plants are modest structures, however, lacking roots, leaves, flowers, and all the features we think of as plantlike today. They were probably confined to moist areas and they appeared together with fossil millipedes about 420 million years ago in late Silurian times, only "shortly" before the oldest amphibians and other land arthropods—spiders and scorpions—made their appearance.

Plants changed Earth's barren, rocky surface forever, slowly spreading a mantle of decaying vegetation that became soil. The spread of swamps and, later, forests and prairies not only opened up new habitats for animals but also modified Earth's climate and influenced its atmosphere. Perhaps rotting vegetation led to the spread of stagnant conditions in estuaries, where oxygen thus became restricted, contributing indirectly to the subsequent migration of animals to the land.

Though the oldest plants were no more than three of four inches high and show no division into roots, stem, and leaves, they do have xylem, a vascular structure that provides support and carries water throughout the plant, and a surface cuticle that allows both transpiration and gas exchange and prevents drying out. Other fossil plants of slightly younger (Early Devonian) age are taller (reaching a foot or so) and bear sporangia, carrying asexual reproductive spores (as do contemporary plants) and showing the first traces of leaflike spines on the stems. Well-preserved faunas of small arthropods—mites, spiders, and other arachnids—are associated with some of the fossils. These modest plants may well be the ancestors of the later giant club mosses and other towering lycopods and tree ferns that formed the chief vegetation of the coal-forming swamps that spread across the continents. It is from their remains that much of the world's energy has been and still is derived.

The growth and spread of early algae and other plants literally transformed Earth's atmosphere by photosynthesis, in which chlorophyll is transmuted into carbon and oxygen in the presence of sunlight. These first lowly plants changed

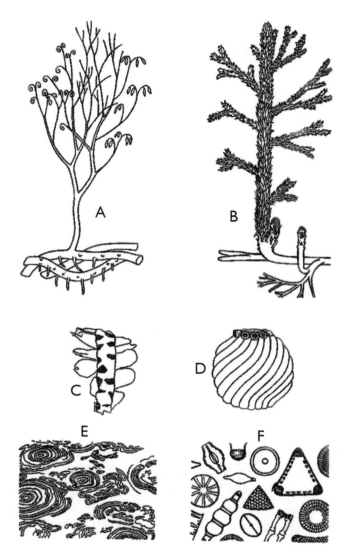

12.2 Paleozoic plants: A–B, Devonian land plants; C, Carboniferous bryophyte *Hepaticites*; D, fossil *Charaphytes*; E, calcareous algae Cryptozoan; F, variety in living diatoms. From Frank Rhodes, *Evolution of Life* (New York: Penguin, 1976). Used with permission.

the world. Our presence on this planet today is the direct consequence of their ancient alchemy. Without these simple plants and their steady, unspectacular work of photosynthesis, the lands of this beautiful planet of ours would still be uninhabitable. (Figure 12.2.)

Although the earliest plants were small and their asexual style of reproduction limited them to damp, moist environments, the appearance in mid-Devonian times of some plants with two sizes of spore indicates the development of sexual differentiation, and the later appearance of seeds—with their built-in food and water supply—allowed plants to colonize the drier upland areas.

Jaws, Fins, and Backbones

The oldest known vertebrates are represented by two kinds of fish reported from the 530-million-year-old marine Cambrian rocks of China. In later (430–490-million-year-old) rocks from Russia and Colorado are bone fragments from what were probably small, bottom-dwelling marine fish, which lacked true jaws and "real" fins. In 450-million-year-old rocks in South Africa, small, soft-bodied, marine fish have been found with eyes, scales, stomach, liver, and bronchial openings preserved. It is only later, in late Silurian and Devonian times, that we get a better glimpse of these early fish. Most were heavily armored, jawless fish (Agnatha), many having bony head shields, but others were jawed and finned placoderms (spiny sharks), which flourished in Devonian times. Most of these were small creatures, no longer than ten inches or so, but some placoderms grew to thirty feet in length and were the most formidable carnivores of their day, developing massive bony skulls some six feet long, armed with large, shearing teeth.

Both groups developed a rich variety of forms and relatively soon invaded freshwater streams. Agnatha vanished at the end of Devonian times, being represented today only by the jawless lamprey and hagfish. Placoderms became extinct at the end of the Paleozoic. These groups were replaced by their descendants—sharks and bony fish, with strong fins and jaws, which now dominate life in the oceans and, in the case of the bony fish, in rivers and lakes as well. (Figure 12.3.)

These two major groups of living fish—sharks and bony fish—are supremely successful. Sharks and their allies, the rays and skates, stand at or near the top of the ocean food chain. They have a cartilaginous, rather than a bony, skeleton and either sharp slashing and cutting teeth (sharks) or blunted crushing teeth (rays and skates), and they have developed highly refined sensory systems. Because of their soft cartilaginous skeletons, sharks and rays are rarely fully preserved as fossils, except for their resistant teeth, which show them to have changed little over the 400 million years since they first appeared, their very constancy indicating their fitness for the active predatory lives they pursue.

It is the other great group—the bony fish—from which we ourselves were subsequently to develop. This group includes the countless ray-finned fish, from tuna, swordfish, and sturgeon to herring, plaice, and sardines, and from garfish to minnows, which, with such evident abundance, fill every niche of Earth's watery envelope. Their light, flexible skeletons and thick scales have been adapted to an endless variety of forms and structures. But bony fish—Osteichthyes, as biologists describe them—include two other less conspicuous, related groups: the lungfish,

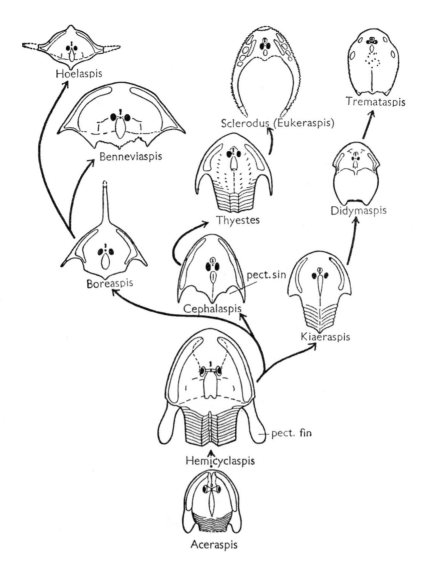

12.3 Variation and development of head shields in cephalaspid ostracoderms. From Frank Rhodes, *Evolution of Life* (New York: Penguin, 1976), after W. K. Gregory, *Evolution Emerging* (New York: Macmillan, 1951). Used with permission.

or Dipnoi, known only from Africa, Australia, and South America; and the crossopterygians, or lobe-finned fish. Both have strong, stumpy fins, quite unlike the fan-shaped, flexible fins of their more abundant ray-finned relatives.

Living lungfish are well adapted to life in rivers and warm, moist areas of seasonal rainfall. Here they protect themselves in times of seasonal drought by burrowing in the mud of riverbanks, breathing by means of their lungs, and so reducing their level of bodily activity in a process not unlike hibernation. Devonian

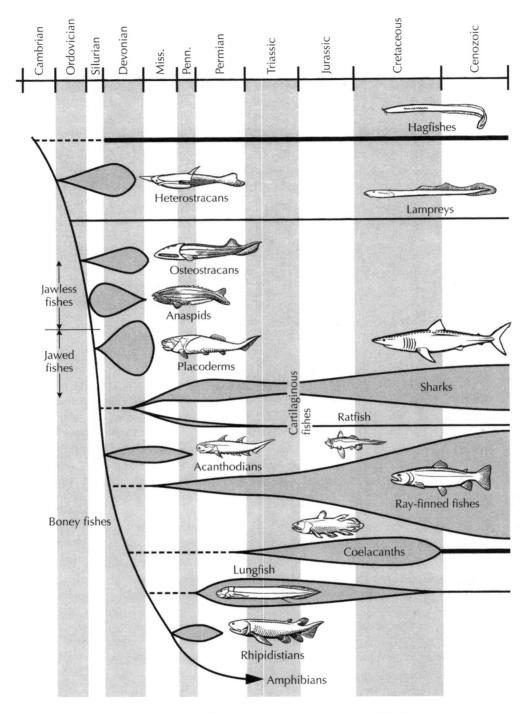

12.4 Evolutionary history of fishes and other early vertebrates. From Donald R. Prothero, *Evolution: What the Fossils Say and Why It Matters* (New York: Columbia University Press, 2007). Drawing by Carl Buell. Copyright Columbia University Press. Reprinted with permission of the publisher.

fossil representatives resemble them, and their persistence with almost no change over more than 350 million years since Devonian times qualifies them as true living fossils. It would be easy to conclude that here in the lungfish we have found our ancient ancestors, but this seems unlikely. Their successful and long survival reflects their specialized, passive, and limited way of life. (Figure 12.4.)

Their relatives, the crossopterygians, on the other hand, were powerful carnivores, actively pursuing their prey, chiefly in freshwaters but later also in the sea, where they are represented now only by the rare, deep-sea dwelling coelacanth—"Old Fourlegs"—of the Indian Ocean. In the freshwaters they were supreme in their time, their distinctive upper jaw articulation making them formidable predators and their stout, trunklike fins able to support them as they moved through weed-choked lakes, rested in shallow streams, or perhaps basked on riverbanks. It was from these creatures—now, apart from the elusive coelacanth, quite unknown—that we and all our kind developed, for their strongly built fins later gave rise to the sprawling limbs on which amphibians were able to clamber ashore.

Crawling Ashore

In the summer of 1933 two young Swedish geologists, Gunnar Säve-Söderbergh and Erik Jarvik, were working across the ice-bound mountains of Celsius Bjerg in east Greenland, climbing slowly across a steep cliff face. They were inspecting the dark-red rocks, bed by bed, and layer by layer, searching for fish—fossil fish—that had lived there in late Devonian times, about 360 million years ago. They were also searching for fossil amphibians, the oldest of which had been discovered by Säve-Söderbergh in Greenland a few years earlier. The stratified rocks they studied were formed from the sand and silt that had accumulated in an ancient river system that once covered what was then an ice-free tropical or subtropical region. In fact the region then supported a lush cover of lowly vegetation—mostly creeping, low-growing club mosses and other simple plants. That was a time when land plants, even simple, low-growing ones, were still a novelty on the planet: there were no forests, no trees, no grasses, no flowering plants. Even the amphibians and fish for which Säve-Söderbergh and Jarvik were searching were still relative newcomers in those distant days. They were the only vertebrates in a world of invertebrates—scorpions, insects, and snails—slowly making a laborious transition from the turbulent stability of the waters of the oceans to the turbulent instability of the waters and shores of the land.

The fossils for which Säve-Söderbergh and Jarvik hunted were scarcely spectacular: small fishlike creatures entombed and thus preserved in what were once the soft muds and other sediment at the bottom of the water that had once been their home. But the other vertebrates for which they searched were altogether more substantial. Up to a meter in length, they were fishlike creatures but had four short, stubby legs and feet; they were part fish, part amphibian, tied to the water, but capable of clambering onto the land. These fossils—creatures of lazy river meanders and warm swamplands, appearing in the rocks of what are now the frigid Arctic wastes—had to wait until 1952 before making their first public appearance: World War II intervened and Säve-Söderbergh died in 1948. But when their descriptions were published, they were widely recognized as typifying the ancestors of all the later legions of air-breathing amphibians, reptiles, birds, and mammals. They were not fully described until 1996, by which time Jarvik was eighty-nine years old. (Figure 12.5.)

These remarkable fossils—called *Ichthyostega* and *Acanthostega*—are not our earliest direct ancestors: they are a little too late in time to be that. But they must be broadly similar to what our distant ancestors looked like. They were a jumble of fish and amphibian characteristics, with fishlike bodies but well-formed walking limbs; fish-patterned skulls but amphibian-like proportions and eye placings; fishlike tails but distinct fingers. They were a yard or so long, but part fish, part amphibian; part aquatic, part terrestrial. And there were significant differences between the two forms Säve-Söderbergh and Jarvik found. Jenny Clack and Michael Coates described additional material in 1987 and showed that *Acanthostega* was a fish with legs. It retained its internal fishlike gills and had such a weakly constructed skeleton that it probably never left the water—in spite of having four limbs (with eight digits on the forelimbs and seven on the hind limbs) and a strong pelvic girdle. *Ichthyostega,* in contrast, seems to have had sufficiently strong forelimbs to lift the front part of its body. It appears to have been "more terrestrial" than *Acanthostega,* but still capable of only crawling across land surfaces.

Ichthyostega lived 360 million years before humans first appeared on Earth, but it had one essential feature that links it with every member of the human race: it was an air breather, capable of living outside the water from which it emerged, and to which it returned to lay its eggs.

Ever since the days of *Ichthyostega,* and the few of its relatives more recently discovered, Earth's lands have been inhabited. Animal life on the land was the direct result of the earlier growth and spread of lowly algae plants that transformed Earth's early atmosphere by photosynthesis, in which chlorophyll is transmuted into carbon and oxygen in the presence of sunlight. It was these early plants that made the lands habitable for animals.

The transition from life in the water to life on the land involved profound changes in all organisms—not only plants but also animals—that moved ashore. For animals it meant the development of new supporting structures to make allowance for the loss of buoyancy provided by water; the ability to withstand drying out; the means to take advantage of new food sources; and new methods of respiration, reproduction, and movement, as well as adaptations for seeing and hearing in a new environment. The development and mutual interdependence of these profound changes within such different animal groups as vertebrates and arthropods is an extraordinary process. To describe it as "an invasion of the land," as many writers have, is to see it only in long retrospect and to imply a degree of organic self-determination that was surely lacking. Remarkable as such change is, we can trace at least some of the exquisite transitional changes in fish by which it came about, in part, it seems, as a response to life in increasingly harsh freshwater environments, perhaps in areas of seasonal climate change.

Living amphibians are confined to moist and relatively warm environments, and the fact that their earliest ancestors are found in Greenland reminds us of how little the climate and geography of the present resembles those of the past.

The early amphibians we have just described—the ichthyostegids and related forms—appeared in late Devonian rocks deposited about 360–370 million years ago, and several different genera are known from Scotland, Latvia, Russia, Australia, Brazil, the United States, and China, as well as Greenland. Not all these seem to have been land dwellers, their strong limbs assisting, presumably, an active carnivorous life in the water. Some were clearly freshwater dwellers, but others seem to have inhabited coastal wetlands and brackish deltas, all of which supported a lush growth of vegetation by late Devonian times. They were presumably, like their present-day descendants, the frogs, toads, and salamanders, confined to moist and warm environments, where their distinctive reproductive structure of fishlike eggs, external fertilization, and lack of protection against drying out required them to return to the water to breed. Unlike later tetrapods, the icthyostegids had seven or eight toes on each limb, but, in detailed skull pattern and fundamental limb structure, they show many similarities to the crossopterygian fish from which they arose.

These early amphibians developed into a striking variety of later forms. Many belonged to a group of large, heavy, squat carnivorous forms known as labyrinthodonts (from the infolded wall structure of their hollow teeth), while others were known as stegocephalians (from their heavy, broad skulls). They were worldwide in their distribution and survived for some 100 million years. Many inhabited the widespread coal-forming swamps of the late Paleozoic, and some displayed structural changes that better fitted them for life on the land: changes in the nostrils, ear structure, and vertebral column, for example. A few forms grew to a length of fifteen feet. For some 120 million years, amphibians were the dominant "land" animals.

It was from such forms that living amphibians—frogs, toads, urodeles, and caecilians—later developed, about 150–200 million years ago, and it was also from these early amphibians that the reptiles—in all their exuberant variety—arose. But in one sense, though some of its members were large creatures, with worldwide distribution, the amphibians were imperfectly adapted to life on the land. Because

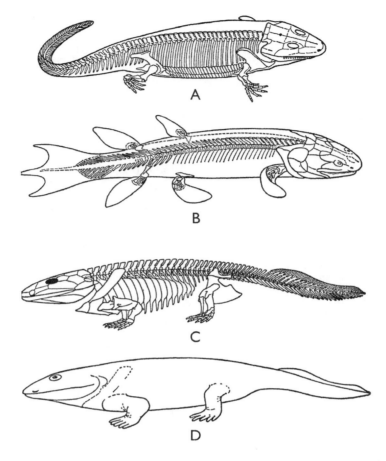

12.5 The origin of the amphibia: A, *Diplovertebron*; B, *Eusthenopteron*; C–D, *Ichthyostega*. From Frank Rhodes, *Evolution of Life* (New York: Penguin, 1976). Used with permission.

they were tied to the water to reproduce, they were—as their name implies— amphibian. Some developed into formidable terrestrial carnivores, while others— seemingly perversely—returned to life in the water from which their ancestors had slowly and recently emerged.

The changing geography of Earth exercised a profound influence on the development of life. Continental fragmentation, for example, produced varied, isolated environments and rich diversity in animal groups, while the aggregation of land into "supercontinents" tended to reduce organic variety. Several of the great episodes of animal extinction that marked the course of Paleozoic time seem to reflect changing patterns in continental geography, leading to profound changes in climate and vegetation. For example, most of the great lycopod trees and the varied amphibians of the Carboniferous coal swamps declined as the climate became more arid in early Permian times. Among the marine invertebrates, too, the end of Permian times saw a great dying— a sweeping wave of extinction in what was, perhaps, the most devastating episode of mass extinction in the history of the Earth.

The Reptilian Dynasty

The amphibians can scarcely be said to have "conquered" the land, but they did establish a significant beachhead. It was from this that their descendants, the reptiles, were to become the first fully independent, terrestrial animals, and the key to their success was the development of an egg produced from internal fertilization and enclosed in a tough protective membrane. This amniotic egg, providing both protection and nourishment, allowed the infant reptile to develop so that, by the time it hatched, it was already more or less fully developed.

The earliest known fossil reptiles are represented by a handful of specimens from Cheese Bay in Scotland, and so are named *Casineria* ("Cheese Bay"). These come from rocks of early Carboniferous age, deposited some 340 million years ago. Small lizardlike creatures, preserved in tree stumps and logs, are found in slightly younger, coal-bearing rocks from Nova Scotia.

From such lowly ancestral forms a veritable host of later reptilian forms developed, exploiting new diets, reflecting new patterns of life, exhibiting new structural forms, and colonizing new environments. Though "cold-blooded" like living reptiles, they were presumably able to exert a significant degree of control over their body temperature by orientation and movement and by choosing their particular location.

By the end of the Paleozoic era, some 100 million years after the appearance of the oldest reptiles, seven distinct reptilian orders had developed.

Some of the terrestrial forms were ten-foot-long active carnivores, such as the sail-backed pelycosaurs and the broadly similar, but vegetarian, edaphosaurs, in both of which highly adaptive tooth structures reflected their dietary preferences. Their distinctive sails, formed by vertically elongated vertebrae, probably represented some kind of thermoregulation device. These creatures, and the therapsid mammal-like reptiles that arose from them, seem to have occupied more extensive upland areas, demonstrating the gradual invasion and exploitation of environments far beyond the marginal swampy lowlands from which their amphibian forebears emerged. (Figure 12.6.)

Other reptiles returned to the water, invading not only the freshwaters but also the seas, where dolphin-shaped ichthyosaurs, long-necked plesiosaurs, and savage mosasaurs, as well as giant turtles, spread across the oceans.

One reptilian group—the pterosaurs—invaded the air, some of their gliderlike members attaining a wingspan of fifty feet. On the land, the period of

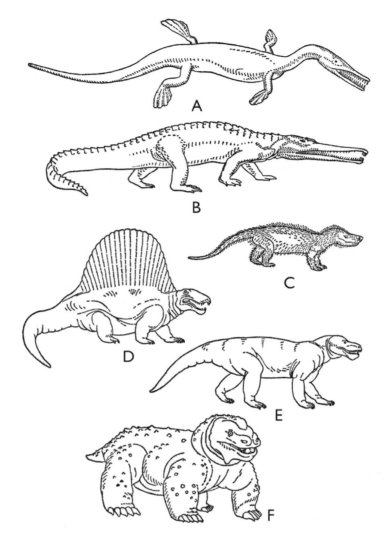

12.6 Early reptiles: A, *Mesosaurus*; B, *Rutiodon*; C, *Oligokyphus*; D, *Dimetrodon*; E, *Cynognathus*; F, *Pareiasaurus*. From Frank Rhodes, *Evolution of Life* (New York: Penguin 1976). Used with permission.

Permo-Triassic times—some 190–280 million years ago—saw a veritable explosion of reptilian form and adaptation. During this period, more "modern" groups replaced these earlier dominant mammal-like reptiles. The thecodont, "stem reptile," gave rise, not only to the pterosaurus and groups related to living crocodiles, lizards, and turtles, but also to the ancestral dinosaurs, whose varied descendants were to dominate life on the land for the next 160 million years. Many of the earlier reptiles were squat, sprawling, slow-footed creatures, but later ones became lighter in structure, more agile in build, and bipedal in posture.

The dinosaurs remain creatures of fascination. Two distinct groups developed: one—the ornithischians—characterized by a birdlike hip structure, the

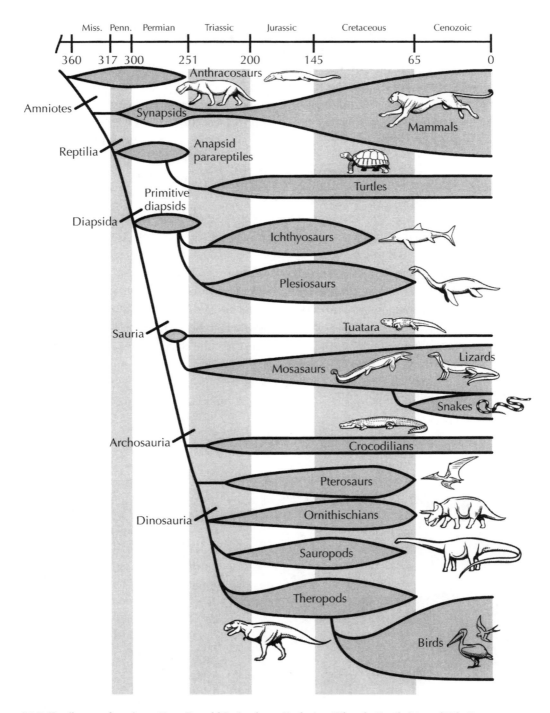

12.7 Family tree of amniotes. From Donald R. Prothero, *Evolution: What the Fossils Say and Why It Matters* (New York: Columbia University Press, 2007). Drawing by Carl Buell. Copyright Columbia University Press. Reprinted with permission of the publisher.

other—the saurischians—by a lizardlike hip. The saurischians included both bipedal carnivores, theropods, ranging from the slender, six-to-ten-foot-long *Coelophysis*, to the forty-foot-long *Tyrannosaurus*, and the huge, four-footed, herbivorous sauropods, the largest of which probably weighed some fifty tons. *Brontosaurus*, the "thunder lizard," is typical of the group.

The other great dinosaur group, the ornthischians, was vegetarian and its members exhibited extraordinary variety in later Mesozoic times. They included armored ankylosaurs, horned ceratopsians, plate-backed stegosaurs, duck-billed hadrosaurs, and many more, becoming worldwide in their distribution and prolific in numbers. Some appear to have lived in herds, and others are known to have used "nests" of sand in which to lay their eggs. (Figure 12.8.)

The upright posture of dinosaurs was quite distinct from the sprawling gait of the ancestral reptiles from which they evolved. Some have suggested they may have been warm-blooded, though conclusive evidence is lacking.

Overshadowed by the variety and success of the dinosaurs, one other inconspicuous group coexisted with them: the mammals. But throughout the long reign of the dinosaurs, mammals filled a lowly role and occupied a restricted niche.

For all their lengthy and exuberant history, the decline of the reptiles was as dramatic as their extraordinary development. For over 150 million years, they dominated life on the land, in the air, and in the seas, but by the close of Cretaceous times, 65 million years ago, most of their members had vanished, and they are represented today only by a remnant handful of lizards, turtles, crocodiles, and the lowly New Zealand tuatara (*Sphenodon*). What happened? How can such

12.8 Cretaceous dinosaurs with flowering plants. Detail, Rudolph Zalinger, "Age of Reptiles," Peabody Museum of Natural History, Yale University, New Haven, CT. Reproduced with permission.

a mighty race undergo almost total decimation. What can account for this great dying? For it was not just dinosaurs that vanished, striking though that was: it was also giant marine reptiles, flying reptiles, and even the ammonite cephalopods, which had dominated invertebrate life in the seas. And as these groups declined, flowering plants blossomed into dominance on the land. There seems as yet no simple explanation for change on so profound a scale. Meteorite or asteroid impacts, climate change, volcanic outpouring, continental fragmentation, rapid advance and retreat of the oceans over the continental margins, lethal ultraviolet radiation, mammalian predation—all these and more have been argued with conviction and force. The most probable of these, an asteroid impact that produced profound climatic change, is the current favorite, but though each hypothesis has its advocates, none so far seems capable of providing a wholly satisfactory explanation for the relative rapidity, geographic extent, ecological range, and biologic inclusiveness of so great a decline. (Figure 12.9.)

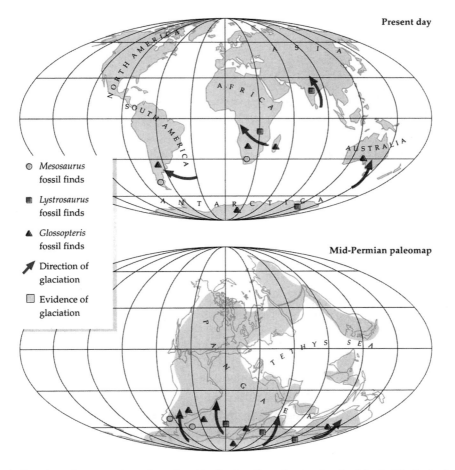

12.9 Permian paleogeography and reptilian distribution. From Stephen Jay Gould, ed., *The Book of Life: An Illustrated History of the Evolution of Life on Earth* (New York: Norton, 1993). Copyright 2001, 1993 by Michael Benton. Used by permission of W. W. Norton.

Feathers, Flowers, and Fur

Any thumbnail sketch of the history of life is likely to overlook the worthy in favor of the conspicuous, and this account is no exception. So it's worth remarking that, for all their continuing attention in the media, the ruling reptiles were not the only significant players on the Mesozoic stage. First conifers and then flowering plants came to dominate Earth's green mantle.

By the end of Cretaceous times, trees and shrubs, as well as plants, developed a reproductive structure that allowed growing colonization of increasingly varied environments, from the equatorial tropics to the fringes of the tundra regions. All animals ultimately depend on plants as a source of food, and the great reptilian expansion across the Mesozoic landscape reflected the spread of new kinds of plants. The earlier plants—from modest club mosses to tree ferns and giant-scale trees— were seedless and were limited by their reproduction to damp environments, such as the great coal swamps of late Paleozoic times. The dawn of Mesozoic times saw the emergence of seed-bearing plants, capable of colonizing a much wider range of areas. The oldest representatives were the so-called gymnosperms—"naked seeds"—in which the seed is not protected but is naked, as in pinecones. This group, which includes conifers, cycads, and ginkgoes, became worldwide in distribution in early Mesozoic times and must have provided an important source of food for herbivorous reptiles. Gymnosperms are well represented as fossils and are still abundant in many areas.

The second great group of seed-bearing plants—the angiosperms—includes the flowering plants, represented by some quarter million species today. In them, pollen fertilizes the egg, which develops into a seed, whose hard covering protects it against drying. Flowering plants, which appeared in Cretaceous times, came to depend on animals to assist in pollination, and their rise was accompanied by a marked expansion of herbivorous dinosaurs and pollinating insects. With their rapid expansion, the Earth literally broke into bloom. This blossoming of the land was later to play a decisive role in the rise of the mammals, by providing a rich source of food for the newly developing herbivores. Side by side with this blossoming of Earth, in mutual interdependence, insects in countless varieties and of exquisite form fertilized and fed on these plants.

It is perhaps no coincidence that the remains of the earliest known birds are found in rocks of Mesozoic age, though the bird in question—*Archaeopteryx*— was, for all its feathers, clearly a carnivorous predator, with sharply pointed teeth and claws reminiscent of the carnivorous dinosaurs from which it developed. Birds are warm-blooded vertebrates with highly developed senses and, of course, flying ability of unique refinement. But they are also poor candidates for fossilization, their fragile skeletons and avian habits making them only rarely preserved. In fact, after the appearance of *Archaeopteryx*—known only from five specimens from one locality in Germany—there is a gap of 50 million years in the fossil record of birds. By the time we meet them again, in rocks of late Cretaceous age, we find them represented by seabirds, some of gull-like form and others of diverlike form, with webbed feet. A little later in time, giant, flightless carnivorous birds appear, indicating not only what remarkable adaptability birds represent but also how little we know of the detail of their remarkable history.

Flowers and fliers—whether insect or bird—make up much of life's economy in our present world, but there is one other group in which we have a significant

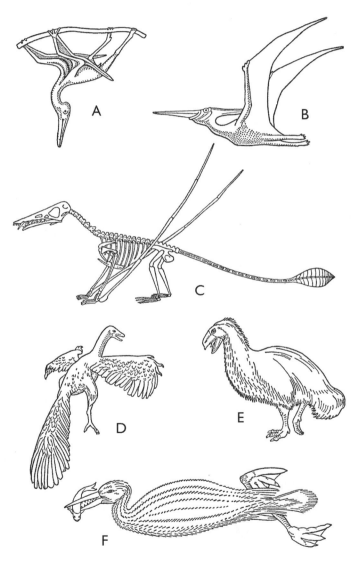

12.10 Flying reptiles and birds: A, pterosaur's probable method of walking; B, *Pteranodon*; C, pterosaur skeleton; D, *Archaeopteryx*; E, *Diatryma*; F, *Hesperornis*. From Frank Rhodes, *Evolution of Life* (New York: Penguin, 1976). Used with permission.

interest. With the great decline of the dinosaurs in late Cretaceous times, the lowly fur-bearing creatures underwent explosive expansion, exploiting the niches once dominated by their reptilian forebears. From that distant expansion, we ultimately arose.

Profound changes in global geography and climate during late Mesozoic times may well have influenced the decline and extinction of the reptiles and other groups. The small mammals of that period—insectivores and others—persisted but gave rise to a great expansion of new forms, which rapidly filled the environments once dominated by the dinosaurs. Hoofed vegetarian mammals (ungulates) developed rapidly into an astonishing variety of horned and hoofed creatures providing the forerunners of such living ungulates as horses, cattle, elephants, deer, camels, giraffes, and rhinos. Some rhinolike forms attained huge size and most developed specialized teeth for biting and chewing vegetation. Both browsing and grazing forms are known, and the latter underwent great expansion after the widespread development of upland prairies and grasses in Miocene times, between about 5 and 25 million years ago.

Alongside and preying on these herbivores were the carnivorous creodonts— hyena-sized, long-tailed, slender, clawed, active creatures, with strongly differentiated teeth—which appeared in the Cretaceous. From weasel-like members of this group the ancestors of living cats, dogs, and bears developed. Other mammals, including the ancestors of web-footed seals and walruses, replaced the great host of marine reptiles in the oceans. Closely related cetaceans (whales and dolphins) appeared in Eocene times. The niche once occupied by flying reptiles was later filled, not only by a greatly diversified range of birds, but also by flying mammals, such as bats. (Figure 12.11.)

The path of evolution between different genera is particularly well illustrated by such groups as horses and elephants, where successive forms display adaptation to changing environments and plant life.

Because terrestrial mammals are influenced in their distribution by ocean barriers, the changing faunas and migrating patterns of some groups—such as the mammals of North and South America—can be traced in exquisite detail.

Human Origins

Among the less conspicuous of the early mammals were small, insectivorous, arboreal shrew-sized creatures, which were contemporaries of the dominant dinosaurs some 85 million years ago. From such small creatures it seems likely that the later primates developed. This group—primate means "the first"—includes living lemurs, tarsiers, monkeys, apes, and humans. For all their differences, all the primates share several distinctive features, including large brains, acute vision, hands

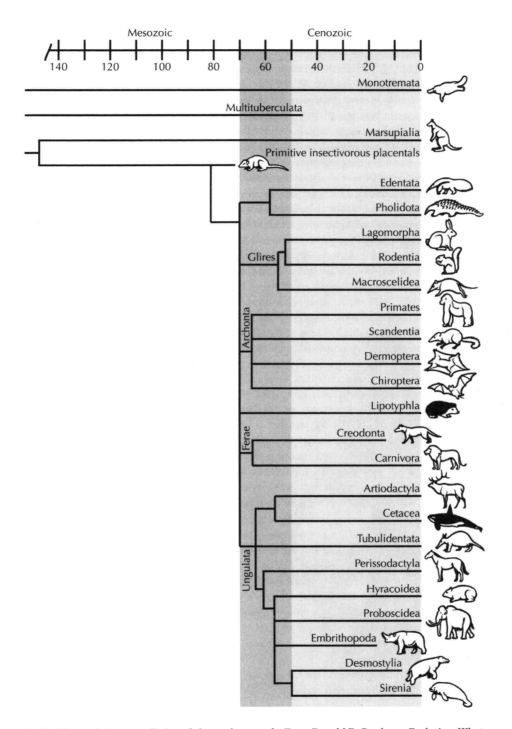

12.11 The evolutionary radiation of placental mammals. From Donald R. Prothero, *Evolution: What the Fossils Say and Why It Matters* (New York: Columbia University Press, 2007). Drawing by Carl Buell. Copyright Columbia University Press. Reprinted with permission of the publisher.

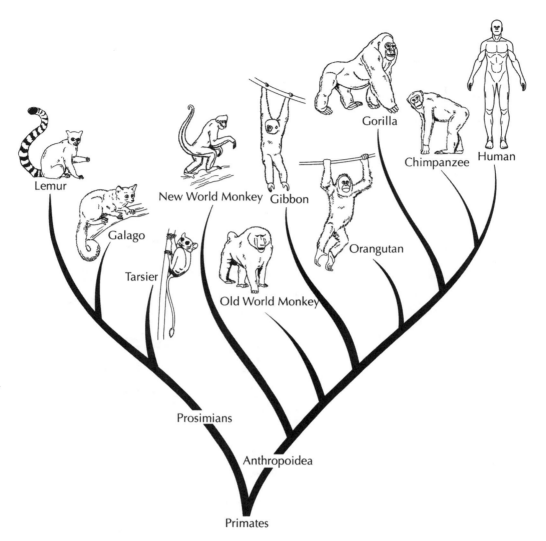

12.12 Family tree of the major groups of living primates. From Donald R. Prothero, *Evolution: What the Fossils Say and Why It Matters* (New York: Columbia University Press, 2007). Drawing by Carl Buell. Copyright Columbia University Press. Reprinted with permission of the publisher.

adapted to grasping, slow growth and development, rather long life spans, few offspring, membership in social groups, and displays of social behavior. All seem to have developed from a common ancestor some 60 million years ago. Primates tend to be relatively rare as fossils, largely because of their generally arboreal habits, to which their large eyes, stereoscopic vision, and grasping hand structure have made them superbly adapted.

One group of primates—the hominines, or hominoids—includes both apes and humans. Some living apes are chiefly arboreal—the gibbon and the orangutan, for example—and are exquisitely adapted to life in the trees; others—the chimpanzee and gorilla—are chiefly ground living. All lack the tail typical of monkeys. Ancestral fossil forms are widely known from rocks of Miocene and Pliocene age. In fossils, humans are distinguished from apes chiefly by their distinctive semicircular dental pattern, as opposed to the rather quadrate pattern of apes, as well as by their larger brain size and their toolmaking abilities.

One of the problems in tracing human origins is defining the term *human*. Our species is unique not only in its anatomy but also in its degree of intelligence, creativity, conceptualization, and communication. The human brain is large. Fossil "humans" are generally defined by their tool-making habits and skills. Direct evidence of human use of language goes back only about five thousand years, but the extensive use of language—as opposed to calls, expressions, or gestures—makes humans unique.

The oldest stone tools are some 2.6 million years old, created not only from conveniently, naturally formed stones, but from flaking, by striking one stone on another. The most basic tool—the hand ax—appears some 1.5 million years ago.

The oldest fossils that fit these tool-making criteria are the australopithecines (southern apes), who may well be ancestral to our own species, *Homo sapiens*—"wise men," as we choose to name ourselves. These creatures, and their immediate forerunners, *Ardipithecus,* were widespread in Africa about 5 million years ago. They were about four feet tall, upright in posture, with heavy brows, rather protruding jaws, and a brain capacity about half that of living humans. Four fossil species have been identified, including both earlier "gracile" and later, closely related, "robust" *Paranthropus* forms. Footprints that are 3.7 million years old, found in volcanic ash at Laetoli in Tanzania and presumed to have been made by *Australopithecus afarensis,* show these creatures to have been bipedal. They lived from some 1.2 to 4 million years ago and overlapped the earlier members of the genus *Homo.* Stone tools and butchered bones found in association with the later species *A. garhi* show them to have been carnivores.

The precise origin of humanoids remains a matter of debate. *Sahelanthropus,* a hominid discovered in 2001, appears to be about 6–7 million years old, while genetic studies of both humans and chimpanzees suggest a later divergence and origin about 5 million years ago.

Our own genus, *Homo* ("human" or "man"), is represented by only a single living species, but five extinct species are known, stretching back some 2 million years in time, and these are found in Africa, Asia, and Europe.

Modern humans appeared in Africa some 250,000 years ago, descended, it seems, from *Homo erectus*, a form widely known from Africa, Europe, and Asia, or *Homo ergaster,* a closely related species.

Homo erectus has all the bodily characteristics to suggest that it (he? she?) is directly ancestral to our own species. Intermediate in skull structure, brain size, and posture between *Australopithecus* (with whom "he" seems to have been a

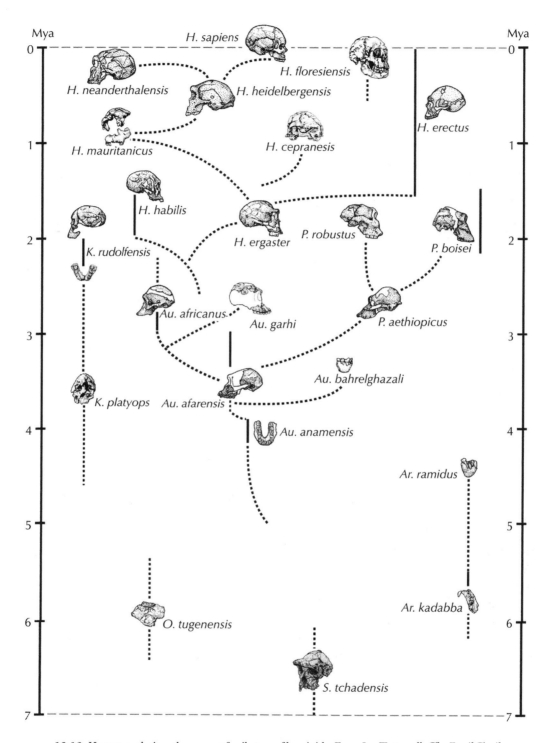

12.13 Human evolution: the current family tree of hominids. From Ian Tattersall, *The Fossil Trail: How We Know What We Think We Know about Human Evolution*, 2nd ed. (New York: Oxford, 2009). Used with permission of the author.

contemporary for some 2 million years or so) and *Homo sapiens, Homo erectus* was a skilled hunter, toolmaker, fire user, and probably a cannibal, and spread widely over Africa, Asia, and Europe.

These patterns of human development are based on fossil remains, but stone tools from ancient deposits provide additional evidence of lifestyles of these early forms. The oldest known tools are conveniently—but apparently naturally—shaped pieces of stone, found associated with remains of some 4-million-year-old australopithecines. The oldest known manufactured tools are some 2.6 million years old and consist of hand-sized pebbles that have been split by striking (see chapter 8, figure 8.1). Crudely chipped and fashioned stone tools have continued in use until our own times, but successively younger ones show increasingly fine workmanship (chipping, flaking, polishing) and increasingly specialized use (hammers, axes, scrapers, slicers, and so on). Decorative jewelry, statuettes, and cave paintings appear at a later date.

Neanderthal people—generally regarded as an early race of our own species rather than a distinctive species of our own genus—appeared some 350,000 years ago, during a warming episode of the glacial period. They included one group that inhabited areas of warmer climate, whose remains are associated with now extinct elephant, rhino, and elk species, and another, younger group from colder areas, associated with remains of now extinct wooly mammoth, rhino, reindeer, and cave bear. Neanderthals were cave dwellers, with marked brow ridges, a brain size comparable to that of contemporary humans, and males who had an average height of some 5.5 feet. Neanderthals buried their dead with care and buried tools with them, implying, perhaps, some belief in an afterlife. They were widely distributed throughout Europe, Asia, and Africa. Later forms were contemporaries of, but probably not directly ancestral to, our own species, *H. sapiens sapiens.* Some recent DNA studies suggest that they may have interbred with our own species, but this conclusion is controversial.

Neanderthals were replaced by a race known as Cro-Magnons, who were, to our anthropomorphic senses, finely built, tall, muscular, large-brained humans, shapers of carefully crafted tools, not only of stone, but also of bone and ivory. Their remains are associated with cave paintings and carvings of great sensitivity and their dead were buried with ornaments.

The oldest known sewing needles are some 28,000 years old, though it is unclear whether they were used for sewing skins or cloth. The oldest pottery is found in Japan and dates from some twelve thousand years ago. The technology to form ceramic containers was developed some two to four thousand years later in Africa and the Middle East and later still in Europe.

Farming seems to have developed some ten thousand years ago, as the last glaciers retreated from Europe and Asia. Established communities and settlement followed and, with them, new skills and new crafts.

With the coming of modern humans a new force entered into Earth's long history. Widespread in distribution, unspecialized in structure, versatile in ability, these people—as we may now call them—crafted tools, first of polished stone and then of bronze and later of iron. Evidence of tool manufacturing sites and trading routes, of the growth of shelter and communities comes later in time. But so also does evidence of rapid growth and population spread, growing impact on the environment and the rapacious character by which modern humans, in spite of their brief span of existence, have already had a significant effect on the environment from which they sprang. We are no longer just one of the many species that share this parent planet. We are unique among all living things in the degree to which we exploit and pollute it. Every other species, of course, also exploits the planet. That's the way life works. But our very success, our explosive population growth, our burgeoning consumption, and our spread to every habitable corner of the Earth all extend our impact and increase the demands we make and the stresses we impose, not only on the planet but also on the countless creatures with whom we share it.

CHAPTER 13

The Warming Planet

"Stop climate change," exhorts the bumper sticker. It's a laudable ambition but an impossible task. Climate change has been an ongoing process for as long as the Earth has existed. That's 4.6 billion years. Earnest as these pleas and warnings about global warming undoubtedly are, they can be addressed only if we understand their context. Earth's temperature has always varied. The only times Earth has not been warming are when it has been cooling. It is never constant over any significant time span. Nor can it be. The dynamic system that climate reflects is the same system that sustains us. An unchanging system, nondynamic and dead, would mean a dead planet: a Moon or a Mars. Earth's priceless gift to us is its dynamic systems, and, of these, climate is the most dynamic. It is not just local weather that changes day to day. It is also the more general climate that changes year to year, decade to decade, century to century, millennium to millennium. Without that variety, we should not be here.

And these changes involve not only the experiences of a lifetime—"Ah, we had real winter, real snow, in my younger days"—but also the longer trends preserved in records of local weather, rainfall, and temperature. Historical global records based on direct measurements go back only about 150 years, but there are somewhat longer local historical records, including temperature estimates based on records of harvest yields, crop prices, and other indirect indications of climate, which go back further.

We shall explore these in a moment, and trace the patterns that they show, but it is worth making one observation before we do. Many of these records reflect natural variations in our climate. The earlier ones occurred long before humans appeared on the scene. Some of the later occurred while human numbers were so small and human influence so limited as to make it unlikely that the changes reflect the effects of human intervention. This does not imply that present human influence is unimportant. It is not. But it can be understood only as an added influence, imposed on the underlying pattern of natural climate change that is part and parcel of our planet.

Earth Past: Paleoclimatology

Paleoclimatology is the study of ancient climates; through such study some sense of the climate of the past can be developed. One of the most useful questions we can address is whether Earth's past climates have included times that were similar to those that are now projected for the future and, if they did, how well Earth and its inhabitants responded. We need to know what the physical results might have been. And it will also help us if we are able to reconstruct the timing of such changes and understand whatever we can about their natural causes. The great potential benefit of this knowledge is that it will then allow us to compare present climate projections—with all their admitted uncertainty—with the events of the past, and so test and refine our responses to them.

First, the historical evidence: reliable local records of weather are limited in time but they can be extended back by making use of additional biological and physical evidence—known as proxy records—that reflect past weather conditions. These indirect records serve as a proxy for direct readings because they record the responses of animals, plants, water, and air to our changing climate. Tree rings, for example, can provide an annual record of climatic conditions that extends back as far as ten thousand years. Beyond that, the distribution of fossil wood from ancient forests can be used to indicate the distribution of distinctive coniferous and deciduous species, which in turn allows us to reconstruct ancient climatic zones. Pollen, preserved in sediments and deposited in ancient lakes, can be used to reflect the mix of characteristic plant species, and thus of vegetation zones of earlier times. Studies such as these can be used to reconstruct some aspects of the climate and geography in particular regions.

On a more global basis, two additional methods are available. First, cores obtained by drilling through the sediments of the ocean floor contain the fossil remains of microscopic plankton, animals, and plants that float in the surface waters of the oceans. The study of the ratio between two oxygen isotopes (^{16}O and ^{18}O) of the minute plankton "skeletons" has proved to be a reliable indicator of ocean

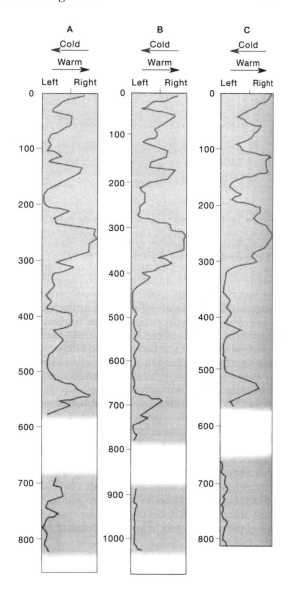

13.1 Proxy records of temperature based on distribution of temperature-sensitive *Globorotalia truncatulinoides*. From Harold L. Levin, *The Earth through Time*, 9th ed. (Hoboken, NJ: Wiley, 2009), fig. 15–42, p. 492. Used with permission of the publisher.

temperature at the time these microorganisms lived. Similarly, on land, deep drilling in the ice sheets of Antarctica and Greenland has produced ice cores, drawn from depths as great as two miles down. Ice, like wood, forms in annual layers and retains an imprint of the conditions under which it formed. Studies of the hydrogen and oxygen preserved as ice allow the reconstruction of temperatures over a two-hundred-thousand-year time span, while bubbles of air trapped in ice provide a proxy record of earlier atmospheric composition, especially the concentration of the greenhouse gases (CO_2, CH_4, and others). In practice, the older ice at lower levels in these ice cores becomes so compressed that the earlier (i.e., lower) record tends to be less reliable than the later. These land-based ice records can often be correlated with those of the ocean sediments by the presence of thin bands of volcanic dust, which settled both on the land and on the surface of the oceans as clouds of ash from particular volcanic eruptions reached the stratosphere and drifted across the planet.

Other direct measurements of some surface features also provide an indication of recent global temperature change. The rate of melting and retreat of some European glaciers has doubled in the last twenty years, for example, providing direct proof of the impact of observed temperature change.

These various reconstructions can be pieced together to give a broad portrait of climate change over the past millennium. But it is a portrait, not a movie record. Like projections of future climate, it has its own limitations. But, though exact comparisons with both present and future are not possible, it is still a useful portrait.

Reconstructions of temperatures of the earlier years of the past millennium indicate, as we shall later see in more detail, a warm period over Europe from the tenth to the fourteenth centuries—the so-called Medieval Warm Period—recorded in historical documents. This was followed by a drop in temperature that has been described as "The Little Ice Age," when, as recorded over much of northern Europe from the mid-fourteenth century to the mid-nineteenth, global temperatures may have been some 1 to 1.5°C (1.8–3.3°F) lower than today.

Earth's Distant Past

Over a longer period, temperatures have shown far greater variations than those of the recent past. For much of its history, Earth has been warmer than it is at present, with higher sea levels and higher CO_2 concentrations than now. It has also, at times, been considerably colder. The Earth has experienced a total of at least four major periods of global glaciation in the past. Northern Europe and North America, for example, are blanketed with distinctive deposits of sand, gravel, clay, and huge boulders that are uniquely characteristic of the debris found in association with existing ice sheets and glaciers.

Oxygen isotope records suggest there have been fifteen to twenty "ice ages" in the last 3 million years, the earlier ones lasting some forty thousand years and the later some hundred thousand years. At least four of these are recognized as major glacial episodes, as continental ice sheets, over two miles in thickness, advanced southward to cover much of the Northern Hemisphere. In North America, the locations of such cities as New York (where Central Park still bears the erosional scars of glaciation) and Saint Louis were covered by ice. In Europe, the sites of present Copenhagen and Edinburgh were buried. Between these major advances

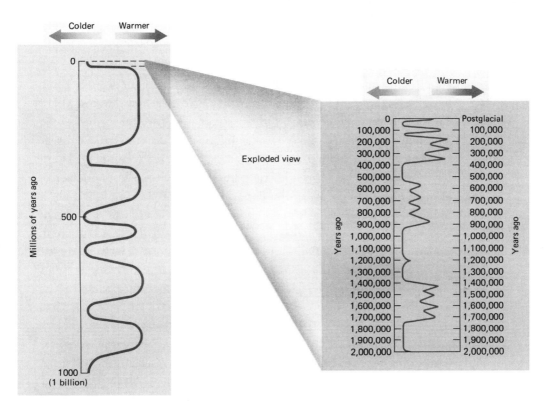

13.2 Average global temperature variations during the past one billion years. The times of lowest temperature are thought to coincide with ice ages. The right-hand scale is an expanded view to show temperature fluctuations during the Pleistocene Ice Age. From Graham R. Thompson and Jonathan Turk, *Earth Science and the Environment (with EarthScienceNow and InfoTrac)*, 3rd ed., copyright 2005 Brooks/Cole, a part of Cengage Learning, Inc. Reproduced with permission, www.cengage.com/permissions.

of ice—each lasting several hundred thousand years—were warmer interglacial periods, each ten to twenty thousand years in length, marked by the northward migration of plants and animals, including humans during later periods.

The global pattern of cooling leading to this most recent episode of major glaciation (the so-called Pleistocene) began some 35 million years ago, though the glaciation itself began some 3 million years ago, and ended "only" some ten to eleven thousand years ago (figure 13.2). We now live in a period (the so-called Holocene Period) that may be either an interglacial interval, before the onset of the next glacial epoch, or a postglacial period, marking the close of this recent episode of prolonged continental glaciation. Which of these our present period is, we do not yet know. We do know, however, that the presence of significant volumes of permanent polar ice is an unusual feature in the history of the Earth.

Although it was in such unusual glacial conditions that humans migrated across Eurasia, glacial episodes such as that from which we have recently emerged have been rare in Earth history. In only a handful of earlier times—in the Carboniferous-Permian rocks of the Southern Hemisphere, deposited some 260 to 340 million years ago; in Ordovician rocks of North Africa and elsewhere formed 450 million years ago; and in the late Precambrian rocks formed some 550 to 850 million years ago ("Snowball Earth," or the late Proterozoic glaciation)—do we find evidence of huge ice sheets spread across the continents. There may also have been one or two earlier ice ages, but the evidence is less conclusive.

Each of these extended ice ages—the Permo-Carboniferous lasted 80 million years—was marked by repeated glacial and interglacial episodes. The interglacial episodes that separated recent advances of the ice sheets tended to last only some ten thousand years or so before the return of cooler conditions. The most recent switch from the peak of glaciation to the start of warming probably began about fifteen thousand years ago, and the temperature has probably risen 4° to 5°C (7.2° to 9°F) during that period.

Three questions emerge. First, what was the broader impact of this recent glaciation beyond the areas buried by ice? There were several changes that were particularly significant. Sea level fell by about 350 feet, as water became locked up in continental ice, while on land, drainage was blocked, lakes were formed, and river courses were modified. The broad continental shelves emerged above sea level and allowed migration of mammals, including humans, across the Bering Strait from Eurasia to North America. In addition to this impact, climatic belts were displaced and weather patterns severely disrupted. Areas that are now arid—the Sahara, for example—enjoyed significant rainfall. Vegetation belts were displaced north–south and oceanic circulation patterns most probably were influenced.

Second question: What sort of temperature variation do these glacial episodes represent? We cannot yet be sure. But in the second-to-last interglacial period (the Eemian), temperatures may have been 1° to 2°C (1.8–3.6° F) higher than those at present, though this is not certain, and sea level was probably five to twenty feet higher. Records of the ancient concentrations of greenhouse gases show a strong correlation of high concentrations with warming episodes. But that may not necessarily mean that greenhouse gas emissions are the *cause* of global warming. Cold water holds more CO_2 than warm water. Increasing CO_2 concentration may, therefore, be a product of temperature increase, caused by such other means as increase in solar activity.

Third question: What is the cause of such extreme, natural climatic fluctuation within a comparatively brief span of time? In a brief period of little more than 2 million years, four glacial advances buried the Northern Hemisphere. How can we account for such changes? We do not yet know, and each mechanism offered seems to present obstacles to its acceptance. Many observers now believe that the broad onset of glacial conditions may be caused by particular disposition of the

13.3 An ice sheet covered most of northern North America 18,000 years ago. It was thickest near what is now Hudson Bay and from there flowed outward in all directions, as shown by arrows. From Graham R. Thompson and Jonathan Turk, *Earth Science and the Environment (with EarthScienceNow and InfoTrac)*, 3rd ed., copyright 2005 Brooks/Cole, a part of Cengage Learning, Inc. Reproduced with permission, www.cengage.com/permissions.

continents near the poles and distinctive patterns of oceanic circulation. Such geography may also have been accompanied by other effects—for example, changes in the incidence and distribution of solar radiation on Earth's surface may be a significant factor, and that, in turn, could be influenced by slight changes in the Earth's axial tilt, orbital eccentricity, precession (wobble), and orbital inclination around the sun. These changes occur within a regular but complex cycle, the Milankovitch cycle.

But the repeated temperature swings that mark ice ages are not easily explained by any one such mechanism, though it is possible that pre-Pleistocene temperatures were too high for them to have been influential until a tipping point was reached. Another possible extraterrestrial cause is variation in solar activity, and thus in solar radiation. We know from satellite observations that the intensity of solar radiation does vary over an eleven-year cycle, which is correlated with the development of sunspots. Here again, this mechanism alone seems inadequate to create the major temperature variations experienced on Earth, being able to account for no more than 0.5°C variation. There may, of course, be longer-term solar cycles of which we are not yet aware, and solar influences may possibly combine with those from other sources.

Other suggested mechanisms are of terrestrial origin: intense volcanic activity, or complex and perhaps chaotic interactions of oceans and atmosphere that may trigger climate change or changes in oceanic currents, but here again the scale of change involved seems inadequate to create a glacial episode.

Over longer time frames than that of the last 1 or 2 million years, other factors have undoubtedly had an influence on climate: the changing number, positions, and nature of the continents, the distribution of mountain chains, and changing patterns of solar radiation, volcanic activity, and oceanic and atmospheric circulation, for example, have clearly had significant influences on climate change.

Over shorter time frames, by contrast, though the climatic trends are clear, the mechanisms that produced them are not. Global warming, for example, though clearly established during the last thirty years, was preceded for a quarter century or so by global cooling. This suggests that human influence may be insufficient by itself to account for the present phase of global warming, and may be superimposed on an underlying pattern, produced by natural causes. At times one will counteract the other: at other times, they will reinforce one another. That may be both a help and a hazard in our uncertain future.

Unknown in its relationship to these various possible causes is the exact degree of impact of greenhouse gases. The general range of temperatures we have experienced and, at least in part, produced in the last 250 years or so seems broadly comparable to the temperature estimates that marked previous warm interglacial episodes. From those episodes, Earth emerged, as did our earlier ancestors. That conclusion may offer us only limited comfort. We know nothing, as yet, about the complex interactions of these possible mechanisms. There may well be other causes of which we are now unaware. And there may be a tipping point, at which the dynamic equilibrium that marked the stability of the earlier part of an extended warmer phase, such as the present, is disturbed, to produce a quite new, sudden, and unpredictable climatic change. And there is one sobering comparison. During the last 650,000 years, atmospheric CO_2 concentrations did not exceed about 300 parts per million, even during interglacial periods. They have now reached 390 ppm as of 2011.[1]

---

1. Nicholas Stern, *The Economics of Climate Change: The Stern Review* (Cambridge: Cambridge University Press, 2007); "Trends in Atmospheric Carbon," NOAA Earth System Research Laboratory website, November 2011, Dioxidehttp://www.esrl.noaa.gov/gmd/ccgg/trends/global.html.

Helpful as they are as indicators of the impact of past climatic change, and as proxies to test the validity of present models against past trends and events, we cannot conscript the data of the distant past as a comprehensive model for the present and the future. The very factors that influenced past climate change—the geography of many earlier times, as well as the elevation and detailed form of the continents, the location of mountain ranges, the pattern of ocean currents and climatic belts, and so on—were all sufficiently different from those of the present as to undermine precise predictions based on close comparisons. We know, for example, that large dinosaurs lived in what is now Antarctica and Alaska during earlier times. We can conclude, in general terms, that Earth and life have shown remarkable resilience to intense climate change. But even that requires clarification. We—mammals though we are—are in some respects less adaptable than our mammalian or even reptilian or invertebrate forebears. More inventive we may be, but more mobile we are not. We cannot move Chicago three hundred miles south, or Calcutta one hundred miles north to escape climate extremes. We could, of course, rebuild such cities, but we should not underestimate the huge costs and implications of such climatic displacement. For all our creativity, our patterns of settlement have made us more rooted and therefore less mobile than many of our fellow creatures.

Earth's natural climatic changes have been both significant and continuous. Earth's temperature reflects both incoming radiation from the sun and the reflectivity (albedo) of Earth's surface. Periods of climatic stability are rare. On these changing natural patterns we now impose our own. It is to that discussion that we now turn.

The State of the Planet

In the context of the past, what are Earth's present vital signs? How significant is climate change? How real is global warming? What is the probable outcome of whatever trends we can determine? What is the likely impact, both on the planet and on us? And in what time frame is all this likely to take place? These are the practical questions on which any effective policy has to be based. We shall later have to address the additional questions of what, if anything, we can and should be doing about it. In the meantime, one question at a time.

So, what is the present situation? Let's summarize: longer term, we are in a natural warming trend, which began about fifteen thousand years ago. Some ten to eleven thousand years ago, this warming led to the start of melting and the retreat of continental ice sheets that had covered much of Eurasia and North America, when temperatures were perhaps 4° to 5°C (7.2–9°F) lower than those at present. This more recent warming trend has not been constant. Cooling spells over a span of both several decades and several centuries have taken place within the general warming trend of the last millennium.

From 2000–600 BC, for example, it was about 2°F colder than it is today; from AD 200–600—the warm Roman period—about 1°F warmer. The Dark Ages (AD 600–900) were some 2°F colder, the Medieval Warm Period (900–1300) some 1°F warmer, and the Little Ice Age (1300–1850) some 2–3°F colder. Since 1850 we have been in a Modern Warm Period, but as recently as the 1950s to the early 1970s, the general warming trend has been reversed. These reversals seem sometimes to have been abrupt, rather than occurring over an extended time span, perhaps reflecting changes in oceanic salinity or circulation. Temperatures during the 1980s and '90s, though warm, were relatively stable. There is some evidence we may now be in the early stages of a new cooling trend.

But what of the broader picture? Compounding this recent broad natural warming trend is the additional impact produced by an increase in greenhouse gas emissions. GGEs include water vapor (H_2O), carbon dioxide (CO_2), methane (CH_4), nitrous oxide, and other gases that absorb and retain outgoing solar radiation reflected by Earth's surface, thereby acting as an atmospheric blanket and so increasing global temperatures. Studies of air bubbles trapped in ancient ice indicate that concentrations of these gases have been steadily increasing since the time of the Industrial Revolution (about 1750). Direct readings show that carbon dioxide concentration in the atmosphere has increased some 30 percent since the mid-nineteenth century, and continues to increase by more than 1 percent every year. Methane and nitrous oxide concentrations have also increased over the same period. These increases in greenhouse gases are the clear result of human activities, reflecting the growing emissions produced by increased industrialization, transport, deforestation, intensive agriculture, urbanization, and growing population. And the gases have long atmospheric lifetimes: nitrous oxide over a century, for example. Because records of high greenhouse gas concentrations show strong correlation with higher temperatures in ice cores, these increases are assumed to compound natural temperature increases. There is no significant disagreement in the scientific community on this point. The debate, rather, is about the degree of impact and future scale and timing of human influence on global warming. The United States and Europe, for example, are likely to be able to adapt to modest increases in global temperature, but other regions face serious problems.

If these temperature increases prove to be small and take place only slowly, most countries could probably adapt to them. The problem is that because of growing GGEs from industrialization, deforestation, and transportation, a number of projections suggest that such temperature increases are unlikely to be slow and could produce temperature change of some 2.5°C (4.5°F) over the next century. To initiate that degree of potential warming would be, as some have remarked, to conduct a whole Earth experiment in real time.

That warming may well be compounded by other unpredictable events: changes in the pattern of oceanic circulation, for example, or the possible release of methane gas, now trapped in the frozen tundra or on the ocean floor.

Perhaps the most striking evidence of climate change comes from the shrinking coverage of the polar ice. On the Antarctic peninsula, the Larsen Ice Shelf is breaking up, with an area the size of Delaware disappearing over the last five years. In Greenland, the outer portions of the ice caps are melting rapidly, as are some permafrost areas. In Africa, the glaciation of Mount Kilimanjaro has decreased by 80 percent since 1912, and it is projected to disappear within a few decades if present trends continue. This shrinking ice cap may, however, also reflect increased warming and decreased snowfall, produced by changes in land

13.4 Global average temperature and carbon emissions from fossil fuel burning, 1950–2009, and atmospheric concentrations of carbon dioxide, 1960–2009

Year	Carbon Dioxide (parts per mill. by vol.)	Temperature (degrees Celsius)	Emissions (bill. tons of carbon)
1950	n.a.	13.87	1.63
1955	n.a.	13.89	2.04
1960	316.9	14.01	2.58
1965	320.0	13.90	3.14
1970	325.7	14.02	4.08
1975	331.2	13.94	4.62
1980	338.7	14.16	5.32
1981	339.9	14.22	5.16
1982	341.1	14.03	5.11
1983	342.8	14.25	5.10
1984	344.4	14.07	5.27
1985	345.9	14.03	5.43
1986	347.2	14.12	5.60
1987	348.9	14.27	5.73
1988	351.5	14.30	5.95
1989	352.9	14.19	6.07
1990	354.2	14.37	6.14
1991	355.6	14.32	6.23
1992	356.4	14.14	6.10
1993	357.0	14.14	6.10
1994	358.9	14.25	6.23
1995	360.9	14.37	6.40
1996	362.6	14.25	6.55
1997	363.8	14.40	6.68
1998	366.6	14.56	6.67
1999	368.3	14.33	6.51
2000	369.5	14.32	6.64
2001	371.0	14.47	6.82
2002	373.1	14.55	6.95
2003	375.6	14.52	7.25
2004	377.4	14.48	7.57
2005	379.8	14.62	7.97
2006	381.9	14.55	8.22
2007	383.7	14.58	8.36
2008	385.5	14.44	8.51
2009	387.3	14.57	8.39

Source: Data compiled from GISS, BP, IEA, CDIAC, DOE, Scripps Institute of Oceanography, and Earth Policy Institute.

use. Meanwhile, in Europe, Alpine glaciers have shown significant rates of retreat, and milder winters now produce daffodils, magnolia, camellia, and apple blossom in full bloom in mid-February in some areas. In contrast to those warming trends, global temperature in 2008 was 0.4°F colder than 2005, and 2008 was the coldest year since 2000. Satellite images suggest that global sea ice in 2009 is slightly greater in extent (19.9 sq km) than it was in 1980 (19.7 sq km). Some of the changes described carry their own feedback mechanisms that may, in turn, accelerate temperature trends. Ice core studies show, for example, that at the end of the cool phase, known as the Younger Dryas, some 11,500 years ago, the rate of warming was more than four times greater than at present. The possibility of such an acceleration in warming happening again cannot be discounted.

Most changes in global temperature have been slow and gradual, but Earth's temperature is known to have changed abruptly and dramatically in the past, largely, it seems, because of abrupt reorganization of the ocean conveyor currents. Others have suggested that abrupt climate change could be triggered by the albedo effect of Earth's ice or cloud cover. Ice, for example, has a higher albedo, or reflectivity, than land or water, and so a slight increase in the ice cover reflects more solar radiation back into space, thus reinforcing the spread of ice or cloud cover. Any such change, if it does occur, is liable to be not only abrupt but also uneven in its effects on population, and may pass a tipping point beyond which positive feedback reinforces the intensity of the effect.

Present greenhouse models do not fully explain such variations, but most paleoclimatologists accept the implications of the correlation between greenhouse gas concentration and global warming in the geologic past and project that, even if we are able to cap GGEs today at their present levels, global mean temperatures would still increase by several degrees centigrade by 2050. Natural cyclic perturbations might increase or decrease that estimate, but probably not by more than about 0.5°C either way.

Earth's Recent Past

The most comprehensive review of the impact of climate change and present trends is given by the recent (2007) seven-hundred-page report of the Stern Commission,[2] a panel under the chairmanship of Sir Nicholas Stern, former chief economist at the World Bank, appointed by the British government to review the evidence for climate change and assess its economic implications. Such commissions are by no means infallible, but it is worth reviewing their major conclusions.

---

2. Ibid., 74–76.

I quote the conclusions of the Stern Commission at some length because they are among the most comprehensive and representative of what has become the conventional wisdom concerning global warming. Though they have involved thoughtful scientific judgments, the generalizations and conclusions are inevitably based on short-term direct observations, limited data, and significant assumptions. This is especially true of the modeling of future climate projections. The need for continuing refinement of observations and projections is illustrated by the significant quantitative changes in the commission's 2001 and 2008 reports. For that reason, I have also included some of the comments and concerns of other informed observers working on the same problems.

After a careful documentation and review of the body of scientific studies, the commission drew several conclusions:

- "An overwhelming body of scientific evidence now clearly indicates that climate change—largely caused by human activities—is a serious and urgent issue."
- "Human activity has had a major influence over climate change for at least the past 50 years." There is continuing debate on the extent of this influence, but there is general agreement that burning fossil fuels, as well as deforestation and industrialized agriculture, have a significant influence on climate.
- "Since the beginning of the Industrial Revolution (about 1750) atmospheric carbon dioxide concentrations have increased by about a third, to 380 parts per million (ppm) today." Other greenhouse gases, especially methane and nitrous oxide, have also increased, as has water vapor, the most pervasive greenhouse gas of all. But exactly how the effects of other greenhouse gases are related to water vapor is not fully known.
- "Global mean surface temperatures have risen by 0.7°C since 1900." This warming has not, however, been uniform across the planet, nor has it been without reversals. The period from 1940 to 1970, for example, was marked by falling temperatures, and there is some evidence that we may now be entering another cooling period.
- "Over the last 30 years, global temperatures have risen rapidly at around 0.2°C per decade, reaching what may be their highest level in the last 400 years." The commission noted that others regard present global temperature as the warmest level in the past twelve thousand years, and that nineteen of the hottest twenty years on record have occurred since 1980. It should be noted, however, that temperatures were higher during the "Holocene optimum" (five to seven thousand years ago) and the Medieval Warm Period than they are today.

- "The causal link between greenhouse gas emissions (GGEs) and temperature is now well established and allows climate modeling for a given level of atmospheric greenhouse gases"; and "if annual greenhouse gas emissions (GGEs) continue to increase at their present level, by 2100 their concentrations would be more than treble pre-industrial levels, creating a temperature increase of 3–10°C."

- "Temperature increases themselves may trigger and amplify further warming, by reducing the capacity of plants and soils to absorb carbon dioxide and possibly releasing methane now trapped in permafrost." Such positive feedback mechanisms could lead to an additional rise in temperature of 1–2°C (1.8–3.6°F) by 2100.

- "The impact of past GGEs has yet to be fully realized. Even if global emissions stopped immediately, the fact that 84 percent of the total heat increase is taken up by the oceans means that Earth would continue to warm by 0.5–1°C over the next few decades.... If GGEs continued at their present

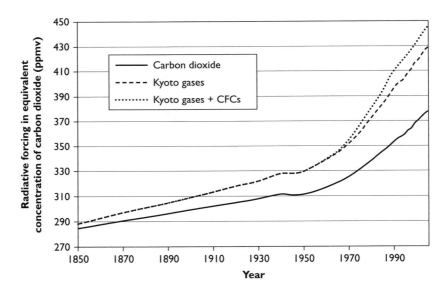

13.5 Rising levels of greenhouse gases: This figure shows the warming effect of greenhouse gases in terms of the equivalent concentration of carbon dioxide (a quantity known as the CO_2 equivalent). The bottom line shows the value for carbon dioxide only. The middle line is the value for the six Kyoto greenhouse gases (carbon dioxide, methane, nitrous oxide, PFCs, HFCs and SF6), and the dotted line includes CFCs (regulated under the Montreal Protocol). The uncertainty on each of these is up to 10%. The rate of annual increase in greenhouse gas levels is variable year-on-year, but is increasing. Source: *The Stern Review Report*, copyright British Crown 2006, *The Economics of Climate Change: The Stern Review* (Cambridge: Cambridge University Press, 2007), based on L. Gohar and K. Shine, Dept. of Meteorology, University of Reading.

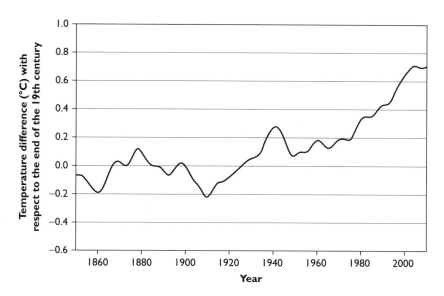

13.6 Global average near-surface temperatures, 1850–2011. Data supplied by the Met Office, British Crown copyright 2006, the Met office. Redrawn with permission.

levels, by 2050 GGEs would reach almost double pre-industrial levels and temperatures would rise by 2–5°C or more."

- "Temperature increases will be significantly greater in higher latitudes and continental areas. Polar regions will experience still higher temperature increases." Furthermore, "increasing global temperatures will increase the frequency of extreme climatic events, such as heat waves, hurricanes, droughts and flooding, and may also create significant change in rainfall patterns. They may also trigger sudden major changes in oceanic circulation and the pattern of global climate."

- "Sea level is now rising globally at 0.3 cm a year and the rise has been accelerating. The inherent time lag means this rise will continue for several more centuries, with a rise of up to one meter already 'built in.'" The impact of rising temperatures will also increase snowfall and cause some glaciers (such as those of West Antarctica) to grow.

- "The cost of tackling global warming now would be some 1% of global GDP. If nothing is done, the eventual cost would be at least 5% of global GDP and perhaps as high as 20%.".

While the conclusions of the Commission have been broadly embraced by the scientific community, a small but significant minority of informed writers conclude that there is no linear relationship between increases in atmospheric carbon dioxide and global mean temperature change. Global mean temperatures, it is argued, have both increased and decreased during the rise of atmospheric carbon dioxide. Other anthropogenic factors—deforestation and industrialized agriculture, for example—are said to be far more significant than CO_2 emissions. Water vapor is by far the largest influence on the greenhouse effect (some 95 percent), with total carbon dioxide—both man-made and natural—making up only 3.6 percent of the total contribution

So Kyoto-type calls for mandatory carbon dioxide reductions of 30 percent from developed countries would have a negligible impact on climate but a serious economic and social impact.

Other authors, while accepting global warming, argue that we should concentrate on adaption, as well as mitigation, and devote funding to health, nutrition, social, or other programs, as well as to longer-term climate research.

The Impact of Climate Change

If the planet is warming, what is the impact likely to be on the lives of the world's peoples and their subsistence, as well on Earth's wild places and wildlife? The effects of global warming will be unevenly felt in different regions and will vary with the amount of temperature increase. Modest increases (in the 1–2°C/1.8–3.6°F range above preindustrial levels) could benefit some regions, extending growing seasons and increasing crop yields. Above that level, there would be a rising toll on many species, with potential extinction for as many as 15–40 percent, and great hardship for many of the world's poorest people, especially those in south Asia and sub-Saharan Africa. Because the impact of climate change will be global in its extent, widespread in its influence, and long term and intergenerational in its effects, it creates complex policy and ethical issues. Consider one recent summary of the possible impacts, assuming a 2–3°C (3.6–4.5°F) rise over the next fifty years:[3]

- Melting glaciers will increase flood risks and reduce dry-season water supplies for one-sixth of the world's population, especially those of India, China, and the Andes.
- Declining crop yields will afflict many areas, especially in Africa.
- Increasingly acidic oceans will be created by rising CO_2 levels in the atmosphere.
- Rising sea levels will threaten many areas, especially in Southeast Asia, and small islands, as well as the world's large coastal cities, which will

---

3. Intergovernmental Panel on Climate Change, *Climate Change 2001: Working Group I; The Scientific Basis* (GRID-Arendal, 2003), http://www.ipcc.ch/ipccreports/tar/wg1/index.php?idp=338.

face massive population displacement. A 3°C (5.4°F) temperature increase above preindustrial levels could permanently displace some 200 million people.

- Disease from heat stress, malnutrition, and other causes will increase.
- Ecosystems will be under severe stress.
- Continuing temperature increase will become disproportionately more damaging, with possible triggering of rapid shifts in climate patterns and the creation of extreme weather events.
- Climate change will affect all areas, but the people of sub-Saharan Africa would be especially hard-hit.

Serious as such changes as these would be, temperature cannot be controlled by us, or by any other species. Nor will it remain constant for any significant time. Earth's dynamic systems inevitably create climate change. Existing temperatures already involve hardships, shortages, and displacements of populations. Falling temperatures would bring their own set of problems. Our task will be to accommodate these changes and adjust to them as well as we are able, optimizing the support and preserving as best we can the systems of the benevolent planet that supports us.

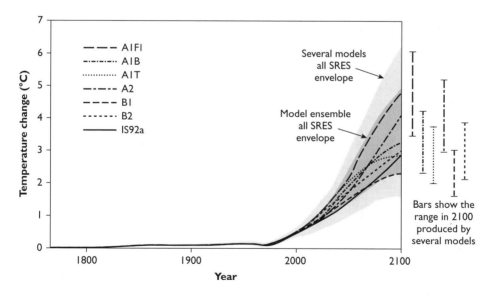

13.7 Global mean temperature change from pre-industrial levels and projections for the future. Global temperature in the twenty-first century will depend on natural changes and the response of the climate system to human activities. The graph shows various emissions scenarios. For further explanation of the scenarios see the IPCC Third Assessment Report "Climate Change 2001." Courtesy of UNEP.

Global Concerns and Policies

The 2007 report of the Intergovernmental Panel on Climate Change (IPCC) suggests that it is "very likely" (greater than 90 percent probability) that recent human activities lie behind recent climate change. This is a stronger conclusion than that reached by the IPCC six years earlier, when it concluded such a responsibility was only "likely" ("between 60 and 90%" probability). Furthermore, the actual increases in temperature and sea level over the last ten years have been higher than estimates that had been made in earlier studies by the same Intergovernmental Panel. The report of the panel is the result of the work of 2,500 scientists from sixty nations and represents six years of scientific study. It is probably broadly representative of the views of most people working in the area of climate studies.

We should also note that there is a significant body of literature that challenges what are regarded as the excesses of "bombast and exaggeration" used to describe the "vision of hellish climatic catastrophe" projected by some political leaders and climate specialists. These authors generally accept that there has been global warming over the last century, to which greenhouse enhancement has contributed, and that this warming may continue, but they challenge both the magnitude of future warming and the accepted extent of the impact of the human contribution. They also question the relative benefits of what they regard as the crippling costs of such programs as a carbon treaty, including the Kyoto Protocol, which would reduce greenhouse gas emissions by what are argued to be trivial amounts.

Global average surface temperature is currently about 15°C (59°F). While it is true that at various times in the geologic past it may have been as high as 27°C (80.6°F) and as low as 7°C (44.6°F), this natural variation in Earth's climate has now been reinforced by the effects of human activity and especially by the increase in the "greenhouse effect." Now, the role of the greenhouse effect is critical to life: without it Earth would be too cold for habitation. And all the greenhouse gases existed and have modified climate long before humans appeared on the scene. The concentration of some of these gases—such as water vapor—has remained virtually unchanged over historical times, but sharp increases in the concentration of other gases reflect the influence of burning fossil fuels, changing land use, and the intensive levels of mechanized agricultural production and factory farming required to feed a growing population. Precise measurements of CO_2 atmospheric concentration over the last decade, for example, show a steady annual increase of about 1.9 parts per million (ppm), reaching 390.19 ppm by volume in 2010. Now those measurements may not seem much to be concerned about, but extended measurements back to 1800 suggest CO_2 concentration has risen by some 30 percent in two centuries. And equally careful measurements indicate temperature is still rising: it increased by about 0.6 + 0.2°C in the twentieth century.

This rise in temperature has been accompanied by a rise in sea level of four to eight inches, largely from expansion of the warmer ocean waters and partly, perhaps, from glacial melting. It is projected to continue to rise throughout the twenty-first century and will continue to rise for at least a thousand years into the future. If this present warming trend continues, the Stern Commission projects that temperatures will probably increase by 3.2 to 7.2°F or higher by the end of the twenty-first century, with related sea level rises of 28–43 cm (0.92–1.41 ft). This will almost certainly involve longer and more severe periods of drought and heat waves.

These temperature increases will also affect the frequency of extreme weather events, with increasing numbers and intensity of heat waves and tropical storms. To recite such projections and describe such storms in abstract terms should not disguise the global havoc and misery they will create. Millions will be displaced from their homes, and the ecological balance will be profoundly disturbed. Such projections are not, of course, infallible, and both higher and lower temperatures are possible. For comparison, five years earlier the same Stern IPCC panel projected increases over the same period as 2–11.5°F.

In spite of continuing uncertainty in the complex computer modeling that lies behind such projections, one additional threat lies in the inertia built into the Earth system. Even if global CO_2 emissions were immediately to be capped at present levels, global temperature would still continue to rise for many decades to come, as a result of the heat capacity of the oceans.

Any solution to the problem of greenhouse gas emissions must be a global solution, but it cannot be effective without the active participation and leadership of the United States, Europe, China, and India. The United States, for example, has about 5 percent of the world's population but produces about a quarter of all GGEs. Though China now slightly exceeds the U.S. total in emissions, it has a population over four times that of the United States, and its level of GGEs is growing rapidly.

The impact of extreme climate change, if it develops, would be serious for all the world's peoples. It could be potentially disastrous for some of the world's poorest people. But projections are not destiny. Although some continuing climate change is already unavoidable, being built into the system by existing temperature increases of the oceans, we are not powerless to respond. We have the capacity both to limit future greenhouse gas emissions and to reduce future harmful impacts, even though both actions may offer the prospect of only limited relief. And countermeasures such as cap and trade will be costly, perhaps far more costly than now estimated, and they will impose their own degree of hardship. That is why another thoughtful minority of climate scientists asserts that we should "wait and see": that "hasty response is premature, or even ill-advised." "We should," it is argued, "wait for more evidence," before indulging in costly countermeasures.

Now it is true that more evidence will certainly refine our projections and improve our understanding of the prospects we face. Our knowledge is incomplete. It always will be, and there are many aspects of the Earth's systems involved where our knowledge is still sketchy. We must make improved scientific understanding of climate a priority. It is also true that countermeasures will be costly and will fall disproportionately on the world's poor. The 1 percent of GDP suggested by the Stern Commission may prove to be an underestimate. It is also true that there is no "silver bullet" solution to this problem as there was for, say, CFCs or acid rain. The processes that produce GGEs are the foundation of the world's economy. We should not—we dare not—decrease GGEs by destroying the economy that supports us. But inaction could well prove far costlier than action. The prudent course is to develop economically sound remedial measures, as well as improving energy efficiency and developing alternative energy sources and favorable economic policies that will provide benefits, whatever the future temperature trends may prove to be.

In the midst of the present debate the overall situation we face is clear. Although there is not unanimity, most reputable scientists agree that a continued increase in GGEs will produce significant increases in global temperatures. The debate concerns not the fact of global warming but the relative impact of GGEs and the extent and timing of future increases. The debate also involves the interpretation of climate records of the last two centuries and the reliability of very complex computer models of future trends. We must also remember that if existing paleoclimatic reconstructions are accurate (figure 13.2), Earth has, for most of the past 550 million years, been appreciably warmer than it is at present. That is why a combined, economically sensitive "no regrets" and "inherent benefits" program represents the most prudent policy response.

But we must also develop a far more sophisticated foundation of knowledge on which to base that response. An integrated reaction has to include intensive efforts to improve our scientific understanding, as well as investments in new energy technologies, improved methods of conservation, mitigation, and adaptation, and decreased deforestation. And all policies will need to be developed in a partnership between the authority of governments and the creativity of open global trade and economic markets. Developing these policies will require far more deliberate debate around the scientific, technological, and economic questions than is now taking place, as well as probing review of climate projections. Such policies will have to be adaptive, inherently beneficial in their own right, and open to ongoing correction and revision. Responsible leadership will require a careful calculation of the costs and hazards of future risks—not only climatic and agricultural, but also economic, personal, and social—and the costs and benefits of adaptation, mitigation, remediation, reduction, redesign, alternative energy sources, and improved efficiency. That is a topic to which we turn in chapter 24.

CHAPTER 14

The Polluted Planet

On June 22, 1969, the residents of Cleveland, Ohio, were startled to find the Cuyahoga River on fire—again! The river, which passes through the city on its winding one-hundred-mile way to Lake Erie, had long been heavily polluted from the cumulative effects of waste dumping, sewage disposal, industrial contamination, agricultural runoff, growing urban development, and dam construction. But pollution was one thing; a blazing river was quite another. It was, wrote one reporter, "the river that oozes rather than flows."

That fire, and the press attention it created, not only generated a torrent of legislation—local, state, and national—but also revived a sense of public awareness and outrage that contributed to the birth of the environmental movement.

For the burning river—spectacular though it was—was only part of a much larger problem. Lake Erie was already known as the most heavily polluted of the "Late Great Lakes." Its shallow depth, warm temperatures, large surrounding population, heavy industrialization, and adjacent agricultural use were known to have had a damaging effect on its wildlife and limited its recreational use. Fish with conspicuous tumors, birds with birth deformities, frogs with three legs and other abnormalities testified to its sinister influence. Choked with algae and weeds, it was officially pronounced "dead." It became the topic of popular songs and stories, including *The Lorax* by Dr. Seuss. It became a byword for the despoiling of the urban landscape.

Legislative Impact of Pollution

The flurry of legislation the Cuyahoga River fire produced—the creation of the EPA in 1970, the Great Lakes Water Quality Agreement of 1972, the Clean Water Act of the same year, for example—led to real improvements over time, with marked reductions in levels of water contamination and the healthy reestablishment of some native species.

Of course, the Cuyahoga River fire wasn't the biggest or the last environmental disaster—many worse were to follow—but it caught the nation's attention. With various degrees of militancy or moderation, the public concluded we had to "Stop Pollution." The slogan had a good ring to it and a virtuous motive.

But what was it, exactly, that had to be stopped? To pollute is to "make foul, dirty, or unclean." Surely every living creature pollutes to some degree; does not our very existence create pollution? The concern was a matter of degree. If the amount of disturbance, the degree of pollution in Earth's natural systems becomes such that it renders the environment unfit for its natural use (such as providing air for animals to breathe, water for fish to swim in, and soil for plants to grow in), then pollution becomes a matter of public concern. And that concern can never be purely local. Whether a particular point source of pollution or a regional catastrophe—the Chernobyl disaster of 1986, or the great wildfires of Africa and South America, for example—pollution can rarely be contained. The accident in the nuclear reactor at Chernobyl produced radioactive iodine in milk in Scotland and radioactive fallout in Sweden and other countries. The particulate matter from the African wildfires drifted across the Southern Hemisphere as far as Australia. Worse yet, many environmental stresses reinforce one another. For example, population growth magnifies the impact of groundwater contamination. More than a billion people still lack clean drinking water, and 2.1 million die each year from diarrheal diseases associated with this contamination.

Some pollution, of course, is inevitable: a part and product of natural processes. Dust particles from the explosion of Krakatau (Krakatoa) volcano in 1883, for example, affected weather throughout the world, leading to the "year without a summer" and grain shortages across the breadth of Europe. There's not much we can do about that kind of natural pollution. But other things we can do something about, especially when natural factors expose or reinforce the damaging effects of human pollution. The "Great Stink" of 1858 in London was produced by untreated sewage discharged into the River Thames, and its severity led to the construction of one of the world's first comprehensive sewage systems. The "Great Smog" of 1952 led to four thousand respiratory deaths in a week in London, and twice that number again in the next few weeks, but it triggered public concerns that led to comprehensive legislation for smoke abatement.

The Effects of Pollution

The disastrous consequences of the explosion on April 20, 2010, of the BP Deepwater Horizon drilling rig in the Gulf of Mexico have alerted us to the far-reaching hazards of pollution. Pollution can affect air, water, and soil and the most dire of its effects often fall most harshly on the world's poor. Air pollution, for example, is one of the world's most pervasive problems. With the growing rapid industrialization of the world's most populous countries, outdoor air pollution is estimated by the World Heath Organization (WHO) to lead to 3 million deaths a year, in addition to the deaths of another 1.6 million from indoor use of solid fuel. In some areas of the Gulf of Mexico, agricultural runoff has produced offshore areas that are described as dead zones. It is tempting to conclude that these tragedies involved villains and victims. But in one sense, each of us is both villain and victim.

None of these tragedies is necessarily the result of villainous behavior, corporate greed, technical carelessness, government neglect, or societal irresponsibility, though some or all of these factors may be involved. But the agriculture that pollutes the Gulf supports the people who live around its shores. Oil from the Gulf fuels the nation's trucks and automobiles. The stoves whose fumes overwhelm their users cook the food and boil the water that sustains their lives. The people of the developing world share the burdens of the cumulative degradation of two centuries of industrial activity in the developed world, whose members are, in turn, affected by the growing environmental impact of the developing world. And all of us share the consequences of the stark fact that pollution is the by-product of the world's peoples making a living. Our goal, presumably, is for that process of living to continue. So we need to develop ways in which to achieve it at much less expense to the benevolent planet that provides our sustenance. And we have—somehow—to work together to achieve it. It's not helpful to insist that because Cairo is the most heavily polluted city in the world—which it is—the "cleaner" cities of the developed world should wait for it to "catch-up" before further reducing their industrial emissions: we have only one atmosphere. We cannot wait for a perfect calibration of previous contributions, present impact, and future liability. If we are to address pollution, we must start now. And we shall have to assume that the shared benefits of present and future reductions will outweigh the supposed inequalities of allocated costs. So where do we start? Perhaps the best way is first to identify the causes and sources of pollution.

Sources of Pollution

Water is our most basic resource, and one of our most vulnerable. Direct pollution may contaminate streams, rivers, lakes, groundwater, or the oceans, but the linkage of each of these into a single interacting system—the water cycle—adds to our vulnerability. It is estimated that waterborne diseases are responsible for 80 percent of the illnesses and deaths in developing countries, killing about six thousand people a day.[1]

Sources of pollution may be thought of as representing three broad kinds of contamination:

- *Discharge or disposal* of polluting agents
- *Surface runoff* of soil, fertilizer, pesticides, or other harmful substances
- *Accidental leakages or spills* of hazardous materials

The effects of each of these may sometimes be magnified by the particular pattern of drainage systems or the form of underground aquifers.

The *discharge* or *disposal* hazards may involve inadequately treated sewage, industrial or chemical waste, heated or contaminated water from power or industrial plants, and acid rain caused by discharge of sulfur dioxide (SO_2).

Surface water runoff may contain agricultural manure and other wastes, as well as soil or silt from forestry and farming (which may in turn interfere with light's penetration of water), pesticides, fertilizers, industrial solvents, detergents, and petroleum and other products. Runoff may also carry mine tailings and wastes into drainage systems.

Accidental leakages and spills may involve any of the liquid items listed above, especially petroleum products. Several giant oceangoing tanker spills (the Exxon Valdez, for example) and the contamination from the explosion of the BP Deepwater Horizon drilling rig have tended to generate high public concern. Leakage of radioactive materials can be particularly harmful if allowed to contaminate water supplies.

Pollution of the soil comes largely from the agents listed above, but it tends to be more site specific than water contamination. Mining and industrial operations can, however, have blighting effects on broad areas if not adequately protected and contained.

Atmospheric pollution is, by its nature, more global in its impact, though, because of atmospheric dispersal and dilution, it is often less severe in its immediate effects. The Bhopal disaster in India in 1984, however, shows how devastating local effects can be. In that tragedy, an accidental leakage of industrial gases at a Union Carbide plant killed two thousand people and left some 150,000 injured. Though each of the activities we have already mentioned has some atmospheric impact, combustion exhaust gases and particulates from power plants,

---

1. NIKA Water Company, "The Crisis," http://www.nikawater.org/the-crisis/.

manufacturing, transportation, deforestation, and agriculture—especially slash-and-burn farming—are by far the most serious. The World Health Organization has calculated that 2.4 million people die each year from causes directly attributable to air pollution.

All pollution is ultimately global. Though local communities will suffer most from particular abuses, we share our wastes and our hazards with all humankind. Groundwater, for example, supplies more than half our water supply worldwide. Local contamination of groundwater, though it will pollute the local village pump, may also spread its harmful effects over a much larger region. That's why local improvements, vital as they are, will not solve our problems unless they are linked with reinforced regionally and globally designed policies and practices. The greenhouse effect is experienced worldwide, even though it is created by the cumulative effects of uncountable local incidents and practices. Our rapidly increasing global population—increasing 1.3 percent per year—compounds and complicates the severity of our problem.

Remedies

We have already seen that some degree of pollution is an inevitable by-product of living, and that all pollution is ultimately local. We should not overlook the fact that if we are to thrive as a global people, a critical part of our adaptive strategy will have to be local. The most destructive pollution can be reversed. Despoliation can be turned around. The Cuyahoga River no longer burns. Lake Erie has been cleaned up. These efforts involved state and federal programs, but they depended on local initiatives for their success. Enlightened self-interest is still a powerful driver of social policy. That's why local interests and efforts remain essential to environmental success. The Farmington River in Connecticut, for example, was heavily polluted for a century, but local voluntary efforts by the Farmington River Watershed Association, based on encouragement rather than enforcement, led to the return of salmon to the river in 1960 and the growing use of the river for recreation. And all this, ten years before the Environmental Protection Agency came into existence. The restoration of the Lower Swansea Valley(discussed in the following section) illustrates the same powerful influence of local concern and initiative. No national or international policies are likely to succeed without such local support.

There is a school of thought that argues that we can still continue business as usual: dumping sewage in the oceans, piping industrial waste into lakes and rivers, burning any fuel that suits us, in whatever way we need to. The "solution to pollution is dilution," the old slogan argues. Alas, there are limits to dilution; there is nowhere available to dump our garbage except in our own backyard. Dilution means putrefying streams and rivers, lakes and oceans. It is true that Earth is a resilient planet; it can absorb such insults. Its physical systems will adjust. But our species cannot adapt so easily, cannot adjust without drastic change. Nor can many of the species with which we share this planet. Reaccommodation for many of these will involve decline and, for some, extinction. For living things, the mechanism of accommodation is natural selection. Only the fit, only those suited to or able to adapt to the changed, stressful environment survive. That harsh reality is what we already begin to see.

To project in the coming decades a continuation of the rapid expansion in population that has marked the last two centuries would be quite unrealistic. In 1800 the global population stood at less than 1 billion. By 2000 it had reached over 6 billion, and was increasing so rapidly that it had doubled in the previous forty years. But that rate of increase cannot be sustained. It took place on a foundation that no longer exists. It depended on a pristine landscape with a soil bank accumulated over the unusually long and stable interglacial period that preceded it; it inherited a virtually untapped reservoir of coal, oil, and gas, representing the decay and accumulation of more than a billion years of plant and animal remains; and it covered a time span during most of which the much smaller, less efficient population placed much less demand on Earth's resources.

Today, the landscape is widely defaced and is under growing stress from urbanization. The soil bank is now much depleted and, though coal is still relatively plentiful, oil is now approaching peak production. Nor can either the soil bank or our store of fossil fuels be rapidly replaced. It takes an average of five hundred years for an inch of topsoil to be formed and it is estimated that we are losing 25 billion tons of topsoil each year. Land covers only about 29 percent of the globe, and of that, some 50 percent is occupied by desert, polar regions, or mountainous areas where temperatures and elevation limit productive soil formation. Of the small remainder of Earth's total area—some 12.5 percent—about 40 percent is too wet, rocky, or steep for effective cultivation of crops. So it's only the remainder of Earth's surface—about 7.5 percent—that can be used for agriculture, and even this limited area must also provide space for the growing spread of urban development, manufacturing, and transportation networks that support our burgeoning population.

How can we begin to address this problem? There are, I believe, four complementary approaches that offer promise.

Redesign mining, industrial, power generation, agricultural, transportation, commercial, and domestic appliances and practices, both to increase efficiency and minimize harmful emissions, discharge, leakages, and runoff. This will require cleaner and more efficient devices in every area of life as well as the development of alternative fuels, including both renewable and nuclear. The goal is to reduce consumption and harmful emissions, as well as to recycle and to reuse materials.

Create incentives to reward efficient designs and good practice, and *regulations* and requirements to strike a responsible balance between demand, production, cost, efficiency, consumption, and environmental impact. Regulations will work best with strong incentives. Heavy-handed regulation without meaningful incentives for compliance can produce not only burdensome bureaucracy but also widespread public resentment, evasion, and resistance.

Reduce deforestation. In the presence of sunlight, trees and plants absorb CO_2 and water by the process of photosynthesis and convert them into sugar and oxygen. So trees are a major carbon sink. But deforestation and loss of existing forests account for some 17 percent of greenhouse gas emissions responsible for climate change. That's more than all methods of transportation combined, including rail, road, and air. Deforestation continues unabated in many of the world's poorest tropical countries, so that any program to reduce it will probably require some incentives, such as a carbon credit scheme.

Restore and remediate. This is costly work, but it needs to be done. There are many sites on each continent (including Antarctica) where once heavily polluted areas have been restored to almost pristine conditions and where rivers—once the sewer of the industrial world—have been refurbished to the point where native species have returned and recreational use restored. One early example is the Lower Swansea Valley site in South Wales.[2]

In the early eighteenth century the once-verdant lower valley of the River Tawe became a center for industrial production, as ships brought in ore supplies from across the world. It was a perfect site, with a sheltered, deep-water harbor, an abundant supply of high-grade local coal, and a skilled workforce. Copper smelting in the early years was followed by the rapid growth of zinc, nickel, iron, steel, and tinplate production. By the mid-twentieth century, all these industries were in decline, and the site had become an industrial wasteland. What had once been "the metallurgical capital of the world" became the largest area of industrial dereliction in Europe, marked by dilapidated mills, abandoned factories, decaying spoil heaps, and such widespread pollution of its once-fertile valley floor that scarcely a blade of grass could be seen.

Careful scientific analysis and ambitious remediation plans were undertaken in the late 1960s, and over the next twenty years the entire community became engaged in land restoration. The area now houses a sports complex, a restored dock and maritime quarter, museums, restaurants, hotels, a shopping center, an enterprise zone, and wooded recreational areas and trails. What was once a blot on the landscape has become a source of civic pride and economic renewal.

To assume that the Lower Swansea Valley Project can serve as a practical model for the restoration of other polluted areas in the world may perhaps be too

---

2. K. J. Hilton, ed., *The Lower Swansea Valley Project* (London: Longmans, 1967).

simplistic. But it does provide an inspirational example of the power of a voluntary partnership in which the local community, its regional government, its university, and its industrial enterprises were able to transform one of the world's most largely despoiled and environmentally devastated areas into one of restored natural beauty and blossoming economic enterprise.

The Lower Swansea Valley Project is a classic example of the influence and power of local community concern in environmental matters. A small, determined band of academic scientists, concerned civic leaders, local government officials, and enterprising corporate executives joined together to create an almost miraculous transformation of one of the world's most desolate industrial wastelands. The neighboring communities were far from wealthy and faced many other pressing and competing concerns, but they concluded that the restoration of their blighted and polluted region would serve as a catalyst and inspiration for progress in other areas. And so it has proved to be. Clean air, clean water, and uncontaminated soil not only benefit all; they are all humanity's fundamental birthright. Without them, all other rights are diminished.

CHAPTER 15

The Crowded Planet

We share our benevolent planet with some 7.0 billion other members of the human race. By 2050 that global population is projected to increase to some 9.6 billion, before declining and leveling off at about 8.5 billion. If these projections prove to be correct, that will mean feeding about 50 percent more people than we now have.

Does that imply looming catastrophe, as some argue? Or is it, rather, a manageable challenge and a solvable problem?

Some conclude that this is a manageable task while others believe we already face a serious crisis. Thus, David Pimentel et al. argue that 70 percent of the world population is already malnourished, the largest number ever in history.[1] And malnourishment is the leading cause of death in humans. This debate began some two centuries ago as a long-running, but amiable, father-son debate. The father, Daniel Malthus, an eighteenth-century English country gentleman, was a friend, confidant, and enthusiastic follower of David Hume and Jean-Jacques Rousseau, who argued for the innate goodness of humankind and for the perfectibility of society. The son, Thomas Robert Malthus (1766–1834), would have none of this. He regarded humanity as anything but good and society as far from perfectible. The younger Malthus, the second son of eight children, was educated at Cambridge,

---

1. David Pimentel et al., "Will Limited Land, Water, and Energy Control Human Population Numbers in the Future?" *Human Ecology* (2010), http://www.populationmedia.org/2010/09/23/will-limited-land-water-and-energy-control-human-population-numbers/.

where he was a contemporary of Coleridge. He subsequently became a fellow of
Jesus College, entered the ministry of the Church of England, and in about 1796
became curate of the parish church of Albury, a hamlet near his father's estate in
Surrey. Encouraged by his father to put his ideas into writing, in 1798 he published a slim pamphlet with the title *An Essay on the Principle of Population as It
Affects the Future Improvement of Society*. Malthus argued that, left unchecked,
growth in human population would always outstrip its means of subsistence,
since growth increases in a geometric (exponential) ratio, while subsistence increases only in an arithmetic ratio. In practice, he argued, population growth is
restrained only by such "positive checks" as war, starvation, and disease that lead
to premature death, and by the influence of misery and vice. To these he added,
five years later, "preventative checks" that reduced the birthrate by what he called
"moral restraint"—things such as the delay of marriage and sexual abstinence:

> Let virtue, knowledge and civilization be advanced to the greatest height
> that these visionary reformers [his contemporaries William Godwin and
> the Marquis de Condorcet] would suppose; let the passions and appetites
> be subjected to the utmost control of reason, ... the principle of population
> will still prevail. . . . The number of mouths to be fed will have no limit; but
> the food that is to supply them cannot keep pace with the demand for it; we
> must come to a stop somewhere.[2]

Malthus concluded that public support, social welfare, and individual relief for
large families only added to the misery and compounded the problem of overpop-
ulation. His writings produced an outcry of vituperation against the quiet clergy-
man. "For thirty years it rained refutations," wrote his biographer. One receptive
reader, however, was Charles Darwin, for whom Malthus's views on "the struggle
for existence" struck a chord that provided a major element in his evolutionary
thinking. Darwin later wrote:

> In October 1838, that is, fifteen months after I had begun my systematic
> inquiry, I happened to read for amusement Malthus on Population, and
> being well prepared to appreciate the struggle for existence which every-
> where goes on from long-continued observation of the habits of animals and
> plants, it at once struck me that under these circumstances favourable varia-
> tions would tend to be preserved, and unfavourable ones to be destroyed.
> The results of this would be the formation of a new species. Here, then I
> had at last got a theory by which to work.[3]

---

2. *An Essay on the Principle of Population as It Affects the Future Improvement of Society* (1798), in *The Works of
Thomas Malthus*, vol. 1, ed. E. A. Wrigley and David Souden (London: W. Pickering, 1986).

3. *Autobiography of Charles Darwin*, edited by Nora Barlow (London, 1958), 130.

Nor was Darwin the only one impressed. By one of those great historical coincidences, Alfred Russel Wallace, Darwin's codiscoverer of natural selection, also happened independently to read Malthus's work, and was greatly influenced by it.

Others were less impressed. William Cobbett railed: "How can Malthus and his nasty and silly disciples . . . look the labouring man in the face?" William Hazlitt concluded, "Mr. Malthus has written nonsense." When Malthus wrote his first pamphlet in 1798 the world's population is estimated to have been between 600 and 670 million. Today, as we have seen, it stands at 7.0 billion. By 2050 it is projected to be 9.6 billion.

But humanity continues to survive. Was Malthus wrong? Or did he overlook something? Well, for one thing, "every extra mouth produces another pair of hands." But more importantly, there have been changes since the time of Malthus that have created a huge jump in agricultural productivity and provide cheap energy and materials. Today we also enjoy the benefits from industry, free international trade and commerce, and major improvements in public health. These have greatly increased the "means of subsistence," while the rise of family planning and female influence and choice, especially in the developed world, have reduced the growth rate of population in some areas. Many today are as critical of Malthus as his contemporaries were. "Will Malthus Continue to Be Wrong?" asks the title of an article published in *Science* in 2005, and reprinted in 2006.

But was Malthus wrong? Or was he perhaps only premature in his forecasts? A report published by the World Wildlife Fund in 2006 concluded that if current trends continue, by 2050 we should need *two* Earth-sized planets to sustain

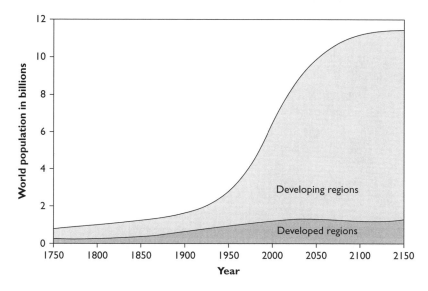

15.1 World population growth since 1750 and future projections. Courtesy of Population Reference Bureau.

us.[4] This report suggests that Earth's natural resources are being used 25 percent more rapidly than the planet can renew them, and that we have been living beyond our (natural) ecological means for the last twenty years. This is the equivalent, in economic terms, of living off our capital, rather than spending only the interest and appreciation.

The report suggests that our "ecological footprint," based on measurements of crop yields, CO_2 emissions, fishery depletion, and deforestation, has more than tripled in size in the past twenty years. "Large-scale ecosystem collapse" is seen as likely by the middle of the present century. How did we get to this point? Looking back at the history of the human race, Joel Cohen has observed that three hundred years ago the human population was a little over 600 million. It was not until about 150 years ago that our population reached the billion mark. It is now 7.0 billion, so the world's population has grown more than tenfold in just over three hundred years. It took from the dawn of humanity until about 1927 to reach the first 2 billion, less than fifty years (1974) to add the next 2 billion, and twenty-five years (1999) to add the next 2 billion. In the last forty years, the human population has doubled.[5] Never before has humankind seen such explosive growth. The biggest reason for this growth has been unprecedented, worldwide decreases in the death rate, reflecting improved technology, health care, and food production.

But it's not just the rate of growth in population that has increased: it's also the location of that growth. Until half a century ago, most of the increase in population was in Europe and North America, while Africa and Asia grew only slowly. That pattern has now reversed. Africa is now the center of population growth, and it is the poorest countries that are growing most rapidly.

Urbanization has also steadily increased. In 1800, only some 2 percent of the world's population lived in cities. By 2000 more than 40 percent did, their migration driven largely by technological changes. In 2000 there were nineteen cities of 10 million or more, only four of which were in the industrialized world.

There is, however, one other change. Although population is still increasing, the most significant trend has been in the recent reduction in the rate of population growth, which rose steadily to a peak of 2.1 percent per annum in the late 1960s, and has been declining since then, falling to 1.2 percent in 2002. In the fifty-year period from 1955 to 2005 the fertility rate has fallen from five children per woman to 2.7 children per woman.

In addition to large variations in regional growth patterns, there will also be major differences within population structure, with the number of elderly people showing major increase. The proportion over age sixty is likely to increase from

---

4. "The Living Planet: Facts and Figures," *BBC News*, October 24, 2006, http://news.bbc.co.uk/1/hi/sci/tech/6080074.stm.

5. Joel Cohen, *How Many People Can the Earth Support?* (New York: W. W. Norton, 1995).

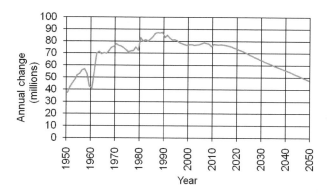

15.2 Annual world population change and projection, 1950–2050. Courtesy of U.S. Census Bureau.

the present 10 percent to around 34 percent by 2100. That leveling off involves significant differences between nations, not only between the developing and developed world, but within them. The U.S. population, for example, has grown from 200 million in 1967, to 300 million in 2006, and is projected to reach 400 million by 2043, largely as the result of immigration. No other "rich country" shows such growth. Japan and the European Union are projected to *lose* almost 15 million over the same period. The fertility rate of U.S. women is now 2.1. In contrast, European women have an average fertility rate of only 1.47 children, a figure well below replacement value.[6] Prosperity, and the freedom that comes with it, allow family choice, and that leads, it seems, to smaller families. But in developing countries, the fertility rate is much higher: in Niger and Mali it is seven children per woman. Over 99 percent of future population growth in the next half century is likely to be in developing countries in Africa; but as these nations become more prosperous, it is argued, population growth will decline and their populations stabilize. In 1960, only six countries had fertility rates below the replacement rate; by 2000, sixty-four countries were below the replacement rate. The general population, as Cohen has pointed out, will also become more urban, as well as older. The reduction in growth rate reflects the new balance between reduced fertility (reflecting the presence of more women in the workforce, the spread of family choice and contraception, and economic development) on the one hand, and improved survival, resulting from better health care and infant survival, on the other.

Now, all projections require assumptions, involving not only those about fertility, but also about such unknowns as migration and catastrophic epidemics or disasters. And none of these is either fully predictable or unchanging. Very slight differences in assumptions about fertility rates from, say, the "European" model of 1.7 children per woman, to the "Argentine" model of 2.5 children per woman could represent a difference between a total population of 4.3 billion and 28 billion

6. "Now We Are 300,000,000: America's Population," *Economist*, October 14, 2006, 29.

in 2150, with the former declining and the latter still increasing. Major differences now exist and will probably continue to exist between countries, economies, and cultures.

Most demographers conclude that world population will stabilize at some 10 billion, though others project a smaller number.[7] But once world population does stabilize, whatever the number, will that not solve our problems of uninhibited growth and finite sustenance? It might if—and only if—the developing world's appetite for more energy, food, and resources were to remain constant. But it will not. It will increase, as we are now seeing in China, India, and other countries. Some will argue that we should all go back to the "simple life," cutting out global trade and transportation and returning to a strictly local and rural economy, with local food production, limited travel, rural communities, and reduced consumption. But it is too late for that. Even our present population of 7.0 billion will not go back into that local box; still less will 9 or 10 billion people. Even trying to impose that pattern could lead to widespread poverty and famine, and would be likely to generate strife between the "haves" and the "have-nots."

The fundamental question is whether this planet of ours can continue to feed, not only the 7.0 billion of us who now inhabit it, but the added 75 million people born annually. Malthus notwithstanding, if one looks at the recent past, while global population doubled from 3 to 6 billion people in the forty years leading up to the year 2000, global food production not only kept up, but many people were eating better than ever before.[8] Even as population doubled, the proportion of the hungry declined. The proportion of malnourished people fell over thirty years to the mid-1990s from 37 percent to 18 percent. World cereal production has doubled since 1970, and meat consumption has tripled since 1961. The global fish catch grew sixfold from 1950 to 1997. This growth in food production has been achieved partly by the continuing green revolution and the massive use of fertilizer and pesticides. So why can't we just keep going? Is all well? Was Malthus wrong?

Not exactly. The per-capita production of cereals has been declining for the past twenty-four years. The very means that we have used to succeed are now bringing their own problems. It requires seven hundred thousand gallons of water to produce an acre of corn, for example, and water is in limited supply in some of the neediest areas. In some areas intensive agriculture has led to soil erosion and degradation, and insect pests are developing resistance to pesticides. Expansion of cities is encroaching on arable land, while runoff from massive fertilizer use and increasing animal waste from intensive livestock farming are increasing nitrogen

---

7. Cohen, *How Many People?* 14–18.

8. Alex Kirby, "Can the Planet Feed Us?" *BBC News,* November 24, 2006, http://news.bbc.co.uk/2/hi/science/nature/4038205.stm.

concentrations. There is growing pressure on other species that once shared what are now agricultural habitats. We already use some 26 percent of the Earth's usable land area to produce food, displacing one-third of the world's forests and a quarter of its natural grasslands.

And, for all our success, which is real, the proportion of hungry people is now beginning to increase. In 2003, it is estimated that there were 842 million hungry people, one-third of them in sub-Saharan Africa.[9] Studies suggest that 25,000 people a day still die of hunger and malnutrition. And the human loss—great as it is—is compounded by the negative impact on the planet for which each of us is responsible. One calculation concludes that every pound of food now produced involves the loss of seven pounds of soil by erosion, and that each human "uses" about 35,000 pounds of minerals and fuels, excluding those used for construction. Our total human use of materials amounts to some 50 billion metric tons a year. Some of this, like wood, is renewable, and some metals are recycled, but the total is more than three times the amount of naturally eroded sediment carried by the world's rivers into the sea every year.

So in spite of real gains in agricultural productivity, it is clear, not only that unlimited population growth on a limited planet is impossible, but also that presently projected population levels will pose a formidable challenge to our capacity to produce food. One of the best informed and most succinct estimates of the outlook for humankind is given by David Pimentel and Anne Wilson, who argue that because of the very young age of a large proportion of the world's population, estimates of stabilization at 9 billion are questionable; that the rates of increase in grain yields per hectare are decreasing in both developed and developing countries; and that the World Health Organization estimate of 3 billion people undernourished is the largest number ever reported.[10] Meanwhile, the decline in per-capita available cropland as a result of population growth is aggravated by the degradation and erosion of soils, leading to the loss of 1.3 percent of total cropland annually. Water supplies, on which intensive agriculture depends, are under increasing stress in many parts of the world, while rising energy costs will impact every aspect of agricultural production, and especially the costs of irrigation, fertilizer, and pesticides.

Garrett Hardin (1915–2003), who was professor of human ecology at the University of California, Santa Barbara, analyzed more carefully than any recent writer the social and ethical dimensions of this Malthusian imbalance into which we are moving. Continuing increases in human population, he argued, will create environmental damage, deplete resources, and degrade the quality of human

---

9. Ibid., 3.

10. David Pimentel and Anne Wilson, "World Population, Agriculture, and Malnutrition," *World Watch*, September/October 2004, 22–25.

life. Nature will exact harsh penalties if we exceed the capacity of Earth. Human nature is such that we exploit anything that is "free" in order to serve our own advantage—for example, polluting the atmosphere—leaving society to foot the bill. Stern measures would be required, Hardin concluded, to prevent exponential population growth. Allowing immigration from overpopulated countries, while providing aid to them, he reasoned, is equivalent to overloading and capsizing the lifeboat that now sustains us. In a broad sense, Garrett Hardin is Thomas Malthus revisited. Hardin also echoed the grim prognosis of the late Kenneth Boulding, the University of Michigan economist, who in 1956 proposed two theories of economics: the dismal and the utterly dismal. However favorable the environment and however advanced the technology, he argued, human population will continue to grow until restricted by starvation and misery. That's the dismal theory. The utterly dismal theory states that if this is so, then any technological improvement will ultimately increase the sum of starvation and misery by allowing a larger population to share those conditions.

Fifty years on, more recent population trends suggest that both of Boulding's economic theories may be a little too dismal. Growing prosperity, as, for example, in Italy and some fifty other countries, has produced declining birthrates, and declining populations bring their own distinct challenges and problems, as these same countries now demonstrate.

To project in the coming decades a continuation of the rapid expansion in population that has marked the last two centuries would be quite unrealistic. In 1800 the total population stood at less than 1 billion. By 2000 it had reached over 6 billion. But that rate of increase cannot be sustained. It took place on a foundation that no longer exists. It depended on a pristine landscape, with a soil bank accumulated over the unusually long interglacial period that preceded it. It inherited a virtually untapped reservoir of coal, oil, and gas, and it covered a time span during most of which the much smaller, less efficient population placed much less demand on Earth's resources.

Population growth has also depended on a series of technological "revolutions" and on growing global trade to support it. Our future prospects for success will require that we maintain the conditions that nurture discovery and free trade.

But the challenge of feeding another 3.2 billion people remains huge. Today, the landscape is under growing stress from deforestation, intensive agriculture, and urbanization. The soil bank is now much depleted, and, though coal is still relatively plentiful, oil is now approaching peak production. Nor can either the soil bank or our store of fossil fuels be rapidly replaced. It takes an average of five hundred years for an inch of topsoil to be formed and it is estimated that we are losing 25 billion tons of topsoil each year. As we have seen, of Earth's total surface, only about 7.5 percent can be used for agriculture, and even this limited area must also provide space for the growing spread of urban development, manufacturing, and transportation networks that support our burgeoning population. That's why recent policy decisions to divert the use of agricultural land from food production to ethanol development appear to many to be inexplicable.

So how many people *can* Earth support? Joel Cohen, a distinguished demographer, has given the best answer by telling a story of a small boy who wanted to know the sum of one plus one.[11] He first asked a physicist, who replied that if one is matter and the other is antimatter, then the answer is zero. But if one is a critical mass of uranium, and the other is also a critical mass of uranium, then it becomes an explosive question. Faced with the same question, a biologist answered that it depended whether we were talking bacteria, mice, or whales, and how much time was involved. Bewildered, the boy turned to an accountant, who replied, "Hmm. One plus one? Tell me, how much do you want it to be?"

And to answer the question of how many people Earth can support poses the same question for us. How many do we want Earth to hold? No meaningful answer is possible until we specify the social, cultural, economic, and environmental conditions we are willing to accept. Are we willing, for example, to accept gross inequalities in income, living standards, and health between different groups or different regions? How do we weigh the rights of each couple to determine the number of children they will produce against the rights of each person to obtain adequate sustenance? Conversely, unlimited freedom and choice in reproductive behavior now may lead to a severe reduction in personal freedom and societal choice in a few generations if our growing population outstrips Earth's capacity to support it. And how do we balance population size against the quality of the individual lives and health of its members? Is a population of, say, 15 billion living in poverty more desirable than one of, say, 2.5 billion living at the standard now enjoyed by those living in North America?

These questions of moral value and social responsibility are so personal that it is unlikely we shall choose to address them in such stark form. It is more likely, I think, that they will be answered on the basis of societal and personal behavior in response to changing—probably declining—conditions. Or that, in the absence of an answer reflected in formal national and international policy, they will be answered by the heavy hand of natural selection, responding to our indecision by imposing its own harsh limits on our expanding human numbers. That's precisely what determines the numbers of every other species, whether animal or plant. Even at our present numbers, with a still rapidly growing population of 7.0 billion, our biological vigor comes at the price of huge inequalities. Fifteen percent of our global population enjoys 81 percent of humanity's total income. Ten percent of the world's people are obese, while a study by the UN and the World Bank

---

11. Cohen, *How Many People?* 261.

estimates that 800 million are not getting enough to eat. Thirteen percent are undernourished. And even though global poverty fell by 20 percent in the 1990s, over 800 million are reported by the UN Food and Agriculture Organization still to be hungry, and 10 million people die every year from hunger and malnutrition. To bring the needy up to the level of those living in North America, Lester Brown has argued, would require six Earths.[12]

Continuing population growth and consumption patterns will produce global demand for a doubling of food supply in the next twenty-five to fifty years, and the growing problems of some areas may well be aggravated by existing trade restrictions and subsidy systems—such as those of the OECD. The impact of large-scale migration, growing water shortages, and rising temperatures in sub-Saharan Africa and other areas will also have significant effects.

Beyond the difficult questions of policy, there lies the additional question: Who can act in this matter? Can nation-states go it alone? Should they? Should regional groups of nations? Can, should, would the United Nations act? Each choice involves uncertainties and unsatisfactory risks and possible outcomes. The ability of democratic governments to commit themselves to hard choices and unpopular policies is questionable. So also, perhaps, is the ability of authoritarian governments.

There is one additional major challenge in our present situation that arises from the stark differences between different national populations. A typical population passes through three successive phases of development over time: the first is marked by high birthrates, high death rates, and little or no population growth; the second, by continuing high birthrates, falling death rates, and rapid growth. The third stage—marked by low birthrates and low death rates—leads to stabilization in numbers, and with it a growing degree of prosperity and comfort. In an ideal world, all nations would move steadily toward stage three, but, as Brown has pointed out, numbers of countries now in stage two are slipping back toward stage one, with devastating increases in death rates. In sub-Saharan Africa the level of AIDS infection has reached epidemic proportions: in Africa alone, over six thousand lives are lost every day to AIDS. In sixteen countries of Africa, adult infection rates are reported to exceed 10 percent. In South Africa the infection rate is 20 percent; in Zimbabwe, 25 percent; in Botswana, 36 percent.

Though less immediately conspicuous than AIDS, some of the countries having the greatest population growth also face the threat of ecological collapse, especially from shortages of water and arable land. In some of the world's poorest and most rapidly growing countries, water table levels are collapsing at the rate of several feet per year, topsoil is being lost, and deforestation is widespread, leading

---

12. Lester Brown, *Eco-Economy: Building an Economy for the Earth* (New York: Norton, 2001).

15.3 World population, total and annual addition, 1950–2012

Year	Total	Annual Addition
	(billion)	(million)
1950	2.55	37
1955	2.78	53
1960	3.04	41
1965	3.35	70
1970	3.71	77
1971	3.79	76
1972	3.86	75
1973	3.94	74
1974	4.01	72
1975	4.09	71
1976	4.16	72
1977	4.23	72
1978	4.30	75
1979	4.38	72
1980	4.45	82
1981	4.53	80
1982	4.61	81
1983	4.69	80
1984	4.77	81
1985	4.85	83
1986	4.94	86
1987	5.02	87
1988	5.11	86
1989	5.20	87
1990	5.29	83
1991	5.37	84
1992	5.45	82
1993	5.54	80
1994	5.62	81
1995	5.70	80
1996	5.78	78
1997	5.86	77
1998	5.94	77
1999	6.01	76
2000	6.09	77
2001	6.17	77
2002	6.24	76
2003	6.32	76
2004	6.40	77
2005	6.47	78
2006	6.55	78
2007	6.63	78
2008	6.71	77
2009	6.79	75
2010	6.86	77
2011	6.94	77
2012 (prel)	7.02	77

Source: U.S. Bureau of the Census.

to serious reduction in the area of productive land available for each person. In Arizona and California, aquifers are being "mined" 10 percent faster than their recharge rate.

What should responsible tenants do? It is unlikely that we shall ever reach agreement on the answer to this question. It seems more likely that—inadvertently—we shall conduct the experiment to test Earth's carrying capacity, providing whatever incentives of support we can to poorer nations. This is likely to include family planning support for incentives to restrain population growth and continuing international aid and humanitarian relief as we navigate our way through the challenges of the next fifty to sixty years until global population begins to level off.

Any such program will be costly and controversial. But it will be less costly than inaction. More than fifty countries now have an average family size of two children or less. But more than one hundred others—all of them developing countries—continue to have growing populations, with many of the largest families living in the poorest countries. China will add 8 million people to its population this year.

Of course the best efforts may not succeed and we shall still face the problem of feeding an additional 3 billion or so people as global population stabilizes. Can we feed that huge number? The opinions of experts differ, but somehow we must. Only one thing, perhaps, is clear. The Malthus debate continues. Right though he was in principle, Malthus underestimated the huge technological benefits of human creativity and the widespread contribution of free global trade. Those two uniquely human qualities we must nurture and promote if we are to provide a meaningful future for our grandchildren.

PART III

EARTH FUTURE

The Sustainable Planet

CHAPTER 16

The Sustainable Planet

For the last 3.5 billion years the Earth and its systems have shaped the lives and influenced the behavior of all its inhabitants. And the inhabitants, in turn, have had a significant impact on the surface of the planet that gave them birth. Think, for example, of the huge influence of early bacteria in providing the initial oxygen of the atmosphere. Or consider the continuing role of plants in the carbon cycle.

But for most animals, this interchange has been more one-sided. Earlier generations of humans were, like most of their fellow creatures, far more dependent on Earth's bounty and far more influenced by its forces than Earth was by them. For most of our ancestors, every dimension of life and every aspect of day-to-day existence was constrained by the planet. Human life exists by courtesy of the planet: it came into existence in adaptation to Earth's conditions and it has survived only as it was able to adapt its behavior and accommodate its activities to the continuing and changing constraints the Earth imposes.

A little over 250 years ago that simple pattern of indenture began to change. Until then, food supplies were largely local, their character and even their availability heavily dependent on local climate, soils, and crops, and on a handful of easily hunted or domesticated local animals. Implements were manufactured largely from local materials, whether stone for early weapons and implements, iron for tools, or clay for pottery. Building styles and architectural forms were also local or regional in character, reflecting the availability of freestone for construction, clay for bricks and tiles, timber for framing and shingles, slate or thatch for roofing. Patterns of settlement and trading were no less influenced by local geography and convenient water supply. Wider migration and trade, such as they were, depended on navigable rivers, protective harbors, and favorable winds.

With the coming of the Industrial Revolution the age-old influence of these physical constraints rapidly declined. To the energy long supplied by human effort, wind, water, wood, and animals was now added fossil fuel—first coal, then oil and natural gas, and later nuclear power. And this brought a revolutionary increase in human empowerment, and with it, not only dramatic changes in individual lifestyles, but also a growing human impact on both the planet and its other inhabitants. Mining and quarrying, drilling and blasting every corner of the Earth, humans now began to exploit it on a scale scarcely imaginable in earlier times. Burning fossil fuels in a growing array of new devices—from domestic stoves to industrial smelters, from steam locomotives to jet airplanes—manufacturing, agriculture, travel, communications, and human society itself were transformed. Nor was this all, for with this transformation in human power went explosive growth in human numbers, and, with both, an enormous increase in human impact on the planet itself. Rivers were dammed, drainage patterns were modified, forests were felled, coastlines were reconfigured, prairies were farmed, estuaries were drained, and the surface of the planet shaped and sculpted to suit a world of human needs. Above the Earth, the atmosphere, too, was changed and the global climate modified to an extent and with a rapidity that were unprecedented in recorded human history. It is as though, one reflected, we were absentmindedly conducting a whole-Earth experiment in real time.

All these changes our ancient planet will survive. It has repeatedly shown its resilience throughout its long history. So, too, will life. It has survived waves of global extinction long before we appeared on the scene. It is not Earth's sustainability that is in question. It is ours. And our sustainability requires the planet to remain habitable, able to sustain not only our own species but the countless others on which the intricately balanced economy of our present existence depends. It is that sustainability that we now examine.

Sustainability involves Earth's ability to continue to supply our basic needs. But what's our most basic need? What resource is most critical to our survival? Food perhaps? Certainly, food is essential. We can't long survive without it. But food—whether vegetable or animal—requires soil, and soil requires water. Each of us needs about two and a half liters a day just to keep our bodies functioning. Maybe water's equally basic. So, of course, is air—clean air. And so is energy. No energy, no food production. No energy—no heat, no light, no fuel. And minerals—stone, metals, materials. No minerals, no tools, no shelter. So it's not just one basic resource: it's all the above. Put all these together—food, soil, water, air, energy, minerals—and you have Earth and its components. We need them all. They're all connected. It's in the interplay of Earth's endless cycles of interacting systems that these resources on which we depend—Earth's bounty—are produced and sustained.

Now the definition of sustainability requires refinement and clarification. Suppose we say that sustainability is "Meeting the needs of the present generation without compromising the ability of future generations to meet their needs." That definition—often called the Brundtland definition, after the former prime minister of Norway, Gro Harlem Brundtland, whose commission coined it—is broadly ethical. Other, more precise, definitions involve a social equality component, requiring the elimination of hunger as a necessary definition or a certain level of income or food supply, for example. Still other definitions are economic. Thus Herman E. Daly of the University of Maryland gives an operational requirement that the rate of use of renewable resources must not exceed their rate of generation, that the rate of use of nonrenewable resources should not exceed the development and supply of alternatives, and that pollution rates should not exceed Earth's natural restorative capacity.

My colleague at Cornell University, Norman Scott, suggests a more comprehensive definition: "Sustainable development is a process of change in which the direction of investment, the orientation of technology, the allocation of resources, and the development and functioning of institutions meet present needs and aspirations without endangering the capacity of natural systems to absorb the effects of human activities, and without compromising the ability of future generations to meet their own needs and aspirations.[1]

Any workable definition will require some combination of these ethical, social, economic, environmental, and operational components. Two questions arise: one of equity, the other of costs, and the two are closely related. The equity issues loom large in any discussion of responsibility and costs. The average environmental footprint created by a U.S. resident, for example, is twice that of a European, twelve times that of an Indian, and twenty-four times that of a Somali. The United States and Europe are prosperous, developed countries, India a growing, strongly developing country, and Somalia a poverty-stricken, war-torn, famine-ravaged country. Each is a polluter. Who pays, and what should be the relative contributions? The UK's Stern Commission calculated that a restorative national investment now of 1 percent national GDP could avoid a potential 20 percent cost in subsequent liabilities for environmental restoration.

Nor can we isolate these sustainability questions from the broader question of how many humans Earth can sustain. Several authors (see, for example, Cohen, Brown, Pimentel, Ehrlich, and others) have addressed this issue and we have already touched on it in chapter 15.

---

1. Personal communication.

The most basic requirements for human survival have been estimated as follows, expressed as per person, per year:

- four hundred liters of relatively clean water;
- three hundred kilograms of food, mostly from grain; and
- adequate clothing and shelter from freezing temperatures.

Now these are basic, survival requirements. In contrast, the average U.S. resident in a typical state, Rhode Island, consumes each year:

- one hundred thousand liters of clean water;
- one thousand kilograms of food, including significant amounts of meat and imported fruits;
- between five hundred and one thousand liters of gasoline for transportation;
- one to two thousand equivalent liters of gasoline for power; and
- tonnes of metals, plastics, fabrics, chemicals, and construction materials.

Clearly, therefore, estimates of Earth's carrying capacity involve social, ethical, and economic as well as physical and biological considerations, and, because there is no agreement on the relative priority of each, it is scarcely surprising that there is little agreement on Earth's carrying capacity. Add to this uncertainty the speculative nature of cost estimates and prospects for improvements in technological efficiency (alternative fuels, more-efficient energy use, synthetic materials, new crops, and improved water safety, for example) and the problem of social sustainability becomes still more complex. Profoundly personal views and values thus have to be incorporated in any estimate of Earth's carrying capacity: for example, is it better to have a smaller, prosperous, well-fed population, or a larger, poorer, less well-fed population? How does one decide? How do Earth's people decide? What is the balance of personal reproductive choice and social responsibility?

These and other issues influence the widely varying estimates of Earth's carrying capacity, but if one includes social sustainability as a requirement, estimates range from 0.5 billion to 14 billion, depending on the standard of living, the longevity of nonrenewable resources and the adequacy of renewable ones, and the likelihood of technological improvements. The medians of high and low estimates, however, range from 2.1 to 5.0 billion.[2] That means our present population level of 7.0 billion is already unsustainable. That we are able to maintain this imbalance reflects our dependence on nonrenewable energy sources—especially petroleum and natural gas—and the huge social inequalities that our present global population represents.

---

2. Joel Cohen, *How Many People Can the Earth Support?* (New York: W. W. Norton, 1995).

Sustainability projections such as these inevitably involve assumptions about the global framework within which they will apply. Prolonged global conflict, for example, would clearly involve greater stress than a peaceful global society in which creativity and technology were nurtured and encouraged, free trade was embraced, a sense of global responsibility was fostered, and human population leveled off and stabilized. Given such perhaps utopian conditions, sustainability—though difficult—does not seem unattainable. The alternative—human population growth that exceeds a sustainable number, with the loss of the most vulnerable in large numbers, whatever is done—is simply unacceptable to all our deepest instincts and all our finest values.

These are the difficult questions that underlie any discussion of sustainability, and in the pages that follow we'll first examine the requirements for sustainability and then consider these difficult policy questions.

CHAPTER 17

Water as Sustenance

The Role of Water

To Thales of Miletus (ca. 625–546 BC), the first of the Greek philosophers, water was Earth's most basic substance, its most elemental component. It was, he believed, the vast reservoir in which the disklike Earth floated, surrounded by the great hemisphere of heaven. Water, he argued, was the basic stuff of the universe. That's not surprising, perhaps. Humanity, after all, exists within an envelope of water.

The oceans and seas surrounding the continents were once seen as boundless. The surface of the land was itself interlaced with streams, rivers, lakes, springs, and swamps, carrying water in an endless cycle from the land back into the oceans.

So also above the ground: the air we breathe contains water. Below the ground, too, there is water. We are born from water, encased in its protective womb until the moment of our birth. Our bodies are watery, as is the fruit and other food we eat. Without water, we perish.

Nor is that all, for water has the remarkable quality of falling as rain and snow, precipitated from the vapors of the clouds, freezing as a solid—ice—or flowing as a liquid in a Gothic filigree of streams and rivers over the surface of the land. And, though changed from one state to another and back again, it somehow remains itself: water. Over the centuries, Thales' view of water as a single primordial substance was eclipsed by its identification as a remarkable compound of two elements: hydrogen and oxygen. But as one interpretation faded, another insight emerged, for the properties of water seemed no less remarkable than its ubiquity: it appeared almost custom-made to serve its earthly functions. Its boiling point and melting point are "just right" to allow it to exist as three phases within Earth's narrow temperature range. It is an excellent solvent, capable of dissolving a wide range of other elements, without itself being transformed. Its solid ice phase floats on its liquid phase, unlike almost every other substance. It is this difference in density that allows fish and frogs to survive the winter at the bottom of ponds. Its abnormally high heat capacity insulates the land and moderates its temperature extremes. It has the highest surface tension of all common liquids and so forms droplets, wetting and cohering to surfaces with which it comes into contact. This improves its ability to travel through the narrow pores and vessels of the bodies of animals and plants and to flow through the crevices and cracks of rocks and soil.

It was in water that life first arose on Earth and, long after its origin, it was only when living things developed their own internal plumbing systems, allowing them to incorporate water into their bodies, that they emerged as dwellers on land.

We remain watery creatures. Water forms up to 65 percent of our bodies. Our blood is 83 percent water. Our wateriness reflects, in part, our watery origins. Each day we need about two and a half liters (about two-thirds of a U.S. gallon) of water just to survive. The UN suggests a minimum of fifty liters a day for dining, washing, cooking, and sanitation. This excludes water needed for growing food. Water's remarkable ability to dissolve, transport, metabolize, and cleanse provides the basis for life's sustenance. Thales was right, at least in part: water is the great unifying, sustaining substance. It forms the essential component of all living things, Earth's ubiquitous liquid, performing so many functions of our existence and our planet's well-being that it is easy to take it for granted.

But although water is a renewable resource, it is also a scarce resource in some areas where population growth, growing industrialization, spreading urbanization, climate change, and increasing pollution have, between them, all contributed to mounting concerns for future water supplies. But why do we need more water, except for those now underserved? Where does it go? Is it not, like aluminum cans, recyclable? Those are the questions we now explore.

Water was not primordial, not one of Earth's earliest components. As the young Earth cooled, gases and steam vented from its molten surface, hissing from the fractures and fissures of its congealing crust, forming dense layers of impenetrable clouds and ultimately, over millennia, falling as rain on Earth's glowing surface, drenching the land, flowing into the developing ocean basins. Later, the earliest living things gradually emerged. They did so in the water, bathed in its envelope, sustained by its liquidity. Only later did plants and animals gain a tentative foothold on the land, first in shallow lakes, rivers, swamps, estuaries, and deltas and then later still on "dry land" itself. To describe this long development as "a conquest of the dry land" is to give a triumphalist cast to what was, in another view, a continuing search for alternative sources of freshwater as the swamps and shallow lakes in which these early vertebrates lived dried up during the warmer seasons.

Water is our most abundant resource. It is also our most basic, essential to every aspect of life on Earth. Every organism, plant or animal, depends on it. It has shaped the history of life, influenced the course of human history, determined the pattern of development and settlement. It sculpts the land, it forms its rocks, it controls the weather. It envelops the planet and all its creatures in a continuous, benevolent cycle. But, abundant as it is, it is not everywhere plentiful, and one-fifth of Earth's population still lacks access to clean, cheap water.

Most—97.2 percent—of Earth's water resides in the oceans: more than 2 percent is locked up in glaciers and ice caps, and 0.001 percent is held in the atmosphere. So all that water—over 99 percent of Earth's total—is unavailable to us at present, at least at anything approaching a reasonable cost. Less than 1 percent is available for human use, almost all of it in the form of groundwater. Between major regions of the Earth, and between countries, there are huge differences in the availability of water. Even within some countries—the United States among them—there are major differences. So water use, water rights, and water policies loom large in economic, industrial, and agricultural development, as well as in domestic and international policy. That's why water supply is a major issue in our role as Earth tenants.

The Supply of Water

In most of the world, water is not thought of as a critical resource. In fact, it's scarcely thought of at all. Yet each of us, as we have seen, requires a minimum of 2.5 liters (about two-thirds of a U.S. gallon) of water a day, just to keep our bodies functioning. In the United States, each of us "consumes" about 1,500 liters (almost four hundred gallons) a day, to produce the food, fuels, and all the various manufactured products we depend on, as well as the domestic and municipal services we require. And in most of the developed world, this picture of dependence is much the same. You turn on a tap, and out comes clean water, plentiful and cheap. And we are able to do that only because Earth is a well-watered planet. We use, on average, only about 3 percent of all the rainfall that the Earth receives, and that is ultimately recycled back into the Earth through the hydrologic cycle.

But there are parts of the world where the picture is very different. Some 1.2 billion people—almost 20 percent of the world's population—lack access to clean water. This deficiency takes a terrible toll on life: thousands die every day— many of them children—from waterborne diseases. It is estimated that the total is more than 5 million a year worldwide. And even in areas of plentiful rainfall, intensive agriculture, urban expansion, and growing population have led to the depletion of essential aquifers by overpumping; they have also caused the pollution of lakes, rivers, and coastal waters and the loss of vast areas of wetlands that play a critical role, not only in biological conservation, but also in maintaining the long-term stability of water supply and water quality. Overpumping and depletion of groundwater in burgeoning metropolitan areas—from Miami to Cape Cod—have allowed the incursion of saltwater into many coastal aquifers. And as the world's population continues to grow, standards of living are raised, and industrialization increases, these problems will be of growing significance.

Nor is this all. For addressing these problems, and developing responsible solutions, involves not only questions of technology but also a complex interplay of local, regional, and national jurisdictions, and competing economic interests. Many of the Earth's major river systems, for example, are shared by several states or countries, each having its own priorities and interests. Giant engineering schemes to construct dams and reservoirs have sometimes proved a mixed blessing, giving benefits to one area, even as they deny them to another, changing long-established patterns of agriculture and settlement, and sometimes increasing the spread of schistosomiasis and other water-related diseases.

The Origin of Water

We have already seen that water is the only natural substance that can exist in three distinct states—solid, liquid, and gas—within the range of Earth's temperatures. And, so far as we know, Earth is the only planet in the solar system to possess liquid water, though it seems to have existed once on Mars. How did we come to possess this remarkable liquid? Where did it come from? Why is it found only here, as opposed to other planets?

If, as seems increasingly likely, Earth was formed from the accretion of cosmic dust, it would soon have begun to heat up, both from the effects of impacts and from radioactivity, leading both to the differentiation of Earth's core, mantle, and crust, and to the diffusion of lighter materials to form an encompassing atmosphere. The composition of this early atmosphere is not known, but it must have been chemically reducing in character—rather than oxygen rich—and included perhaps methane, ammonia, hydrogen, and water vapor, with carbon monoxide, carbon dioxide, and nitrogen joining them later. It was probably the photodissociation of these gases, as well as steam from volcanoes, that led to the development of some free oxygen in the atmosphere and later to the presence of liquid water and gradually to the formation of the oceans. As early bacteria and algae developed in these waters, photosynthesis added further oxygen to the atmosphere, which in turn added to the development and buildup of water. The existence of 3.3-billion-year-old fossil algae shows that all this must have taken place within the first billion and quarter years or so of Earth's history.

We have seen that less than 1 percent of Earth's water is available for human consumption. The rest is locked up in the oceans (more than 97 percent) and ice caps and glaciers (more than 2 percent), either too salty or too remote for general use. And one-third of Earth's surface is desert or semidesert. In these regions, groundwater is the chief source of supply, supplemented in some areas by dams that control and redistribute the waters of major river systems. In Egypt, for example, construction of the Aswân High Dam in 1970 brought almost 2.5 billion acres of new land into cultivation, as well as providing 10 billion kilowatt hours of electricity each year. In more humid parts of the world, about three-quarters of all water used comes from surface sources—rivers, lakes, and reservoirs—with the rest coming from groundwater.

But all Earth's waters—oceans, ice caps, glaciers, lakes, rivers, springs, and groundwater—are linked together, involved in continuous movement within a cycle of evaporation, transpiration by plants, condensation, and precipitation. Of the 110,300 cubic kilometers of water that falls on the land each year, about a third (37,400 km^3) comes from the oceans and returns to these as it runs off into lakes, streams, and rivers. The rest of our rainfall comes from the land itself, by direct evaporation and from transpiration. Most of this water subsequently returns to the atmosphere by the same two processes, with the rest soaking into the ground. The relative amount of water following these various paths varies greatly, depending on the topography, the amount and rate of rainfall, the temperature, and the vegetation. Between these water sinks, there is a continuous movement of water, with a ceaseless cycle of hydrologic change, driven by the energy of sunlight. (Figure 4.3.)

The world's great ocean basins—"the face of the deep"—are so vast that the waters within them are not greatly changed by the daily evaporation from their surface that drives the hydrologic cycle. Yet every day, so it is estimated, some 280 cubic miles (1,167 km^3) of water are evaporated from the land and oceans, condensed from clouds, and then precipitated back to Earth's surface in the form of rain and snow.

The Distribution of Water

The movement of water within the water cycle influences Earth's climate, plays a major role in creating local weather conditions, shapes the landscape, irrigates the soil, and removes soluble food, nutrients, and other material. But for all its benefits, the pattern of Earth's precipitation is strikingly uneven. Precipitation is highest in a zone around the equator, where rainfall can exceed eighty inches a year in many places; it is low in the two desert belts that encircle the equatorial zone, increasing in the temperate regions and then decreasing again toward the arid areas around the poles.

But these broad latitudinal patterns of precipitation are everywhere modified by local conditions. Mountain chains, for example, have a profound influence on local climate, creating both areas of high precipitation on their windward sides and areas of rain-shadow aridity on their leeward sides. Arid regions, although they may experience seasonal flash flooding, have only ephemeral streams and rivers, their sparse vegetation allowing substantial infiltration of water into the exposed surface rocks, and high rates of evaporation that return liquid water to the atmosphere. Surface water—springs, rivers and streams, lakes and ponds, which are such conspicuous features of the worldwide temperate zones—exists in arid areas only when surface water runoff exceeds the combined effects of downward water percolation and upward evaporation and transpiration by plants.

Water supply for any region reflects not only the availability of water—either surface or ground water—but also its quality. Some surface and spring waters, for example, are so high in dissolved salts as to be unsuitable for domestic water use unless they are first treated and purified. In arid areas, evaporation rates are often so high that water supply from reservoirs and rivers is diminished, and residual salt deposits are gradually formed, with significant loss of arable land over time.

In the temperate zones of the Earth, about three-quarters of the water supply comes from surface water. But over a quarter of Earth's surface is desert and these areas depend on groundwater to supply their needs. Impressive as springs, streams, rivers, and lakes appear as major features of the temperate landscape, they carry only a small amount of water in comparison with the volume contained in the soils and surface rocks of the Earth's crust. This water, replenished by rain and snow, together with a small amount formed when the rocks were formed, filters downward through the pore spaces and crevices in the rocks near the surface until it reaches a zone of saturation, where all the pore spaces of the rock are filled with water. The upper surface of this saturation zone—the water table—varies in depth from place to place but generally reflects the topography of the landscape. In tropical and temperate regions the water table generally lies near the surface; in arid regions it tends to lie much deeper. Where the water table intersects the ground surface, springs, streams, rivers, and lakes are formed. Water within the Earth's crustal rocks is in continuous motion, whether filtering downward from the surface to the zone of saturation or moving within that zone. The rate of movement below the water table varies greatly from place to place, depending on the topography and the permeability of the rocks, but it is typically less than an inch a day. Rock formations that have high pore space (porosity)

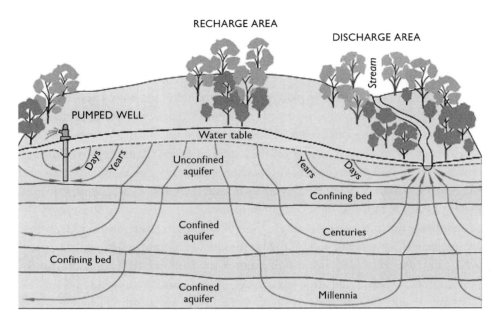

17.1 Water table. Courtesy of USGS.

and allow water to flow freely (high permeability) are termed aquifers. In general, igneous and metamorphic rocks, which cover much of the ancient cores of the continents, are formed from closely interlocking crystals, and so have little pore space to carry water. Sedimentary rocks, especially sandstones and limestones, are high in pore space and fissures, and their high porosity and permeability provide major aquifers, both in arid and in humid regions. Superior aquifers often underlie vast areas. Just two rock formations, the Dakota Sandstone and the Ogallala Formation, for example, which extend from the Rockies across the Great Plains, provide artesian water for much of the agricultural heartland of the central United States.

Subsurface aquifers provide important sources of water supply. At their best, they give a constant, reliable flow of relatively uniform quality. But they are not without problems. Water pumped out of an aquifer must be replenished by precipitation if the supply is to be dependable. For many years, it was required that the rate of extraction should not exceed the rate of natural recharge. That has now changed. In some areas, annual withdrawal of water exceeds recharge rates by as much as one hundred times. The lowering of the water table produced by this "mining" creates not only dry wells but local land subsidence and, especially in coastal areas, the inflow of saltwater into what had formerly been freshwater aquifers. Recharge of depleted aquifers may take decades to achieve, so extraction policies are vitally important in maintaining future water supplies. (Figure 17.2.)

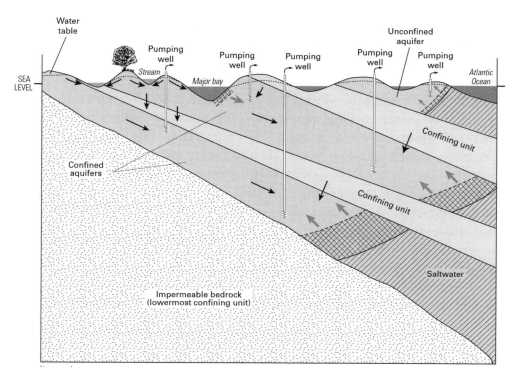

17.2 Saltwater intrusion in a multilayer regional aquifer system caused by groundwater pumping at wells. From Paul M. Barlow, "Ground Water in Freshwater-Saltwater Environments of the Atlantic Coast," USGS circular 1262, 2003. Courtesy of USGS.

Water Supply

In Earth's temperate regions, adequacy of water supply has rarely been a problem, but even in these areas increasing demands for water from industry, agriculture, and growing settlement place a strain on future development. Furthermore, distribution is uneven, significant quantities of freshwater are contaminated by pollution, and recharge and recycling times are often measured in years or even decades.

The water budget of the United States is typical of a major industrial nation. We can think of the overall budget in this way:

- about 29 percent of total annual precipitation flows in streams and rivers directly into the oceans;
- about 71 percent of total annual precipitation falls on unirrigated land and evaporates or transpires back into the atmosphere;
- about 12 percent of total water used is consumed for municipal use;

- about 44 percent is consumed for agricultural use; and
- about 44 percent is consumed for industrial use.

In many other countries, agricultural use consumes 80 percent of the water used. Worldwide water use increased about sixfold in the twentieth century, far exceeding the increase in population growth. Water use has steadily increased in the United States, growing from about 150 billion gallons a day in 1950 to over 400 billion gallons a day by 2000. This pattern of use and significant level of increase is broadly similar to that of other developed industrial nations. These societies, generally located in temperate areas, are not about to run out of water, but the total supply of water potentially available to them at reasonable cost is not unlimited and may be only two or three times present usage. This makes future conservation and the avoidance of pollution and contamination even more important than they have been in the past.

Some regions of the United States, including California, Arizona, and the High Plains, are already "mining" groundwater to such a degree that recharge of aquifers to their earlier state would require many human lifetimes. The Ogallala Formation, for example—the great aquifer of sand and gravel that supports the agricultural economy of much of the High Plains area of eight states—provides, as we have seen, some 30 percent of all irrigation water used in the United States. But overpumping has produced a drop in the water table of some one hundred feet and it would require several thousand years of natural recharge to restore it. One estimate is that the aquifer will dry up within twenty-five years. Nor is this an isolated example of overpumping. The Arabian Peninsula is using groundwater at a rate three times faster than the rate of recharge.

Though groundwater is a renewable source of freshwater supply, such long recharge times show that excessive usage effectively makes it exhaustible in some areas. This problem is not limited to the intensive agricultural use of semiarid areas but is also found, as we have seen, in the increasing salinity of aquifers in heavily populated coastal areas. Although in some of these areas engineering schemes have increased recharge rates and restored water table levels by constructing dams, drainage systems, and recharge wells to allow greater infiltration of surface waters, the continuing growth of population frequently aggravates water shortages by reducing natural wetland and other beneficial recharge areas.

Water Quality

So far we have been concerned with water availability and quantity, but water quality is also of vital importance for potable supplies. In general, water supplied from surface rivers and lakes is more readily polluted than that obtained from groundwater, where the natural pores and fissures of the aquifers act as a filter, removing particles and some bacteria and viruses. But both groundwater and surface water may be contaminated by dissolved materials, some natural, but others arising from municipal waste disposal, industrial and chemical plants, sewage disposal, agricultural fertilizers and pesticides, and other hazardous materials. In a few areas, contamination from dangerous elements—lead and arsenic, for example—is a hazard. Contamination of water supplies by arsenic in Bangladesh now exposes some 50 million people—one-third of the population—to dangerous levels. The gradual buildup of drugs (especially hormones) in the water supply may produce serious harm. Not all contaminants are hazardous. The presence of small concentrations of fluorine (fluoridization) in drinking water, for example, has been shown to assist in preventing tooth decay. Nor is all contamination the result of human activity. Waters in some areas are influenced by the surface rocks. Waters in limestone regions, for example, are "hard" and often need to be softened by the removal of calcium carbonate. Careful monitoring of water quality is standard practice in the developed world and any deficiencies can generally be corrected by appropriate treatment.

All human societies depend on water use: although the minimum required for individual survival is 243 gallons a year, most developed societies consume huge amounts for their sustenance. A ten-minute shower requires fifty gallons, a bath thirty. It takes 250,000 gallons of water to produce a ton of corn; ten times that much to produce a ton of rice. Each gallon of milk requires two thousand gallons of water; to produce a pound of beef requires 3,700 gallons. Aluminum "costs" 350,000 gallons of water per ton, synthetic rubber 660,000 gallons. That is why aluminum, which is highly recyclable and transportable, is refined in places like Iceland, where water is abundant. So as population continues to increase, and industrialization and development continue to grow, the demands for water will be intense.

Nor is that all, for as we have seen, over a billion of the world's people still lack a dependable source of clean water and 30 percent of the population lacks adequate sanitation: disease from contaminated water sources still takes a terrible toll in many areas. The water needs of each country—even those at roughly comparable stages of development—are varied. Mexico and India, for example, use over 50 percent of their water supply for agriculture; the United Kingdom uses some 20 percent. So water-use policy—unlike, say, energy-use policy—can be intensely local in its application and implications. And the problems water shortages bring may vary from village to nation in their impact. Tribal wars have long been fought over water rights. Some have argued that the Six Days' War of 1967 between Israel and the neighboring states of Egypt, Jordan, and Syria was provoked by a dispute over water rights and the proposed diversion of the River Jordan. Although water conservation in one region may or may not be of little consequence or benefit in other areas, the effects of increased water use and constraints of river flow often have profound regional impact. Many of the world's greatest rivers are shared by two or more nations.

Future Water Supply

We need to recognize that, on a global basis, we have a looming water problem of serious proportions. It is estimated that water use will increase by 50 percent in the next twenty years and this will create increasing competition between regions and sectors. Water may well be the most critical commodity for the developing world, where rapid population increase, combined with widespread poverty and growing urbanization and industrialization, add to the problem of increased demand the threat of growing degradation of water resources. In spite of increasing access to treated drinking water, many living in urban areas lack waste treatment facilities. About one-third of Earth's population (2.4 billion people) lack adequate sanitation services. Large areas of arid and semiarid lands still lack adequate drinking-water sources. As many as 1.1 billion people—about one in six of Earth's population—still lack access to improved water supply.

Water Policies

So what should we be doing to fulfill our role as responsible tenants? There are several basic approaches that we need to consider.

1. *Recognize, analyze, and address problems of supply, access, efficiency, conservation, and pollution at the local level.*

   - Water use grew fourfold in the last half century, and with growing population a substantial increase in demand is inevitable. But every water problem is local as well as regional. For the immediate future, growing local populations in densely populated areas of developing countries lacking potable water are likely to depend on drilling new deep groundwater wells, which are less easily contaminated than shallow aquifers. Governmental, international, and voluntary programs are urgently needed to address this aspect of water supply.
   - But this cannot happen without local agreement and community planning. So at the local level—where water supply, water quality, the infrastructure needed to supply water, and local ownership are primary concerns—there

is a need for analysis, planning, and oversight, based on projected use, water quality, and supply. In most of the world this already exists, but in large areas of the developing world, the need for this planning and some financial and technical assistance to undertake it is urgent.

- This planning is urgently needed to confront the growing problems of water supply, conservation, and reuse that are critical to both local and regional solutions. In many parts of Africa, for example, farmers now face crop failures in one out of every three years, and no schemes yet exist to meet their needs. Small-scale projects, involving shallow wells, small pumps, microdams, improved farming practices, and conservation techniques, have been shown to be far more effective in some areas in meeting these needs than many of the large international dam projects of earlier years.

2. *Develop coordinated regional social, economic, and technical policies.*

- Local solutions can provide only for local needs. At the regional level it will be vitally important—but especially difficult—to coordinate these short- and intermediate-term local needs and plans and to hammer out regional agreement on depletion and recycling policies and user charges for large rivers and aquifers. We have already seen that water depletion rates now often exceed recharge rates. Recharge time for major aquifers can require hundreds or even thousands of years, and the areas of natural recharge by rainfall may be far distant from the places of drilling and usage. Water rights and ownership are of major importance in this planning. Ownership of water rights may be represented by municipal authorities, major public utilities, or private companies. All have worked successfully. None is perfect. But this regional planning will need to be soundly based on local plans, adequately funded, and effectively administered.

- But *no* technical schemes will succeed without sound economic and pricing policies. We have assumed for much of our history that water is a free good. Because in any given area the rain falls on us equally, it is easy to assume that it is common property. So we have allowed anyone to drill into an aquifer and use water at no cost; we've also allowed agriculture to do this on a massive scale, aided by substantial public subsidies. And until recently we have allowed both agriculture and manufacturing to pollute surface and subsurface water without any charge or penalty. But if water is a common good, and if overpumping of aquifers and pollution of surface streams diminish the common supply, it is difficult to argue for free access, larger user subsidies, or penalty-free pollution.

- The goal of pricing and supply policies must be to provide improved access, as well as maintenance of water quality, efficiency of distribution, and use and conservation at the optimum social and economic cost. Nor can the supply and cost of water be unrelated to larger economic questions and patterns of trade. The Great Lakes drainage basin agreement between Canada and the U.S. states that border the Great Lakes restricts water use to the Great Lakes drainage basin. But this limitation of "free trade" ignores the importation of agricultural and other water-intensive products from states outside the drainage basin. Any regional legislation of water supply must confront similar anomalies.

3. *Improve the efficiency of agricultural production, industrial use, and manufacturing processes.*

- This improvement lies at the heart of any effective water strategy. Productive agriculture is the foundation of human health and well-being. It is also our single largest user and polluter of freshwater. Manufacturing and industry are the economic basis of modern society, but they also consume huge volumes of water at marginal levels of efficiency, frequently contaminating and polluting it, and often with significant public subsidies.

- Worldwide agriculture consumes about two-thirds of all the freshwater used, and irrigation schemes are estimated to have an overall efficiency of only some 40 percent. Most water therefore never reaches the crops for which it is intended. Huge improvements are possible, by making simple and relatively inexpensive changes in agricultural irrigation schemes. Some alterations include the use of surge irrigation to replace open-flow channels; careful use of valve and timer mechanisms; and replacement of aerial spraying irrigation by low-energy, low-pressure sprinkler heads, drip irrigation, and microirrigation, as well as improvements in the design, distribution, and flow management of irrigation ditches and canals. Pilot studies in many regions show gains in efficiency of 10–40 percent by using these various techniques, and, though the results vary from area to area, the productivity of the land has also generally risen with this increased efficiency. Drip irrigation schemes, for example, have an overall efficiency of some 95 percent, in contrast to only about 60 percent for the more widely used flood irrigation. Improved crop selection, rotation, and management can also contribute to conservation. Together with efficient use comes reduced pollution. Too frequently, runoff from agricultural land is contaminated by residues from fertilizers, pesticides, herbicides, and animal manure. Reduced water use also reduces contamination levels, though it is not by itself enough to prevent it.

- Industrial activities—mining, quarrying, construction, and manufacturing of all kinds—are also major water consumers and sources of pollution. Improved technology has already led to major reductions in water use and in costs. In Japan, industrial productivity has increased its efficiency of water use 300 percent over a period of two decades, with many of the gains coming in the chemical, iron and steel, and paper-manufacturing industries. Recycling and reuse are important elements of these improvement schemes. Overall payback periods for the investment required to produce those improvements range from two to ten years. The economic results of some of these schemes can be spectacular. In one case redesign of the papermaking process and the construction of a new facility allowed production using only 1 percent of the water required at the older plant.

- The incentives for these industrial gains have been the introduction of stricter pollution controls and rational pricing policies. Fortunately, industrial use, in contrast to most farming, allows the recycling and reuse of water up to seventeen times in some industries, thus reducing water consumption, costs, and the possibility of pollution.

4. *Redesign, retrofit, and reuse municipal water supply.*

- Although domestic consumption accounts only for some 10 percent of all water use, population increase and urban spread make this an area of opportunity for improved efficiency. Here again, redesign and retrofitting to allow recycling of domestic and small-business supply schemes and appliances have been shown to be highly cost effective. Managing demand through education and redesign of building and distribution systems has also proved effective in avoiding costly new supply schemes. Comprehensive retrofit— from showers and toilets to heating appliances, dishwashers, washing machines, and metering and audit systems—together with public education, pricing incentives, irrigation management of public parks, and sophisticated leak-detection schemes, have all produced significant economies.

- One promising method of improving the efficiency of domestic water use is the introduction of gray water recycling schemes. Gray water is household water that has been used in sinks, showers, dishwashers, and washing machines but not in toilets. Various schemes are now available to use gray water in irrigation.

- Several metropolitan communities have shown the benefits of this comprehensive approach. The Greater Boston area, for example, with some 2.5 million people, launched a comprehensive conservation and efficiency drive, involving education, retrofitting, and technical advice for consumers, which is reported to have reduced water consumption by 16 percent over four years.

5. *Pursue policies, pricing, and financial subsidies that are likely to be the critical social components of any successful strategy for improved water use. Vital as local initiatives will be, they can be effective only as long as they*

are linked into more comprehensive regional schemes. The reason for this is geological rather than political or economic. Neighboring communities frequently share the same aquifer. Adjacent counties frequently draw water from the same river. Pollution in one county's streams is a serious matter for downstream counties. No water use and no water policy can ever be strictly local. It impinges in some degree on neighboring communities.

- Subsidies play a major role in agriculture in many countries. Some farmers in the United States, for example, pay only 1–5¢/1,000 liters of irrigation water, while the public pays from 30–80¢/1,000 liters for treated water. This pricing provides little incentive for effective irrigation management. Total worldwide government water subsidies in the mid-1990s are estimated to have averaged $80 billion per year.

- The benefits of regional studies of water supply have repeatedly been demonstrated. The Great Artesian Basin of Australia, for example, is underlain by an aquifer that extends beneath one-third of the continent and receives its water supply from the Western Highlands. From the earliest settlement of this vast area, the government required detailed records of water supply, well levels, and production flow. As a result, the system is now in a sustainable steady state.

- Elsewhere, resolution of many of issues of water supply is already proving difficult, even within a single large region or a single country. Once the issues involve international boundaries and interests, they become still more challenging. It requires leadership, time, and understanding to resolve the problems involved. It will also require generous financial aid for the poorer countries of the developing world. Without that, even with technical support and equipment, we have little hope of meeting the water needs of one-third of our fellow humans.

6. *Address pollution.*

- We must deal with known sources of pollution. There are differences of viewpoint among specialists about some aspects of climate change; on others, there is virtual agreement. Consider one: acid rain. Studies in western Europe, especially Scandinavia, as well as in the eastern United States over the past forty years have shown that, whereas normal rainwater has a pH of about 5.7, many lakes and ponds until recently had pH readings around 4.0, with one extreme reading of 2.1, about the same as vinegar. This results from emission of SO_2 and NO_2 by coal-burning power plants, combining with rainwater to produce acids, such as sulfuric acid. Other factors may also be involved in this acidification. While there is still debate over the precise mechanism involved, some fifteen thousand lakes now show the deleterious effects of acid rain and the impact on wildlife has been severe. We need to deal with known sources of pollution, both local and regional.

7. *Develop complementary social and technical solutions.*

- Effective public policy requires not only effective analysis and projection of social needs but also the best scientific and technological assessment of broader regional intermediate- and longer-term climate issues, environmental impacts, and technical remediation. Global climate change is likely to pose major issues of future water policy, for example, not least in the displacement of climatic zones, changes in circulation and weather patterns, and possible rises in sea level, with consequent migration of some populations. Study of and planning for these possible developments should be a high priority at the national and international level.

- At the technical level, the biggest problems involve the present costs of providing clean freshwater in the arid and semiarid parts of the world. Dryland areas cover 40 percent of Earth's land area and support more than 20 percent of the world's population. Despite their use in some oil-rich Middle Eastern states, nuclear-powered desalination plants are scarcely feasible for the poorer nations of the world. There may be useful possibilities for wind power in coastal areas, but in most areas long-term economic development and assistance will be required for such solutions.

8. *Finance known solutions.*

- The development, maintenance, and affordability of simple drilling and supply schemes for rural areas must become a major priority, but this cannot happen without international financial support. And solutions have also to involve commitment to responsive agricultural practices, such as leaving land fallow to prevent soil depletion by salinization. Irrigation, which allows year-round crop production, has already proved unsustainable in some areas.

9. *Reconsider mega supply schemes.*

- Regional studies must increasingly balance the benefits and efficiency of large-scale, "hard" supply schemes—dams, aqueducts, major pipelines, large central plants, and so on—against their social, ecological, and financial costs. These costs arise from the effects of sedimentation behind the dams, the high costs of dam maintenance and repair, the fact that many of the "best sites" have already been used, and the potential population displacement and environmental impact of construction on the remaining sites. Great as the benefits of such schemes have undoubtedly been in the past, attention to less ambitious, complementary, local schemes is likely to become increasingly characteristic of development in many countries.

10. *Integrate local and regional studies and policies.*

- Effective policies will require the integration of local schemes at the national and international level. Among the other technical issues to be resolved are regional problems of water runoff, flood control, dam construction, wetland preservation, groundwater depletion and recharge, water quality, aquifer contamination, saltwater incursion, alkalization, and related irrigation, erosion, and land use problems. Local solutions to these regional problems may do more harm than good. In many cases we have already had sustained study of these supply systems on a regional, national, and international basis. Only in the light of the best available knowledge and technical forecasting can broader questions of social policy be addressed, including such contentious issues as water ownership and price policy, depletion limits, recharge responsibilities, land use limitations, and other comparable issues. To attempt to develop comprehensive water policies in the absence of a sound scientific and technological framework and projections is as likely to compound existing problems as it is to resolve them.

- There is nothing like unanimity among the scientific community about some of these issues, but at present the international framework to address them scarcely exists. Efforts everywhere are piecemeal, often inadequately funded and staffed, and sometimes supported by interested parties to an extent that may threaten their impartiality. Few nations except the largest—China, India, and Russia, for example—can go it alone on this issue. Climate, water use, pollution, and conservation are ultimately global issues in which the behavior of any party has implications for all the others. If ever there was a requirement to act locally but think globally, this is it.

- We need not only the fullest knowledge about local water supply needs but also the will and the agreement on how best to harness science and technology to address water supply issues, both local and regional. We carry the responsibility, not only for our own future sustainability, but also for the health and well-being of the world's poorest people, more than 1 billion of whom still lack clean drinking water, and an additional 1.4 billion of whom lack adequate sanitation. We have yet to develop the will and the means of addressing the huge toll of death and disease that arises directly from this lack. Within national states, the ethical issues of water use and pricing, aquifer depletion, personal initiative, and public constraints will be sharply focused and contested. Within the broader global context they are still largely unacknowledged and ignored. But these are not just "other people's problems": they are the world's problems, and therefore also ours.

Future water supply has been described as the world's silent, looming crisis. Each of the major global trends—population growth, climate change, industrialization, energy availability, pollution—compounds the problem of the adequacy of future water supplies. The "silence" of this threat should not deceive us. It is serious. The good news is that careful planning and prudent policy can resolve the problem.

CHAPTER 18

Air as Sustenance

If Earth is our dwelling place, air is our enfolding mantle. We spend our lives bathed in air, enveloped in its encompassing embrace. It provides our every breath, it supports our every activity. Every morsel of food we consume is a gift of air, a product of the ancient alchemy of photosynthesis, by which plants withdraw one component of the air (carbon dioxide) and so create another (free oxygen). Air encloses the water of the planet, evaporating here, precipitating there, responding in its movements to the radiant energy of the Sun as it drives the water cycle and powers the winds. Air is everywhere—above the ground, on the ground, within the ground—supporting all Earth's living things, participating in all its ceaseless surface changes, acting and reacting endlessly as the chief participant in all the great mediating systems of the planet.

We have already discussed some aspects of the air in chapter 5. We reviewed its layering, its distribution and extent, its nature, and its broad composition. We described its probable origin as Earth's "third atmosphere." And we reviewed its changing temperature through time, including especially its recent warming, and the factors—natural and human—that influence those changes in temperature. We've also considered atmospheric changes that present us with a challenge: the hole in the stratospheric ozone layer caused by CFCs and the spread of acid rain and snow, for example. The attention we've already devoted to the atmosphere in each of these different contexts indicates how significant it has been and still is in the economy of the Earth and its life.

Earth and its atmosphere have developed together over their long history, with the atmosphere evolving in its composition, changing in its temperature, and responding to the planet's slow structural development and changing geographies and also to the influence of its successive inhabitants. It was, for example, simple bacteria that contributed to the development of an oxygenating atmosphere some 3½ billion years ago. Major volcanic explosions continue to leave their atmospheric imprint, injecting particulates so high into the atmosphere that they persist there for a year or more, changing the weather, even influencing the quality of harvests. Sandstorms darken the skies over vast areas on the fringes of the world's deserts to such an extent that they can be seen from space. To these changes—both the long-term natural changes in the Earth and its inhabitants as well as its short-term local perturbations—the atmosphere has responded: evolving, adapting, and accommodating to each.

And in spite of the carelessness of our present patterns of human pollution, Earth will, no doubt, continue to absorb these latest changes and, on its own terms and in its own time, accommodate and adjust to them. The question is not whether the atmosphere is sustainable: it is. The question is whether the atmosphere, changed as it must then become, can continue to sustain us. It is we and our fellow creatures who are at risk from the very changes imposed by our own pattern of living. The hazard we face is whether the atmospheric changes we are inducing may now have become so severe and so rapid that they will overwhelm the atmosphere's capacity to provide a habitable environment for us and for all its other creatures. We have assumed that we could funnel and pipe away all our global gaseous garbage. Building even higher smokestacks for our factories and power plants, fitting quadruple exhaust pipes on our automobiles, we have assumed that, once piped away, our toxic fumes and combustion smoke are gone.

But nothing is "gone." In one form or another, our refuse comes back to us, as acid rain or snow, corroding buildings and bridges and damaging crops; or as contaminating particulates, producing a spreading pattern of pulmonary illness and other chronic ailments. Though some pollutant gases are short-lived, others remain in the atmosphere for a century or more.

And in contrast to other atmospheric hazards—tornadoes, thunderstorms, hurricanes, and so on—we cannot escape indoors for protection from pollution, for, even as we slam the door on the threat, the same atmosphere awaits us within our buildings. In fact, the air within many of our buildings—schools, homes, factories, offices, and even hospitals—can be even more threatening to our health and safety than it is outside. Consider, for example, the discovery in the 1980s of the link between asbestos—a naturally occurring, fibrous mineral, once widely used in building construction and insulation—and mesothelioma, a serious form of lung cancer. Or consider the "sick building syndrome" of more recent years, or the continuing concerns about "radon houses," constructed over areas of natural background seepage of radon, an odorless, tasteless, colorless gas, said to be second only to smoking in causing lung cancer. Like it or not, the atmosphere is everywhere and we are everywhere dependent on its quality for our life support. A century or less ago, atmospheric pollution tended to be local in extent, confined to the neighborhood or village, and thus limited in its effect, because of Earth's capacity to absorb and dilute contaminants. Now, in contrast, the explosion in human numbers and the impact of growing industrialization and urbanization of society make it a far more global threat.

Cities, for example, tend to have more atmospheric pollution than rural areas, but because there is continued growth in the proportion of the world's population who are city dwellers (now approaching 50 percent), urban pollution is a growing health hazard. In spite of differences between one region and another, atmospheric degradation is now a global issue. Sustainability requires a habitable atmosphere for the whole planet.

Sources and Causes of Atmospheric Degradation

The harsh reality is that much of our atmospheric pollution is the product of our basic means of subsistence. The major sources are all too familiar: heavy industry,

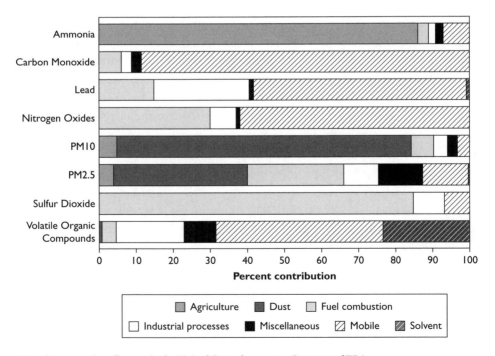

18.1 Common air pollutants in the United States by source. Courtesy of EPA.

power generation, transportation, space heating, agriculture, garbage disposal, and so on. But we can no more eliminate those activities as a basis for modern society than we can cut out eating as a basis for human life.

Nor can we take any action that will guarantee immediate improvement. Consider the atmospheric lifetimes of the various compounds that have been identified as the chief sources of atmospheric degradation. Thus carbon dioxide has an atmospheric lifetime of 5 to 200 years, nitrous oxide 114 years, and hydrofluorocarbons up to 260 years. In contrast, water vapor, though it is the most abundant of the greenhouse gases, has an average residence time of only about nine days. Its changing concentration appears to arise from climatic feedback associated with increasing temperature rather than from industrialization. Its effect seems to be to amplify the influence of other greenhouse gases, though the details are far from clear.

Carbon dioxide (CO_2), the chief greenhouse gas after water vapor, is released from the combustion of fossil fuels, from cement production, and from deforestation. In effect, we are producing this by taking carbon from the Earth, where it is stored in fossil fuels and carbonate rocks, and burning it to produce not only heat but also CO_2, water vapor, SO_2, and traces of other gases. Present levels of atmospheric CO_2 are around 350 parts per million by volume (ppmv), and at current levels of increase this is likely to reach 500 ppmv by 2100. The preindustrial level is thought to have been in the region of 270 ppmv.[1]

Nitrogen oxides (NO_x), including both nitric oxide (NO) and nitrogen dioxide (NO_2), are formed from fossil fuel combustion and the burning of biomass. Though having a residence time of only days in the atmosphere, they contribute to the formation of acid rain and photochemical smog.

Nitrous oxide (N_2O) is formed not only by combustion, biomass burning, cattle lots, and deforestation but also by nitrogen-based fertilizers. An important greenhouse gas, it has a residence time in the atmosphere of 114 years.

Sulfur dioxide (SO_2) is produced chiefly by power plants and factories by the combustion of high-sulfur coal and fuel oils. It is highly toxic and corrosive, and forms, with nitric oxide, secondary pollutants (sulfuric acid and nitrate and sulfur salts) in the form of acid rain.

Chlorofluorocarbons (CFCs) were once widely used as major components of aerosol propellants, as well as in refrigeration and cooling equipment. Now banned, they survive for several thousand years in the high atmosphere, where they break down, releasing chlorine, which attacks ozone in the stratosphere and so damages the ozone layer. CFCs are also derived from some industrial solvents, Teflon polymers, and foam. The industrial use of these products has been greatly reduced by legislation that was passed in the 1980s.

---

1. T. M. L. Wigley, "The Pre-Industrial Carbon Dioxide Level," *Climatic Change* 5 (1983): 315–20.

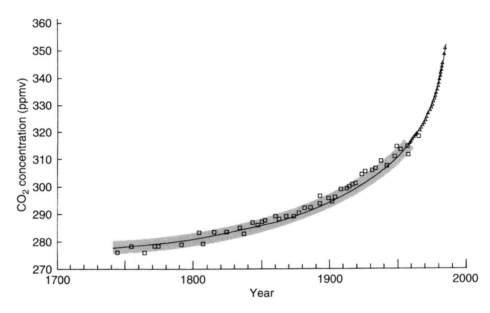

18.2 Level of CO_2 in atmosphere 1750–present. Squares represent values determined from ice cores; triangles represent direct (Hawaiian) atmospheric measurements; the straight line is an approximation to a smooth curve value; and shading represents likely measurement error. From Jonathan Cowie, *Climate Change: Biological and Human Aspects* (Cambridge: Cambridge University Press) and A. Lacey, ed., *Bioresources: Some UK Perspectives* (London: Institute of Biology, 1992), based on NOAA data; used with permission.

Methane (CH_4) is released from combustion of fossil fuels and from rice paddies, animal waste, landfills, and domestic sewage. Its concentration has increased markedly since the Industrial Revolution. Methane also reacts with chlorine in the atmosphere and contributes to ozone depletion. It has a residence time of twelve years.

Not all these gases are harmful, especially in low concentrations. CO_2, as we have seen, is as vital to plants as O_2 is to animals. In fact, as we have seen, many of them occur naturally in the atmosphere. Without their presence, Earth's comfortable 59°F average temperature would be closer to the frigid average—13°F—of the moon.

Studies of direct temperature readings over the last century and a half and proxy records from ice cores from earlier periods reveal the steady increase in concentration of these greenhouse gases and their close, but not perfect, correlation with increasing global temperatures. The balance of evidence strongly suggests a causal link between the two, though there is still some debate on this point. The correlation over the last century between proxies for solar activity and temperature is regarded by some as better, especially in reflecting the temperature decrease from 1940 to 1970.

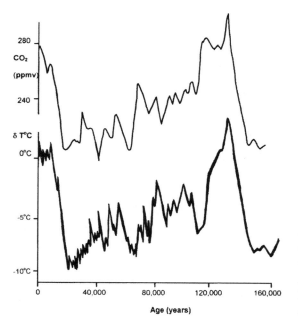

18.3 Temperature over last 160,000 years, with CO_2 plots. Top: atmospheric carbon dioxide; bottom: temperature change. From Jonathan Cowie, *Climate and Human Change: Disaster or Opportunity?* (New York: Parthenon, 1998), based on NOAA data; used with permission.

Now, a further temperature increase of a few degrees would not be unprecedented. We have seen, for example, that the Earth was about 4°C (7.2°F) warmer at the start of the Pleistocene period, 2 million years ago. The previous interglacial period (the Eemian-Sangamon-Ipswichian), which took place about 120,000 years ago and lasted some 20,000 years, was also 2.0–2.5°C warmer than at present, and the still earlier and shorter (Hoxnian-Holstein-Yarmouthian) interglacial, some 220,000 years ago, was also warmer. Within all such glacial and interglacial periods there were many smaller episodes of advances and retreats of the ice in response to smaller fluctuations in temperature.

The last 150 years have clearly been a warming period and GGEs have been increasing. Most informed observers accept that continuation of this climate change poses a serious threat. That is not to suggest complete unanimity in the scientific community about the longer-term consequences of these temperature increases, and especially about the precise future timing, magnitude, impact, rate, and extent of their influence. Nor will all these changes have negative consequences for all areas. They may well prove beneficial for some regions, especially those in mid to high latitudes, by extending growing seasons, although they may also produce more extensive drought. There is also some debate about the extent to which future reductions in GGEs, changes in land use, and other remedial actions can reduce temperature increases that are already "built into" the climate system.

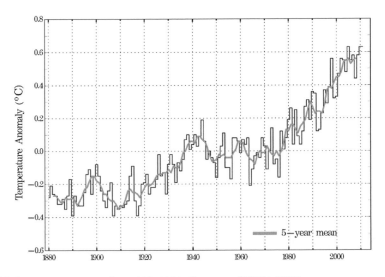

18.4 Global temperature anomalies, 1880–2010. Courtesy of NASA GISS.

Confronting Atmospheric Changes

A major source of debate in all discussions of the impact of climate change is the reliability of the elaborate general circulation models (GCMs) that have been developed as a means of predicting future climate patterns. Alternative methods, such as the use of past warm periods of Earth history to serve as analogues for future climate, have not, as yet, proved satisfactory. This is partly because of the unknown influence of differences between past rates of terrestrial elevation and patterns of oceanic circulation in comparison with those of the present. But even though our predictive skills are limited, the potential threats are so serious that a policy of sensible reduction of system risk would seem to provide the most prudent response. This policy would seek to balance the potential benefits of reducing the various threatening trends with the economic and social costs of their mitigation.

Within such a context, if we are to reduce our vulnerability to the threats presented by atmospheric changes, as to other natural threats, there are two broad categories of response available to us. The first involves *mitigation:* reducing or eliminating the potentially harmful threat. Most of the atmospheric changes we have described cannot now be prevented, because some continuing change is already incorporated in Earth's oceans and systems, but they can be reduced. And reduction can provide great benefits. Mitigation will be costly if we embrace it on a global scale, but we need a careful calibration of those costs against the potential benefits of mitigation. Mitigation in this case would require us to reduce GGEs, develop alternative energy sources, redesign our buildings and communities, improve efficiency of our use of fuels, and reduce our dependence on polluting materials by replacing them with substitute fuels or other materials. All these are worthwhile goals in their own right, quite apart from their influence on climate change.

The second response involves *adaptation*. Although we cannot precisely predict the climate future, we have seen that some temperature increase may already be "built into" the global system and thus be inevitable, and we need to adapt as best we can to its consequences. By implementing programs to reduce deforestation, improve agricultural practices, replace forests, strengthen food defenses, and develop drought-, heat-, and frost-resistant crops, we can make a significant difference. And, again, all these programs are beneficial in their own rights.

The magnitude and extent of programs such as these will require the most extensive scientific and social analysis and financial review. Any plans adopted will have substantial economic and societal consequences. And they will certainly have to be refined and corrected in the light of continuing experience. We shall later review the political context in which such major policy questions will need to be considered.

Even as we wrestle with these problems of greenhouse gases and particulate emissions, the number of coal-fired power plants and gasoline-powered automobiles continues to grow, and it will be decades before they are replaced by some other form of clean power generation and clean transportation. In the meantime, how do we limit further damage to the atmosphere? Air circulation is so pervasive that there is no possibility of remediation on any significant scale. We can't clean up the world's air by remediation. Certainly within a house or a store or a factory we can "treat" the air, dehumidifying, filtering, purifying, heating, or cooling it as we wish. But for any but the smallest point source of outdoor space, that is not possible, as the Beijing Olympics of 2008 demonstrated. The only means of reducing such emissions is the elimination or reduction of the sources of emissions. Unlike soil or water, where contaminants can persist for many years, air composition and quality responds rapidly to reduction at the source of the "polluting" agent. That "solution" has proved to be wholly effective in its outcome, surprisingly rapid in its benefits, but remarkably difficult in its implementation. This may be, in part, because the costs are largely local while the benefits are not only local, but also to a degree regional and global. It is also, chiefly, this same imbalance of costs and benefits that lies behind the inability of the federal government to develop a preventative strategy, in contrast to the agreement of states and regions to enact a common approach. The technical means of reducing GGEs, for example, are demonstrably effective and readily available. The financial incentives and statutory requirements to implement them, however, are not. That means that any effective "clean air" program has to be built on a framework of local, regional, and international partnership. The seeds of just such a program already exist in the Kyoto and other protocols, but we still have a long way to go in their implementation. Most of the signatories to that protocol have not met the goals they established.

Until we can develop such international agreements, the most critical need is to reduce emissions, especially from power plants and automobiles, which account respectively for 40 percent and 33 percent of carbon emissions. We must also improve the efficiency of every machine, appliance, and vehicle that is manufactured, from toasters to tankers, in order both to reduce energy consumption and to limit emissions.

Two major initiatives are now commonly promoted as being particularly promising in this effort. First, emissions of carbon dioxide can be reduced by programs of carbon capture and sequestration (CCS). This involves the injection of CO_2 from large emitters into secure underground reservoirs, rather than releasing it into the atmosphere. It is estimated that the present sum total of human activity is responsible for the emission of some 7 billion tons of carbon dioxide a year: that's about one ton per person. Virtually all of this escapes into the atmosphere, but pilot projects to reduce this pollution are now under way in several countries. Secure underground aquifers in which carbon dioxide may be stored are relatively plentiful. One recent study suggests it should be possible to store all the CO_2 generated by fossil fuel power plants in Europe over the next eight hundred years in a single saline aquifer beneath the North Sea.

Though the cost is likely to be high, the present levels of European carbon taxes could make such sequestration attractive. A possible alternative might be the injection of CO_2 into oil field reservoirs, where the gas can contribute to the secondary recovery of petroleum from depleted oil fields. Recent studies have also shown that injection of CO_2 into underground coal seams can enhance recovery of methane, an important component of natural gas. Coal has an affinity for CO_2, which it absorbs. Other possible methods of sequestration also exist, including injecting CO_2 into the deep oceans to produce icelike clathrate hydrates (crystalline solids in which small gas molecules such as hydrogen, carbon dioxide, methane, and others are trapped inside cages of water molecules) and combining CO_2 with olivine and serpentine-rich rocks to produce carbonates. These are, however, theoretical schemes, which have yet to be demonstrated or tested.

Two factors seem to me to make large-scale carbon sequestration a questionable strategy: present costs and future dangers. The costs of sequestration are likely to be high; they include not only the immediate costs of carbon disposal but also the potential costs of dealing with sequestered CO_2 and possibly leaking reservoirs long after we have finished using fossil fuels. The security of underground reservoirs is not something for which anyone can provide a long-term guarantee.

Nor is there, as yet, any general agreement that carbon sequestration is a safe, necessary, or cost-effective method of reducing greenhouse gases. Some geologists believe the associated costs and hazards are likely to exceed the benefits.

Carbon sequestration is not the only method of reducing carbon emissions. Improved carbon capture in power plants and similar sources of CO_2 emissions can make a major contribution to this reduction. Various effective methods of pre- and postcombustion capture already exist. Most involve additional energy use and thus additional costs to the consumer, but these costs may not be greater than those of electricity derived from other sources, including nuclear plants and other "low carbon" power plants.

The second major initiative is to mitigate the effects of GGEs by encouraging increased forestation and decreased deforestation. Unlike carbon sequestration, this is an established and effective program. It is increasingly constrained, however, by the lack of land available for reforestation, especially as global population continues to increase, together with growing numbers of those who depend on slash-and-burn agriculture for their subsistence.

The remedies proposed to reduce potentially harmful atmospheric emissions may sound simple. They are not. They involve difficult trade-offs and painful priority choices. The recent experience in California, after the governor and legislature agreed on plans to curb carbon emissions by 5 percent by 2020, has highlighted the concerns. Relying on a mixture of cap-and-trade schemes, regulations, and incentives, the program has created sharp divisions in both the business and environmental communities. In the absence of a federal plan, other states are creating their own piecemeal plans, but there is great debate about the effects of regulation on competitiveness, costs, and the overall economy. Some point to the application of the Kyoto Protocol by European countries as a failure from which we have yet to extract the lessons.

Preserving the quality of the air we breathe will be as big a challenge as preserving the land, lakes, and rivers. We are more likely to succeed by a continuing, rigorous balance between clear benefits and real costs (social as well as economic). Redesign and improved efficiency of agricultural, industrial, and domestic practices and products, judicious remediation and restoration, and incentive-based regulation seem to offer our best, if tenuous, hope.

CHAPTER 19

Soil as Sustenance

Time and time again in human history, from ancient Mesopotamia to modern Madagascar, poor animal husbandry and irresponsible farming have led to soil erosion that destroyed the productivity of the once-bountiful land. We are in danger of allowing a similar destruction, not in a single region or on a large island, but over a vast area of continental proportions and global extent.

Soil is Earth's detritus: the weathered, eroded skin of its crumbling surface, mixed with the decaying remains of the vegetation it supports. In some places—mountaintops, high plateaus, the polar regions—soil scarcely exists, for there is little vegetation and few microorganisms, so bare rock forms the surface of the land. In deserts, too, soil is scarce. Rocks crumble to sand, not soil, for here again there is little vegetation and few microorganisms, both of which are essential to the creation of the humus that provides the body and the nutrient of soils. But almost everywhere else, soil is the face of the Earth with which we are most familiar. It *is* the Earth, supporting life and nourishing its growth. For soil is much more than dirt, much more than mineral fragments of eroding rocks. Soil is the place where all Earth's different components interact. Soil is alive, teeming with living things. A single teaspoon may contain over a million fungi, such as yeast and molds, and hundreds of millions of bacteria, several thousand protozoans and scores of nematodes, as well as supporting mites and other creatures. The smaller organisms consume the vegetation that falls on the ground and promote its decomposition, so forming the humus. The work of these microorganisms is supplemented by the

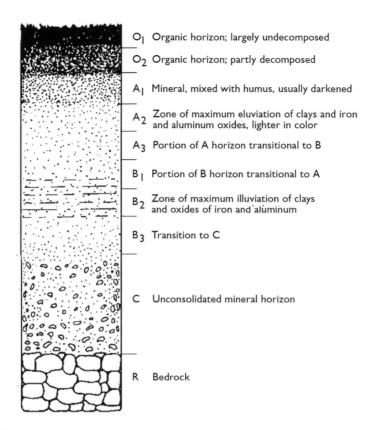

O₁ Organic horizon; largely undecomposed

O₂ Organic horizon; partly decomposed

A₁ Mineral, mixed with humus, usually darkened

A₂ Zone of maximum eluviation of clays and iron and aluminum oxides, lighter in color

A₃ Portion of A horizon transitional to B

B₁ Portion of B horizon transitional to A

B₂ Zone of maximum illuviation of clays and oxides of iron and aluminum

B₃ Transition to C

C Unconsolidated mineral horizon

R Bedrock

19.1 Soil formation. Courtesy of FAO.

work of earthworms and other larger creatures in mixing and aerating the soil. A single hectare of pasture in a humid, midlatitude climate, for example, may contain more than a million earthworms and several million insects.

Each of these various organisms contributes to the development of soil. Some bacteria convert atmospheric nitrogen into nitrogenous organic substances. Others assist in the decomposition of proteins into nitrates, which are vital to plant growth. The nutrition provided by the soil forms the base of the food web, supplying a source of energy from fixed carbon and nitrogen. Most of the organic matter in soil consists of humus, together with roots and organisms. Humus gives the dark color and texture to the upper layers of the soil, allowing it to store moisture and retain nutrients by reducing the downward leaching of fine particles and dissolved salts through the soil. An average soil contains only about 5 percent organic matter. Of the rest, 45 percent consists of mineral particles, with the other 50 percent made of roughly equal parts air and water.

Soils vary greatly from place to place, even sometimes within a single field. The character of a soil depends on the local climate, bedrock, topography, organic content, drainage, and age. These features give soils their distinctive form:

their color, acidity, fertility, and texture. Most soils have a layered profile that results from weathering, the degree of organic decomposition, and downward movement of fine particles and soluble materials. This profile can be readily seen in a shallow trench, dug through the soil. (Figure 19.1.)

The layering is typically about three to six feet deep in most soils from temperate regions. The uppermost layer, the A horizon, consists largely of decayed organic material in its upper portions, together with the more insoluble minerals from the parent rock, especially fine-grained clay minerals and sand (quartz). This is the zone of leaching and of the most intense biological activity. The underlying B horizon contains soluble minerals such as calcium carbonate and iron oxide, leached downward from the overlying soil. There is little organic matter in this layer, which contains in its lower part fragments of the underlying parent rocks. The lowest C layer consists of the weathered portion of the underlying solid rock. The thickness and distinctiveness of these layers vary with the age of the soil and the local climate. The fertile soils of many well-watered areas of eastern North America and Europe, for example, are rich in aluminum and iron but low in soluble calcium carbonate, so farmers regularly have to "lime" the soil. The low rainfall areas of the U.S. West, in contrast, have fertile soils, which are rich in calcium.

All soils change with time: some of the youngest are found on recent lava flows in warm and humid areas where seeds develop in small rain-filled cracks and cavities, gradually forming organic material for future growth.

The local climate plays a large part in determining the mineral components of soils because the particular form of weathering determines the erosional products. In dry, cold climates, mechanical fragmentation, created by repeated freezing and thawing, predominates, producing angular fragments that break down to sand and other resistant mineral grains. In warmer, wetter climates, in contrast, chemical alteration and solution are the dominant agents of erosion, producing dissolved salts and clays. Rainfall, temperature, and time, between them, greatly influence the thickness and character of soils. Near-freezing temperatures, for example, reduce the rate of chemical reactions and inhibit the growth of microorganisms. Low rainfall reduces the downward movement of soluble minerals and also inhibits the development of organisms.

In the tropics, deep red soils predominate. In these latitudes, rainfall produces deep leaching that removes most of the soluble material, including calcium carbonate and even the less soluble feldspar and other silicate salts. The residual "hardpan" soils are rich only in insolubles, typically laterite (iron and aluminum oxides) or bauxite (hydrous aluminum oxides). Though they support lush vegetation, these tropical soils can rarely sustain intensive agriculture.

Some soils are formed in place, lying above the parent bedrock from which they have been derived. Other soils may be made of material transported far from its original source. The boulder clays that cover much of North America and northern Europe, for example, are derived from distant granitic rocks in the north and have been carried southward by moving glaciers. The weathering of these clays has produced fertile soils that are rich in nutrients. It is these soils that support the huge grain harvests by which the U.S. and European agricultural heartlands feed both their own populations and those of much of the rest of the world. Other rich soils are also eroded from sites distant from their present locations. The rich alluvial terraces along river valleys are formed of silt, sand, and gravels derived far upstream from their present locations. The great loess deposits of China and elsewhere are windborne deposits, formed under earlier arid or semi-arid conditions. The sand and silt of which they consist have been eroded often from distant sources and deposited as a thick, unbedded windblown blanket over a wide area, where it now provides fertile and easily worked soils.

Preservation of Soil

Because soil is the ultimate foundation of the terrestrial food chain, it is vital to preserve it. That does not mean, of course, that we can prevent any depletion or erosion. Erosion is part of the natural cycle of change that exists in all landscapes. Without it, there would be no soil. Preservation means, rather, that we must seek to retain the capacity of soil to support life. That requires us to limit the rate of depletion and loss of soil to the rate at which natural processes and human husbandry can replenish it.

In most areas of the developed world, farming practices generally recognize the need to ensure the preservation of soil capacity, but soil is now being diminished in many parts of the developing world, where poor soil husbandry and poor agricultural practices are leading to increased soil erosion and depletion.

Soil Loss: The Quiet, Worldwide Crisis

Some environmental concerns have also become major public concerns—the ozone hole, acid rain, and climate change among them. Others are virtually unrecognized in the media and untouched by public concern. Soil loss—like water supply—is a quiet crisis, profound in its impact, worldwide in its presence. Estimates of the extent to which the world's existing soils have already been damaged by erosion are not easy to obtain. It has been estimated that about 80 percent of all the world's agricultural land suffers moderate to severe erosion, and a further 10 percent suffers light to moderate erosion.[1] A more recent estimate suggests

---

1. David Pimentel et al., "Environmental and Economic Costs of Soil Erosion and Conservation Benefits," *Science* 267 (1995): 1117.

that, even in the United States, about 90 percent of cropland is losing soil above the sustainable rate.[2] The effects of soil erosion have, as yet, been most harshly felt in some of the world's poorest and most populated areas, where the results can be even more serious: loss of soil by erosion in perhaps a third of China's cultivated areas, reduced crop yields in west Africa, famine in Ethiopia, and desertification and hillside soil erosion in India, for example. But even in the developed world, soil loss is a serious and growing problem. Soil loss rates in Europe are reported to exceed the estimated renewal rate by ten to twenty times; in the United States by an average of sixteen times; and in Asia, Africa, and South America by twenty to forty times.

Over half of Australia's agricultural and grazing land has suffered some form of degradation, and well over half of the land area of Argentina has been diminished by erosion, while intensified farming practices in the United Kingdom, supported by subsidies from the European Commission, have had a significant impact on soil erosion. Such rates of erosion and depletion far exceed the natural rate of renewal of soils. It is estimated to take at least five hundred years to form an inch of topsoil under temperate or tropical conditions. Serious as these losses are, the worst effects of soil degradation occur in tropical countries where deforestation, slash-and-burn agriculture, lack of crop rotation, overgrazing, and high rainfall have contributed to the threatening increase and dire effects of soil erosion.[3] And the problems mount, as more and more marginal land is cultivated. These issues will be exacerbated as more land is removed from agricultural use and used for production of biofuel rather than food.

Erosion is caused by the loosening and removal of soil particles by water or the wind. The unconsolidated nature of soils leaves them vulnerable to the impact of heavy rainfall, especially when it falls on bare ground or when it exceeds the soil's capacity to absorb it and thus flows over the surface slopes in sheets or rills and forms gullies. Effective protection against soil loss must first involve reduction in this most destructive form of erosion.

Nor is the problem confined to the loss of soil. Salinization and waterlogging of soils and loss of fertility of remaining soil by removal of finer particles and organic matter all tend to reduce crop yields, while downstream deposition of eroded sediment creates additional damage to fisheries, wetlands, hydropower projects, and wildlife. It can also create severe flooding in some areas.

Even modest erosion reduces productivity because it tends to remove the uppermost soil layers that are rich in nutrients and are most important to water circulation and plant growth. Because erosion control is relatively expensive, it

---

    2. David Pimentel, "Soil Erosion: A Food and Environmental Threat," *Environment, Development, and Sustainability* 8 (2006): 124.
    3. Ibid., 121.

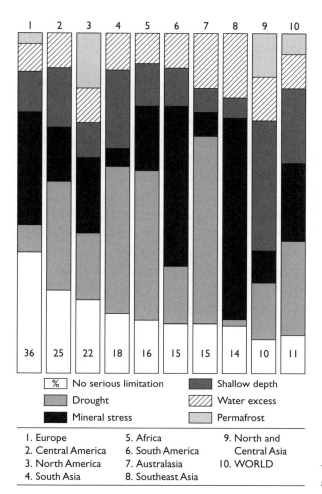

19.2 Regional distribution of soils with or without limitations for agriculture. Courtesy of FAO.

Legend:
- % No serious limitation
- Drought
- Mineral stress
- Shallow depth
- Water excess
- Permafrost

1. Europe
2. Central America
3. North America
4. South Asia
5. Africa
6. South America
7. Australasia
8. Southeast Asia
9. North and Central Asia
10. WORLD

is frequently neglected by individual farmers, not only in developing countries where rapid erosion and the most serious problems exist, but also in developed countries. Social practices often compound the problem. Fragmented landownership and tenancy mean that to avoid serious loss often requires not only regional planning and erosion prevention schemes but also financial support.

How can soil erosion be reduced and productive agriculture be maintained? To answer that question we have to consider both the social issues and human impact involved and the technical steps that can be taken to limit the erosion of soils.

Soil Erosion: The Human Dimension

The soil cover on which we depend is both thin and vulnerable. Rising human numbers, land use changes, and massive technological projects can produce very rapid loss of productive soil. The most direct human impact on soils arises not only from poor crop husbandry, slash-and-burn agriculture, overgrazing, water-logging, salinization, and pollution but also from urbanization, faulty irrigation schemes, and deforestation, all of which can change the character of soils within a generation.

These damaging activities range in scale from very local farming practices to major regional engineering schemes, so that any effective conservation must involve efforts at both levels. Good local soil management, for example, must involve both the education and support of small landowners and tenants—who individually may lack the knowledge and financial capacity to improve their practices, but whose collective impact is great—and the enlistment of the cooperation of global agribusiness. Effective conservation programs must thus combine both local and regional initiatives, and must include education and financial assistance to some of the world's poorest peoples. Only when the benefits of conservation can be demonstrated are small farmers likely to cooperate. The U.S. agricultural extension program—which links the research skills and knowledge of the universities directly to the needs of the farming community—is perhaps the best example of the huge impact on productivity that such educational and financial programs can have.

Conservation Programs

The first priority of soil conservation is the control of erosion, but that is not sufficient by itself to preserve the productivity of soil. Conservation also requires the maintenance of the organic matter, the physical properties, and the nutrients of the soil. And productive agriculture further requires the conservation of enough available water to support plant growth, whether for crops or pasture. We can think of conservation, then, as preventing two kinds of soil loss: erosion on the one hand, and degradation or depletion on the other. Each type of conservation method typically tends also to reinforce protection against hazards of the other type. Let us review them briefly.

Erosion control is especially important in the cultivation of sloping farmland. The most frequently used techniques involve contour planting and terracing and the development of physical barriers to reduce soil loss.

Contour planting is almost as old as agriculture itself. It involves the cultivation of fields by horizontal plowing, following the contours of the land, rather than plowing vertically, down the slope of the land. Because the furrows act as small terraces, disrupting water runoff and allowing time for it to percolate downward, this technique can reduce erosion by up to 50 percent on moderate slopes, and can also conserve water. Water conservation is also improved in some areas by contour ridges and contour barriers—either of grass or of crop residues—which reduce water runoff and prevent soil erosion. In large fields, where mechanized farming is used, strip planting of different crops at intervals of one or two tractor widths is sometimes used to reduce wind erosion.

Terrace farming is used in many parts of the world where steep slopes would otherwise make cultivation impossible. It involves the creation of terraces along contour lines, which are then planted as horizontal seedbeds. Terracing reduces erosion, retains water, and makes small-scale husbandry possible, though it is less suited to larger-scale mechanized agriculture. Terrace farming is often combined with ingenious local water conservation and irrigation schemes.

Physical barriers, culverts, conduits, and water storage facilities, both local and regional, are widely used, both to conserve water and to reduce erosion.

Depletion and degradation prevention involve planting and farming techniques that both protect and replenish the soil. Several techniques are widely employed.

Windbreaks and timber stands are used, both as living barriers to reduce the effects of wind erosion and as a means of improving soil and water conservation. Wind erosion can be especially rapid in areas of increasing desertification.

Remediation is used to treat soils that have been seriously contaminated by industrial spillage, mining operations, or salination. It is costly and may not always be capable of restoring heavily damaged soils to their earlier capacity.

Enrichment of depleted soils by controlling such properties as acidity, salinity, and organic content is an essential part of conservation. The regular use of fertilizer, whether organic or inorganic, is an essential part of this replenishment. Soil enrichment not only provides nutrients but also reduces erosion and encourages downward percolation of water. Many synthetic fertilizers, however, as well as many pesticides and herbicides, are petroleum based and may become increasingly short in supply as the cost of petroleum increases.

Land husbandry involves a range of selective land-use and farming practices that can make a major contribution to soil conservation. Some crops are grown for cover, thus protecting the soil from erosion and excessive evaporation, while others, such as legumes, are grown for their contribution to soil nitrates and the improvement of soil quality. Reduced tillage or no tillage cultivation is common in places where the use of selective herbicides in weed control allows seed planting directly into the soil without the need for plowing. Reducing summer fallow periods is also employed in some areas, while reducing grazing density, using rotational grazing, and protecting streams and rivers are all valuable on range and pasture lands.

Crop selection and soil management involve the appropriate choice and rotation of crop varieties, as well as the selection of particular fertilizers suited to particular crops. These crops can also provide important plant cover for bare ground and improve fertility. The highest rates of erosion—whether by wind or water—occur on bare ground. Early planting, the selection of suitable varieties, and crop rotation all contribute to the prevention of erosion.

Human Impact on Soils

One of the ironies of soil management is that some of the methods commonly used to increase productivity can also create widespread soil damage. Irrigation in semiarid areas frequently produces a gradual increase in soil salinity levels because of evaporation and rising groundwater levels. We have already seen that in some coastal areas of the Middle East and elsewhere overpumping of groundwater has produced seawater incursion into aquifers and thus into agricultural soils. Serious salinity increases have also followed dam construction in some areas (especially the Aswân Dam in Egypt) and are also present in coastal areas of Europe, Japan, and the United States. Almost all irrigated land in warmer areas of the globe is vulnerable to salinization. One estimate concludes that 20 to 30 percent of irrigated land in the United States, Pakistan, Iran, Egypt, and Argentina is now affected by increased salinity. The consequences of increased salinization are serious, making water undrinkable, inhibiting plant growth, and damaging soil structure. As vegetation withers, exposure of soil also contributes to wind erosion. Land impaired by salinity can be reclaimed by various techniques, ranging from improved drainage systems to flushing by freshwater, and from chemical treatment to the planting of salt-loving plants, including those that are edible, such as sea kale and samphire. All these methods are costly, however, and the best method of dealing with salinization is prevention, using better farming and irrigation practices.

David Pimentel and others have calculated that the United States, in spite of heavy investment in public programs for education, prevention, practice, and reclamation, is losing soil from about 90 percent of cropland at a rate greater than its rate of replacement, while some 54 percent of pastureland is now overgrazed and suffering erosion.[4] In some other parts of the world, the problems are even more acute. In China's Huang (Yellow) River, for example, studies show that sedimentation rates—and therefore erosion rates—for the last two millennia were ten times higher than those of earlier times.

Deforestation in tropical and subtropical areas has led to the spread of red lateritic soils, which are also found in the savanna areas of the world where the climate is characterized by warm temperatures and marked wet and dry seasons. These residual soils may reach fifty to one hundred feet in thickness and consist chiefly of iron and aluminum hydroxides, from which the more soluble components have

---

4. Ibid., 124, Pimentel et al., "Environmental and Economic Costs," 1117–1118.

been leached out. Some that are rich in aluminum (bauxite) are valuable ores, but their strong leaching leaves them impoverished in many plant nutrients.

Acidic soils (including peat and podzols), which are widespread in upland areas of Europe, are thought to be the products of both natural processes and long-continued human clearing and intensive agriculture.

Soil Prospects

We have seen that soil and water are the two most essential commodities to human existence. For almost everything else we use—fuels, fabrics, fertilizers, metals, and other minerals—substitute materials can generally be found. For soil and water, there are no substitutes, no alternatives available in suitable quantities, at reasonable cost. But for all their similarity as basic requirements of life, there is one difference. Water is recycled relatively quickly by evaporation and precipitation. Soil is not. Soil, once eroded or depleted takes centuries—millennia even—to recover. It's essentially nonrenewable, at least in the time frames of human history. In addition to the absolute loss of available soil, our growing human numbers mean that we are facing a major loss of soil to support each individual. One estimate is that during the last fifty years of the twentieth century the amount of arable land per person fell by more than 50 percent.

Yet in the poorest parts of the world, soil conservation languishes, losing ground to what are seen as more urgent needs. If we are to preserve a sustainable Earth, we must begin by conserving its soil.

The foundation for any successful conservation scheme has to embrace the need for both responsible local farming practices and free global agribusiness and trade. With that, we can develop fact-based conservation policies, which may well prove to have a relatively low net cost, although the return depends on the specific crop and the particular situation. David Pimentel and others have shown that $1 invested in conservation can return $1.30 to $3.00 in increased crop yields, while each dollar invested in agricultural watershed conservation may save from $5 to $10 in the costs of dredging, levee construction, and flood damage loss. And these benefits are short term, as well as long term. If Earth is to continue to feed us, we need to safeguard its soils. No commodity is more vulnerable and more precious.

The need to address soil erosion is not only important: it is also urgent. Pimentel has calculated that in the forty-year period leading up to 1995, almost one-third of the Earth's cropland (1.5 billion hectares) was abandoned because of soil erosion and degradation that destroyed the productivity of the land. We need to reflect on the lessons of ancient Mesopotamia and modern Madagascar.

CHAPTER 20

Food as Sustenance

In 1944, twenty-nine reindeer were released on St. Matthew Island, Alaska, a thirty-two-mile-long by four-mile-wide island in the Bering Sea. The purpose was to provide an emergency food source for the members of U.S. Coast Guard who were temporarily stationed there. With the end of World War II the island had returned to its earlier uninhabited state: a "reindeer paradise" of thick lichen mats and no predators. By 1957 the reindeer population had reached 1,350, most of them fat and in excellent shape.

By 1963 the reindeer population had swelled to six thousand, though they looked less healthy and the ratio of yearlings to adults had decreased. By 1966 the population had collapsed to only forty-two live reindeer—forty-one females and one male, with no fawns. The island was littered with bleached reindeer bones. By the 1980s, no reindeer were left. What happened? The reindeer grew so rapidly in the early years after their introduction that they overgrazed the island and so exhausted the food supply on which they depended.[1] Almost the whole of the population died of starvation during the winter of 1963–1964. The burgeoning population overwhelmed the carrying capacity of its isolated island home.

---

1. Ned Rozell, "When Reindeer Paradise Turned to Purgatory," Alaska Science Forum, article 1672, Geophysical Institute, University of Alaska, Fairbanks, 2003, http://www.gi.alaska.edu/sourceforum/ASF16/1672. htm.

We too live on an island: we call it a planet. And our human family faces precisely the same constraint. Though we have capacities the reindeer lack—the ability to control our population growth, to modify our environment, to select and change our sources of food, to increase food production and availability—we still face the ultimate constraint of a growing population on a limited land area. Though Earth is a far richer and more varied habitat than St. Matthew Island, it still has limited capacity. For all our freedom, versatility, and creativity, we have only one Earth, and its carrying capacity is finite. And we may be about to test that capacity to support a growing human population at a subsistence level that we would regard as adequate. This will involve a real time experiment and its outcome may not be benign.

For whether we like it or not, Earth's human-carrying capacity—the maximum population size it can sustain indefinitely without reducing its future support capacity—depends on the level of subsistence we select and the degree of equality we demand. Food is a finite resource, and its availability is constrained by the size of our planet.

Earth's Food-Bearing Capacity

Any assessment of Earth's food-producing capacity must begin with the land itself. Earth has a total land area of 150 million square kilometers (about 58 million square miles), about sixteen times the land area of the United States. Of this total area, global land use breaks down as follows:

arable land	10%
permanent crops	1%
forest and woodlands	31%
pasture	24%
unusable	34%

The most striking aspect of this land use is that almost all the potentially arable land is already in use. The only way to increase food supply is to increase the productivity of land use to provide higher yields. That is the pattern that was followed from 1950 to the 1980s, when the green revolution introduced high-yield crops and energy-intensive cultivation to produce a 2.6-fold increase in grain output to match the most rapid population increase ever known. Now, in contrast, grain stocks are at their lowest level in thirty years. During that same period, some regions, such as Asia, fared much better than others, such as Africa, which in twenty years went from self-sufficiency in food supply to producing only 80 percent of the food it consumes. And the amount of available arable land per person in the developing countries is now in steep decline, increasing the dangers of malnutrition and famine.

Although the rate of population growth has begun to level off, the level of grain production has also been falling in recent years, and it may fall even more if we pursue a policy of increasing dependence on biofuels. So two serious challenges confront the world's agricultural industry. First, can the "old" green revolution—which depended on high pesticide, fertilizer, and water use, and high-energy farming patterns—be modified to be economically and environmentally sustainable? Second, can our heavy dependence on high-yield crop varieties of some fifteen plants (wheat, corn, rice, millet, sorghum, etc.) and eight animal species sustain the projected food needs of an additional 3 billion or more people by 2050, when the world will also face a situation of growing urbanization, falling water tables, and rising temperatures? The challenge confronting agriculture is to accommodate these changes, some of which are already built into the system. For example, Joel Cohen has reminded us that when the green revolution began in 1950, the total world population was 2.5 billion. It is now 7.0 billion, and it is estimated that about half of the world's people live in areas where water tables are already falling.

So how are we to interpret our situation? What are the prospects for agriculture? For all its recent successes, food productivity (expressed in production per hectare of land) has not kept pace with overall increases in population, and we have virtually run out of additional arable land and freshwater. Available per-capita cropland has fallen by more than half since 1960 and per-capita production of grains, our most basic food, has been falling worldwide for twenty years.[2] But present policy will take what is now agricultural land out of food production and use it for growing biofuels. And this at a time when the Food and Agriculture Organization (FAO) of the United Nations estimates that at least 800 million people, including, perhaps, one in six of those in the developing nations, are malnourished.

And our rising prosperity is increasing our problems. One important trend in food consumption is what Lester Brown has described as "moving up the food chain." The continuing increases, not only in population, but in wealth, have brought a growing appetite for animal protein: meat, seafood, milk, and eggs. The expression of this particular taste in meat reflects both geography and prosperity. Countries with abundant rangelands—the United States, Argentina, Australia, and Brazil, for example—depend heavily on beef or, in the case of Australia, mutton. More densely populated countries, such as China, rely chiefly on pork. Countries with ready access to the sea, such as Japan and Norway, have diets rich

2. Harrison Wein, "STEP Session Describes Health Challenges of a Growing Population," *NIH Record Archive*, July 1, 2005, http://nihrecord.od.nih.gov/newsletters/2005/07_01_2005/story01.htm.

in seafood. But all involve the greater use of animal protein, and as both prosperity and population increase, especially in Asia, the demand will grow.

With growth in incomes, growth in animal protein consumption continues to rise markedly. The average U.S. citizen consumes the equivalent of four times more grain than the average resident of India does, most of it used to produce meat and dairy products. With overgrazing already widespread, future increases in beef production will have to rely largely on the expansion of feedlots, which represent a costly use of grain to raise animal protein. For this and other reasons, beef production has tended to level off since 2000.

But just as the world's rangelands are limited, so also is one other major source of animal protein: the world's fisheries. A 2002 report by the FAO suggests that 70 percent of the world's fisheries are depleted by overfishing and that some face collapse.[3] We are already experiencing significant decline in seafood catch per person.

In the case of both beef and seafood we may have reached or exceeded the present limits of sustainable production. For other animal protein, the sharp rise of aquaculture ("fish farming") and huge increases in poultry and egg production will provide help, though none of these will remove the larger problems of rising demand and limited supply of animal protein.

No one has described our predicament more forcefully than Lester R. Brown.[4] During the second half of the twentieth century, Earth's population more than doubled, yet somehow, not only did food production increase, but the level of hunger was also reduced. Better still, the global economy increased sevenfold. So, in fifty years, 3.5 billion more people were added to Earth's population, all were fed, and both the proportion and the number of the hungry decreased.

Here, as Brown describes, are other statistics for the period 1950–2000:

- Water use tripled, but Earth's reservoir of water remained essentially unchanged.
- Carbon dioxide emissions quadrupled, but Earth's capacity to absorb it did not increase, leading to growing atmospheric concentrations of CO_2 and rising temperatures.
- World grain production almost tripled from 1950 to 1996. From 1950 to 1984, growth in production exceeded growth in population, but from 1985 to 2000 it fell behind the rate of population growth, requiring the drawdown of grain reserves in 2000.

---

3. Food and Agriculture Organization, *The State of World Fisheries and Aquaculture* (Rome: F.A.O., 2002).
4. Lester R. Brown, *Outgrowing the Earth: The Food Security Challenge in an Age of Falling Water Tables and Rising Temperatures* (New York: W. W. Norton, 2004), 48

- Soil erosion, desertification, loss of cropland, falling water table levels, rising temperatures, and extreme weather events contributed to this decline in agricultural productivity.
- The regional effects of these changes were strikingly unequal, with some of the world's poorest people—especially those in Africa—most harshly affected.
- Shrinking forest areas, collapsing fisheries, rising seas, and disappearing species also imposed "excessive demands on the earth."
- Population steadily increased. We continued to add an extra 75 million people a year to the world's population.
- The number suffering from hunger and malnutrition, though earlier reduced, has now increased, with hundreds of millions of the world's population—many of them children—suffering from malnutrition.

None of these trends show any sign of reversing and therefore improving our prospects of a major increase in food supplies; indeed many now look even bleaker than they did when first reported. For example, many of the pesticides on which past crop increases have depended are losing their effectiveness, as pests develop more resistance to them. The 17 percent of cropland that is irrigated produces some 30–40 percent of all crops, but future water supplies are threatened in many areas by falling water tables. And though the proportion of hungry people is coming down, the actual number continues to rise because of population growth. Hunger and malnutrition are estimated to kill 10 million people a year, exposing their sufferers to diseases; that's equivalent to the death of one malnourished person every five seconds.

But there is yet another major problem: climate change.

Climate Change and Agricultural Productivity

How will climate change affect Earth's productivity? What will be its impact on the production of crops worldwide? To those critical questions there is no simple quantitative answer, but on balance, climate change will create both significant benefits and significant problems. The benefits and the problems, however, will vary from region to region and crop to crop, and will depend on the degree of temperature changes.

Consider the following example. In mid to high latitudes (Europe, North America, Australia, Siberia, and northern China), modest temperature increases of a few degrees could improve crop production by providing longer growing seasons and enhancing the availability of newly arable areas. But that would not be true if temperatures continued to increase significantly.

In contrast to these favorable prospects in some regions, various estimates suggest that on a worldwide basis each increase in temperature of one degree Celsius could produce a 10 percent decline in grain yields. Increasing concentrations of carbon dioxide will also affect plant growth. Because carbon dioxide is essential to plant development, a small increase in temperature of, say, 2° or 3°C would increase production of wheat and rice. Increases of around 4°C or more, on the other hand, are likely not only to reduce yields but also to make many regions unable to sustain their present levels of agricultural productivity.

The biggest problems, however, are likely to be felt in the poorer, tropical parts of the world, especially Africa, parts of Asia, and the Middle East, where crop production is already close to the high end of the temperature range for effective agriculture and where poverty makes it difficult for populations to adapt to changing conditions. The plight of the vulnerable people of these regions is further threatened by their dependence on maize, which has a growth pattern that does not benefit from rising levels of atmospheric carbon dioxide.

These differences in regional effect and crop dependence could produce wrenching social and economic problems. Three-quarters of the world's poorest people rely on agriculture for their livelihood, eking out a living on less than $1 a day. Every day almost a quarter of a million people are added to the planet's population, while per-capita cropland and grain production continue to fall.

In the world's poorer areas, increases in food production are not keeping up with population growth, and there is no more arable land available. Even in those areas of the Earth where increases in arable land are possible, its development can come only at the expense of natural vegetation and forestland, further depleting the level of rainfall because of less rainwater being recirculated into the atmosphere, and increasing the concentration of CO_2 because of the effects of deforestation.

But how serious a challenge do we really face? After all, the global fish catch grew more than sixfold in the last fifty years, while world cereal production has doubled since 1970 and meat consumption has increased tenfold. Global population doubled in forty years from 1960 but food production increased and the proportion of malnourished people was reduced by half. Can we not repeat this success? If we cannot cultivate more land, can we now consume more fish or improve our farming methods?

Increased aquaculture will certainly help but it often uses fish protein or fish oil as food sources, as well as grain. Ocean fishing, in contrast, is already in a precarious position. Recent studies show that if world fishing continues at its present pace, we face the possibility of a "global collapse" in the major fish species.

Then there's agricultural land availability. We now use about a quarter of Earth's land surface for agricultural pastureland and another 11 percent for cropland. Urban settlement takes up 9 percent, and forests—which we clearly need—31 percent. The rest is too cold, too wet, or too infertile for significant plant growth. We've already taken over about a third of the world's forests and a quarter of its natural grasslands for agriculture.

But could we not increase crop yields by using more irrigation or more fertilizer or more pesticides? Unfortunately, each of these presents problems. Water is in short supply in many regions, fertilizer production requires substantial petrochemical feedstock, and petroleum is likely to be in shorter supply and growingly expensive. The World Resources Institute has calculated that half of all the commercial fertilizer ever produced had been applied in a period of only fifteen years, from 1984 to 1999. Pesticides have already had serious effects on species other than those they were intended to restrict, while many of the pesticides used are losing their effectiveness as the targeted pests acquire resistance to them.

Nor, as we have seen, is it only a question of improving crop yields. Growing prosperity has brought with it a worldwide increase in meat consumption, which has tripled since 1961. About 34 percent of all grain produced is now fed to livestock for meat production.

Perhaps the only positive prospect is the possibility of another green revolution. The first one, begun in the mid-twentieth century, allowed food production to keep pace with rapid growth in world population, by the adoption in developing areas of new crop varieties and intensive agriculture, such as the use of pesticides, synthetic fertilizers, and irrigation schemes. The ability of the world to support its growing population over the last half century or so has depended largely on the success of the green Revolution. This transforming movement arose from the efforts of the Mexican government to feed its growing population. With the support of the Rockefeller and Ford foundations, Norman Borlaug, who was subsequently awarded the Nobel Prize, introduced new varieties of wheat, and as a result Mexico, which in 1943 imported half the wheat it used, was able to export half a million tons twenty years later. The program was later successfully introduced into other countries, especially India and other parts of Asia, with improved rice varieties, which transformed India from a land of famine to a major exporter of rice. The new crop varieties (HYVs: high-yielding varieties), together with more intensive cultivation, led to huge increases in grain production. Between 1950 and 1984, world grain production increased by 250 percent. In the developing world, where rapidly increasing population might otherwise have led to famine, cereal production more than doubled between 1961 and 1985.

The green revolution has not been without its critics. Some complain it displaced productive traditional farming patterns, and that—for whatever reasons—Africa has not benefited from the programs. Others argue that it has corporatized and globalized agriculture. Perhaps only one thing is agreed on by most commentators: the revolution required huge energy and water inputs. If we are to support a second green revolution, we will need to develop major new energy sources and water conservation schemes.

If there is to be a second revolution, it is likely to be driven by the use of biotechnology to produce drought-, heat-, and pest-resistant crop varieties. But Europe and some other parts of the world have fiercely opposed genetically modified foods, partly because of fears that they might contaminate traditional crop varieties.

Some authors are optimistic, concluding that "the world can feed twice as many people in twenty years," basing such estimates on sophisticated economic modeling and assuming significant increases in grain production, not only in industrial countries, but also in both developing economies and former centrally planned economies. These projections conclude, however, that Africa presents a special challenge, arising chiefly from political instability, poor economic and agricultural policies, poor farming practices, and drought. Other authors are more optimistic that the world can feed itself if there is a radical change in social and economic policies. They argue for "recapturing" the food system, transforming personal tastes, eating locally and seasonally produced food, supporting local agriculture and business, creating "profound and comprehensive rural change," making "retribution for past injustices," providing living wages for all, ending subsidies to agribusiness, and "supporting a sustainable architecture for food." But such Soviet-style, command economies have repeatedly demonstrated their failure.

Worldwide inequity and poverty are unlikely to be alleviated by government schemes alone, however well intended, just as the global food problem is unlikely to be resolved by our present corporate, industrial-style, multinational agricultural and food-production systems. The most encouraging prospects involve a combination of creative national or regional energy policies and regional development programs and local initiatives. Examples of such proposals are those encouraged by the UN Millennium Project, where local communities use proven low-cost technologies to increase local food production. But trade regimes and subsidy systems (such as those of the OECD) threaten some of these initiatives.

For all our progress, the outlook is not encouraging. Global poverty fell by 20 percent in the 1990s, but the number of hungry people rose by 18 million. Although it is claimed the world produces enough food to feed everyone, economic, political, and social barriers restrict its availability. As we have already seen, malnutrition and its effects continue to kill 10 million people a year: 25,000 every day. And things will become much worse as population continues to increase.

The key to reversing this downward trend is to increase the availability of water and energy. Our planet is awash with both. Our challenge is to find economic ways to conserve and distribute water and harness energy to reverse this terrible trend. It should not be beyond our capacity.

There is perhaps one other major encouraging prospect. Though population is projected to increase to 9–10 billion by 2050, it will then stabilize and, many believe, decline, even perhaps to something at or below its present level. So, like energy, the key problem is how to survive the intervening period. If we are to do so, we will have to come to grips with the intertwined issues of population growth, falling water tables, soil loss, climate change, and energy availability, as well as the economic and social problems of food distribution. Left to their own devices, our governments are unlikely to confront these concerns. But rising food prices and the social pressures they have already produced may induce us to look more urgently at the need for comprehensive policies for sustainability.

CHAPTER 21

Energy as Sustenance

Energy and Society

It is always tempting to identify one particular issue as the most significant issue of our time. "What," we might ask, for example, "is the most critical need for a sustainable planet?" Well, clean water, adequate food, unpolluted air, fertile soils, healthy ecosystems; the list might start with those, and certainly each of them is of vital importance. But if the planet is to have a human population of anything like its present size—let alone a population of half as many again, all demanding mobility and rising standards of living—we'd also have to place energy high on our list.

Every living thing, every animal, every plant is an Earth dependent, an energy user. We are energy machines. That's the basis of life. So we humans are not alone in our exploitation of Earth as energy users. But we outrank all the rest. We ourselves, our bodies, *produce* the energy equivalent of only about one 100-watt lightbulb. But we *consume* far more energy than that. We not only feed off the Earth, but we also rebuild our natural environment, we extend our senses, we transport ourselves, we heat and cool our homes, we manufacture and power our appliances by foraging the Earth for useful materials. Even in the most remote corners of the planet, we drill, we quarry, we mine, we concentrate, we refine, we fashion, we fabricate hundreds of Earth materials to serve our own ends.

If humankind is to continue to prosper, we must provide for adequate, sustainable, safe, and secure energy supplies for the long term, and in doing that there are, as we will see, no easy options.

To sustain our growing global population of more than 7 billion people, and to improve the quality of life of the millions of needy, energy is critical, not only in providing clean water and adequate food, but also in preserving a habitable environment and producing all the materials, objects, resources, and other things we need to live in it. The rise of industrialized societies over the last two centuries has been based largely on the increasing use of energy from fossil fuels and the use of a substantial portion of that energy to extract, manufacture, raise, and grow useful materials—both living and nonliving—from the crust of the Earth. Everything on which we depend comes from the thin crust of planet Earth. And everything requires energy for its growth, extraction, refinement, fabrication, and use.

Perhaps, amid the competing economic, environmental, technical, geopolitical, and social issues involved in energy policies and future energy sources, there is one point that is worth renewed emphasis: the easy options will not work. Those who yearn to re-create in rural communes the simple, energy-limited life of the early nineteenth century forget that the world population was then less than a billion. Add another six and a half billion energy-hungry others—as we now have—and rustic simplicity is not an option.

Nor can we ignore the issue, denying the problem, hoping that, with the help of science and technology, something will "turn up." We are placing growing demands on a dwindling energy supply; we face a future when we will be running short of conventional oil; we are increasing greenhouse gases, polluting the environment, destroying air-saving forests and rich agricultural land in an explosion of new construction and development to house our still-expanding population. Meanwhile, global temperature increases, sea levels rise, the oceans become more acidic, soils are depleted, deserts encroach, and wildlife is endangered.

Earth's carrying capacity is limited. We, perhaps more than any other species, are always pushing the limits. We are very near, at, or perhaps even beyond that carrying capacity if (another big "if") we assume and hope that, given our present energy sources, all the world's peoples can expect to enjoy the living standards of the developed world. Earth's carrying capacity is not a fixed figure: it is a floating number whose value depends on our assumptions about the available resources and quality of life of all the people involved. And that quality will depend in the most direct way on the policy decisions we now must make concerning our future energy supplies. In reaching such decisions, the issues are not simply those of competing national or social interests, or of green versus less green or nongreen. They are those of prosperity versus misery, sustenance versus hunger, stability versus decline, hope versus despair. Energy policy involves not just technology, not just politics, not just economics. It involves survival.

Energy, as we all learned in school, is the capacity to do work. So it is, but that prosaic definition, for all its economy of language, scarcely hints at the centrality of the flow of energy in every aspect of our existence. Reading these words requires energy. Breathing requires energy. Sleeping requires energy. Living requires energy. Every aspect of every activity at every moment in every life requires energy. This energy involves the movement of particles—molecules, atoms, and their components—within and between the bodies of which they are a part. We call this energy, characterized by movement, *kinetic energy*. And the movement is not limited to living things: it's just as characteristic of nonliving things. Earth itself is in motion, spinning on its axis, moving at 65,000 miles per hour as it travels around the Sun, and hurtling through the void as it moves with the Sun and its fellow planets through our galaxy. And on Earth, everything else is in motion: atmosphere, oceans, rivers, land. All of these are in motion: some inconceivably rapid, some unimaginably slow; some conspicuous, some all but invisible. Watch the dancing specks of dust in a ray of sunlight as it falls within a darkened room and see a model of the world in motion. That's energy in action. Every particle of every thing is in motion.

Sometimes the energy of motion is restrained and the energy is *potential* rather than kinetic. Think of water held up behind a high dam. The stored energy of the water has the potential to generate electricity if it is released down an outlet tube to impact blades in turbines at the foot of the dam. This conversion of potential energy to electrical energy is typical of similar energy conversions everywhere in the universe. Energy is never lost. It can neither be created nor destroyed. But it can be degraded or converted from one form to another.

In fact, that conversion, that transformation, is the way we live. We survive by courtesy of the transformation of one form of energy into another. And this same energy transformation also powers the universe. The Sun shines because it is a nuclear power plant, in which deuterium (an isotope of hydrogen) is fused to create a new nucleus, throwing off an atomic particle to produce hydrogen and helium, and releasing energy in the form of heat and light. This conversion of mass into energy was the process that Albert Einstein predicted in his theory of relativity, when he postulated that atomic energy could be converted into heat and light energy. Light energy may itself be converted, as when sunlight falls on green plants and is absorbed by the leaves, where it produces chemical changes by converting water and carbon dioxide from the atmosphere into starch, cellulose, sugars, protein, fats, and other plant materials. Chemical transformation of plant-based biomass produced the world's coal deposits from living forests some 300 million years ago. Burning wood or coal involves the chemical change of oxidation, which releases the ancient energy of sunlight absorbed by the fossil coal-forming plants when they were once growing. The water and CO_2 extracted from the atmosphere long ago by these ancient plants are reformed and released when combustion takes place.

We live within this great web of energy flux, and our lives depend on its effectiveness. The food we eat provides the energy we need not only to work but even to live at all. But the pattern of our lives today—compared with, say, the pattern of our forebears two thousand or even two hundred years ago—reflects something more than a change in diet. It reflects the skill with which we now harness energy to serve our own purposes and needs. Two thousand years ago, nomadic people wandered across the deserts that now represent the biggest oil fields in the world. Today those same lands power the transport and industries of much of the rest of the world. Fifty years ago the north slope of what is now Alaska was good only for caribou hunting, and the deep waters of the Gulf of Mexico only for commercial fishing. Now they produce natural gas that fuels the whole North American continent.

Energy defines our period in history. Its use and availability, and in this past century its affordability, determine the well-being and prosperity of peoples and nations. And the catalyst that converts the capacity to do work into the performance of chosen work is human creativity and inventiveness. Material capacity without invention is powerless. Oil in the ground, untapped, remains oil in the ground. Creativity without resources is barren. An inspired design left on the drawing board remains a great design. But bring capacity and invention together; harness insight with material and inventiveness with purpose, and energy flows. And it flows in directions of human choice, "controlled" or operating by the laws of thermodynamics. That is the secret of the material transformation of society. It has allowed us to increase our human numbers. It has enabled us to transform our planet.

But the development that undergirds civilization depends not only on energy in the abstract but also on useful energy, applied to particular tasks. The Earth is awash with energy but civilization depends on its capture and use for our chosen purposes. Our earlier ancestors provided all their own energy. Then, later on, they harnessed wind, water, and animal energy. In some parts of the world these are still the dominant sources of power. The use of energy is one of the characteristics that shapes and differentiates societies. There is a striking difference in the materials that are used for energy between the developing and the developed countries. In some of the developing countries, it is chiefly biomass—wood and other plant products or animal dung—which is harvested and burned. In the developed world, by contrast, it is overwhelmingly fossil fuels—hydrocarbons and coal—that provide the chief source of energy, amounting now to about 85 percent. Neither of these energy sources is without its problems and its limitations.

So human history is, in part, a history of energy use. Its ready availability and efficiency of use have been critical to human development.

So far, so good. Let the good times roll. Unfortunately, there are a number of factors that now begin to threaten the comfortable pattern of growth we have enjoyed:

- Recent rates of increase in human population, though they are now slowing down, guarantee that our total population will reach some 9 or 10 billion by 2100 before it is projected to level off. That will require us to increase our energy production by at least half as much again as all the energy we now produce just to maintain humankind's present standard of living.
- The growing industrialization of the developing world, especially such large countries as China and India, is rapidly improving the living standards of its people, and in doing so is greatly increasing global demand for energy.
- Some major energy sources and some essential materials are finite and have limited reserves.
- All Earth materials that we use are ultimately recycled, either by us or by natural processes, but recycling times are often long, and the materials typically become so widely dispersed that recycling for many involves very high-cost and high-energy use.
- Energy is not recycled in usable form. Whatever the source of energy, it is ultimately converted to low-grade heat and lost to us in the ambient air and water.
- Present use of the more abundant energy sources—coal, for example—as well as the effects of some essential industrial processes, tend to add to the growing concentration of carbon dioxide in the atmosphere, with its negative effects of global climate change.
- The increasing dependence of much of the developed world on petroleum supplies from politically sensitive areas raises uncertainties about long-term availability of energy supplies.

How, in the light of these constraints, can we support the growing energy demands of our burgeoning population? Petroleum, natural gas, and coal won't last forever. Nuclear power, though growing slowly in public acceptance, is still anathema to many, perhaps to most. Renewable energy, for all its qualities, is at present inadequate to meet all our present needs, still less our future needs. So what's next? What is beyond petroleum? Can we continue to satisfy our ravenous appetite for power? Can we continue to thrive?

The answers to these questions depend, in part, on the answers to three others. First, will the growth in human population level off by the end of this century so that Earth's population stabilizes? Second, will the developing world's use of energy continue to increase so as to approach that of the developed world? Third, are greenhouse gas emissions now seen to be a sufficiently serious problem for world leaders to begin to restrict them?

The balance of evidence suggests that the answers to all three of these questions are affirmative. World population is likely to level off, but it will probably be at a level of 8.4 billion: half as large again as our present global population. The countries of the developing world will continue to increase their energy use. China is now the world's leading energy consumer. Emission caps are already being put in place and are likely to become the norm, though perhaps over a decade or more. And it does not seem beyond our capacity to invent sustainable alternative energy sources, adequate to satisfy the needs of this larger population. Of course, invention is one thing: availability is another. And that will require an open global economy and societies willing both to nurture creativity and entrepreneurship and to embrace technology. No energy policy seems likely to be reached without controversy. The easy options will not work. And we need a long-term policy.

All this implies both a continuing increase in demand for energy through the rest of this century and a much greater emphasis on "clean" energy sources. While no one can predict the quantitative impact of those continuing trends, they will clearly influence both the availability and the choice of energy sources.

With that as background, what are the resources available? How long will they last? What are the relative costs and benefits of each? Can we design a mixture of energy sources that represents a responsible global strategy?

Behind each of the questions we have posed there lies a distinction between two different kinds of energy sources. The most common fuels—oil, gas, and coal, for example—are nonrenewable. They represent the use of solar energy stored over millions of years within the Earth's crust by geologic processes. Once we've used them up, they're gone. While they are available in a range of grades, they are nonetheless limited resources. Using them is like spending down an inheritance. You can use it only once. Oil, gas, and coal are still being formed, but the process of forming usable deposits requires millions of years. The greater part of all the energy we now use—some 93 percent—is nonrenewable, depending on how one "counts" uranium.

So let's start with the nonrenewable energy sources. We'll discuss what they are, how they occur, how they are discovered, extracted, refined, transported, and used. And we'll have a look at their reserves, costs, and future contribution to our needs.

Then we'll consider renewable energy sources. And finally, we'll review where we stand in terms of future needs, future supplies, and the policies we need to sustain our society.

Wood

It began with fire. Perhaps it was a lightning strike or a forest fire that first led to its discovery. Perhaps it was accidental sparks from toolmaking, using flint chipping on flint, that kindled a flame. Once fire was discovered, once its extraordinary usefulness was realized, it had to be treasured. It had to be kept alight. It had to be fed. The oldest evidence of the use of fire is found on the banks of the River Jordan, in charcoal hearths associated with remains of *Homo erectus* that have been dated at about 790,000 years old. Older hearthlike sites and charred bones in South Africa, Kenya, and Tanzania dating back some 1.6 million years are rather less conclusive. Whatever its origin, the realization that fire provided benefits, as well as hazards, brought with it the need for fuel. The first fuel was probably plant material—dried leaves, twigs, branches—and that fuel, known as biomass, is still used today in many parts of the world.

The use of wood as fuel—supplemented by dried animal dung when societies turned from nomadic hunting to settled agriculture—has served humanity throughout most of its history. Wood for fuel meant not only heat for warmth and protection against wild animals, but also heat for cooking, and later for smelting ore and for fashioning metals to form the countless utensils developed to serve human needs.

Wood, though convenient to gather and easy to use, is not a very efficient fuel. It gives out comparatively little heat: not enough, for example, to be very effective in melting metals. In contrast, charcoal gives out about twice as much heat as wood when burned. Charcoal is the residue of carbon left behind when woody or animal material is burned so that the volatile, tarry components are driven off. In earlier times this was done by stacking wood into heaps, with a central vent, covering the stacks with turf, and then burning them. The charcoal produced was used chiefly to produce iron from iron ore, and much later as a constituent for gunpowder.

But wood was also needed for many things other than fuel: for nearly everything, in fact, through most of our history. It was from wood that tools, housing, utensils, furniture, carts, ships, and countless other implements were originally made. And though, in theory, timber is a renewable resource, in practice the extensive use of wood led to the wholesale cutting down of forests, the clearing of land, and the long-term loss of fuel supplies. Wood is renewable only as long as replanting keeps pace with harvesting. On a managed estate or national forest, it can, but in the unregulated slash-and-burn or cut-and-use situations of earlier times, it did not. So forests were cleared all over the world, and that pattern of deforestation continues today in some regions. The forests of Madagascar, with their unique wildlife, are still being clear-cut at a rate that represents a potential ecological disaster. Haiti, with a burgeoning population, has seen its forests wiped out in the endless search for fuel. In contrast, the Dominican Republic, in the same climatic region, has fared much better with enforced protection of its forests.

But wood was the fuel of choice—the only choice—until well into the eighteenth century. Today the continued use of wood as a fuel in some parts of the world contributes to the devastating effects of deforestation, and especially to its impact on the growth of greenhouse gas emissions. We must support programs to provide alternative fuels and increase reforestation.

Coal

Useful as wood had been for centuries as a fuel, the coming of the Industrial Revolution in mid-eighteenth-century England led to its general replacement as a primary fuel by coal. The invention and widespread adoption of the steam engine as a means of providing power led to the demand for a major increase in steel production, and with it a demand for an increased supply of an efficient heating fuel and carbon source. Coal provided both. In England, major coal deposits occur within easy distance of the sites that provided iron ore, as well as limestone, so that these resources provided the foundation for the spread of the Industrial Revolution.

Coal is a brown or black sedimentary rock formed from the remains of ancient plants by their burial, decay, and "metamorphism." Coal varies greatly in character, appearance, and hardness, reflecting the history of its development and the extent to which it has been heated and hardened by the conditions of its burial. Modern peat bogs are often regarded as coal in the making, and *peat* itself is soft, brown, crumbly fuel, full of plant fragments. It burns with much smoke and little heat. *Brown coal* or *lignite*, having undergone deeper burial and higher temperature, is harder, and has a smooth surface, fewer impurities, and better burning qualities than peat. Most coal is known as *bituminous coal*, which, though it comes in several grades, provides better heat than lignite, burns easily with a steady yellow flame, and still produces appreciable smoke. *Anthracite* is the hardest coal of all. It is brittle, breaks smoothly, and burns with a clear blue flame. These varieties of coal reflect variation in composition, with the "lower"-grade coals having more volatile hydrocarbons and producing more smoke, and the "higher" anthracitic and bituminous coals containing more-fixed carbon and less water and oxygen. The depth of burial of these ancient plant remains increases the temperature—and this, plus time, determines the development of the grade of coal.

Coal is widely distributed throughout the world, and the nature of its occurrence indicates that it was formed from the remains of ancient forests growing in extensive low-lying swamps and preserved by later burial in stagnant water, whose low oxygen content prevented their rapid decay. The Great Dismal Swamp of southeastern Virginia and northeastern North Carolina is often seen as a present-day example of a coal-forming swamp. The world's greatest coal deposits were formed during the geological period known as the Pennsylvanian or Later Carboniferous, which lasted from about 280 to 310 million years ago. This was a time when great trees, many more than one hundred feet high, formed extensive forests across the face of the Earth. The vast coalfields of North America and Europe represent this ancient period, as do those in parts of Asia and South Africa. In many areas of the Southern Hemisphere there are coals of slightly younger, Permian, age. Other extensive coal deposits are found in still younger rocks of Cretaceous and Tertiary age. The United States, Russia, China, Australia, and Germany together contain some 69 percent of the world's coal reserves.

Coal is extracted either by surface quarrying—known as opencut or open-pit mining—or by underground mining. Neither is without problems. Surface mining, though less costly and less dangerous than underground working, often scars the landscape and frequently disrupts and pollutes the local drainage. More recent techniques involve the stripping and preservation of topsoil, rolling it up like a carpet, and then returning it after the extraction of coal, so leaving the landscape little disturbed. Underground mining is hazardous and costly and can also occasionally lead to longer-term problems of subsidence and pollution. The landscapes of Pennsylvania, West Virginia, and Ohio, as well as those of Wales and the Rhine, show, in places, the longer-term impact of these mining operations.

Coal has been used since ancient times. There are records of its use in China some three thousand years ago. By AD 1300 it was widely used by blacksmiths and others in England. The development of the steam engine in England in the late eighteenth century, and its widespread adoption and use—both in powering the locomotives of newly constructed railroads and in providing power for the newly created factories of Europe and North America—prompted a huge expansion of coal mining activity. In the United States coal production doubled each decade between 1880 and 1910.

For much of the nineteenth and early twentieth centuries, mining was thus a major component of the production and trade of many of the industrialized nations. In Britain, for example, on the eve of World War I, there were over 1.1 million people employed in some three thousand coal mines, producing over 21 percent of the world's coal. Earlier, in the nineteenth century, it was coal that opened up the American West, fueling the locomotives and smelting the iron needed for that great westward migration.

Coal proved to be a versatile fuel. It was used to produce coal gas, widely used to light streets and buildings and to cook, into the early twentieth century. Coal was, and still is, used as the basis for the production of a wide range of chemicals and plastics.

But from the 1920s, the relative use of coal in the industrialized countries began to wane, partly because it was replaced by other fuels—especially oil and natural gas—and partly because of concerns over the atmospheric pollution it generated. London's traditional "pea soup" fogs, which claimed thousands of lives from bronchial illnesses, were the product of coal burning in open fires for domestic heating, and the fogs ceased abruptly after a ban was imposed on such use.

From employing over 1.1 million coal miners in 1913, Britain now employs only a few hundred.

Coal is still, however, produced in significant volume, largely as a fuel for power plants and for use in the production of iron, steel, and cement. China and India both use large quantities for this purpose, as do the United States and some European countries. Pollution remains a problem, though this is much reduced by avoiding the burning of sulfur-rich coals and by "scrubbing" the emissions of coal-fired plants. At present, coal still plays an important role in overall energy production. It is cheap, widely distributed, easily transported, and reserves are substantial.

Oddly enough, the long decline in the use of coal may be reversed by the future increases in the cost of producing oil and gas and the more limited life of their reserves. Technology is now available to gasify coal underground where it occurs. This involves no mining, limited pollution, and no handling of solid fuel. Under a system devised over a century ago, underground coal seams are ignited and the gas piped back to the surface. Germany used coal in this way during World War II. South Africa successfully used this technique to provide fuel in the 1980s when an oil embargo was imposed on it. The vast low-grade coal deposits of the U.S. West would be ideally suited for this generation, but the economic costs of coal gasification or liquefaction are still relatively unattractive. The much-heralded $85 billion synfuels project of the Carter administration in 1980 was aimed to produce 20 percent of the nation's primary energy, but not a single plant was built and the program was abandoned in 1986. Similarly, the Athabasca oil sands in Canada, in spite of their vast extent and huge potential as a source of energy, at present produce only about 1.2 million barrels a day. Output continues to grow, driven by increasing demand and the high cost of conventional crude oil.

In 1900 coal provided 95 percent of the world's total primary energy supply, but by 2000 it had fallen to 23 percent. Although earlier projections have predicted its rapid demise, it remains important, providing the same proportion of the world's total primary energy as natural gas. The United States, for example, generates 56 percent of its power from coal, and China derives about 70 percent, as does India. It is widely projected that the use of coal will increase by 60 percent in the next twenty years.

In fact, coal production increased by some 18 percent in the last twenty-five years of the twentieth century. And this, in spite of its highly polluting combustion and the dangers and high costs of producing and transporting it, relative to its energy content. The reasons for this are its broadly stable and reasonable cost, its vast and widely distributed reserves, and its versatility—for example, its use as a base in petrochemical manufacturing.

Clean Coal

The biggest problem with this growing use of coal is the emissions produced by its combustion: CO_2, nitrogen and sulfur oxides, and ash particles. In total, it produces about twice as much CO_2 as natural gas does for each unit of power generated. It is responsible for almost a quarter of global CO_2 emissions. Several types of filters and scrubbers are already used to reduce ash, SO_x, and NO_x emissions, while "supercritical" high-temperature power plant turbines provide improved efficiency. Limited use of renewable biomass as supplementary fuel in cofired plants reduces carbon emissions, but major retrofits will be limited to about 10 to 20 percent. Several schemes exist for the capture of carbon dioxide emissions, and, while all are effective, all are also costly. There will need to be either efficiency improvement or public caps, taxes, and subsidies to make them economically attractive.

Public opposition in the United States to the construction of new coal-fired generating plants is strong. Scores have been canceled or put on hold since the turn of the century, though there have been capacity expansions and life extensions at existing U.S. plants. But the newest U.S. coal plants are both more efficient and less polluting than older models and they will, in many cases, replace older, less efficient, and more polluting models. The proposed new eight-hundred-megawatt Cliffside plant about fifty miles west of Charlotte, North Carolina, for example, which will lead to the shutdown of four, older, less efficient plants, will reduce SO_2 emissions by 80 percent a year, and nitrogen oxide by 50 percent, and will increase electricity-generating power.[1] Yet it still faces widespread public opposition.

The major concern is not with plants of modern design, such as this, but rather with construction in other parts of the world of many new plants that lack adequate environmental safeguards. China, for example, intends to build eight hundred thousand megawatts of coal-fired plants over the next eight years: more than two and a half times the size of U.S. installed coal-fired capacity. One estimate suggests that China is opening one or two new coal-fired plants every week, many of which may lack environmental safeguards.

Reserves of coal are sufficient for the next couple of centuries, but it will have a long-term future only to the extent that cost-effective ways can be developed to deal with the pollutants produced by its combustion. Apart from tapping these at the source—which will continue to be very important—the most effective method is likely to depend on the widespread use of carbon capture and sequestration (CCS): the capture and storage of carbon dioxide. This involves the separation

---

1. Cliffside Steam Station, Duke Energy, http://www.duke-energy.com/ power-plants/coal-fired/cliffside.asp.

and removal of CO_2 and its storage in a safe location. Pilot or commercial plants now exist that provide storage in "safe" underground reservoirs, of the type associated with oil and gas accumulation. Properly selected and managed, such sites could provide both physical containment and geochemical "trapping" for hundreds and probably thousands of years. The present costs of CCS are high but are likely to be reduced by further technical developments in separation, transportation, and storage. This is discussed more fully in chapter 18.

Oil and Natural Gas

1859 was a revolutionary year. On November 21st Charles Darwin published *On the Origin of Species.* The first printing of 1,250 copies sold out on the day of publication. The publication of his book was a world-changing event—scientifically, ethically, and socially—and the debate Darwin began still continues to our own day.

Three months earlier, on August 27, 1859, Edwin L. Drake had struck oil at 96½ feet, in a borehole drilled along the banks of Oil Creek, near Titusville in northwest Pennsylvania. That, too, was an event of revolutionary proportions. If Darwin's discovery shaped our view of the development of life, Drake's discovery shaped the world in which we live today.

Oil from this site had been known of for centuries by the Seneca Indians, who collected and traded it for medicinal use. Oil, in fact, was frequently encountered in seeps, boreholes, and brine wells drilled in Pennsylvania, New York, and adjacent states. In the early part of the mid-nineteenth century this oil—"rock oil" as it was then called—began to be used for domestic lighting and lubrication, largely because whale oil, which had been the chief source until that time, became more scarce and more expensive. The problem with rock oil, however, was that it burned with an excessively sooty flame and had a heavy odor. This made it unpopular, until a method of refining it was invented. Once this was developed, demand for "refined carbon oil" quickly grew, and in 1854 the Pennsylvania Rock Oil Company of New York was organized, later becoming the Seneca Oil Company. The young company purchased land around the family sawmill of one of its owners, where oil seeps were conspicuous.

In 1858 Edwin Drake was hired as the company's general agent and sent to Titusville to exploit the oil. Drake was born in Greenville, New York, in 1819 and grew up on the family farm near Castleton Corners, Vermont. After leaving the farm at age seventeen, Drake headed west and filled a succession of different jobs, including hotel clerk, salesman, and railroad agent, until becoming a conductor on the New York and New Haven Railroad. Ill health forced him to retire from this position at a young age and he was hired by the Seneca Oil Company to exploit the oil for which it had been created. Drake studied the methods by which brine-well drillers operated and he hired an experienced brine-well driller to undertake the search for oil. Together, they began to pound their way down, using not the rotary drill of today but a cable carrying a weighted chisel-like tool, which was alternately raised and dropped in the hole. Drake encountered a nagging problem of the hole filling up with groundwater, a problem that had earlier frustrated him in attempting to dig a well in the Oil Creek area. He solved this problem, which caused the hole to collapse, by "casing" the well, driving down a cast-iron pipe to contain the drill string. Drake's successful well, which produced up to ten barrels a day, led to the drilling of scores of wells in the area, and ultimately to today's vast worldwide petroleum production and refining industry, the backbone of the developed world.

All of us are children of the age of oil, living off its legacy to us. We are all the beneficiaries of Edwin Drake.

Drake was not the first to find oil: he may not even have been the first to produce it by drilling. The Chinese are said to have drilled for it over 2,300 years ago. Drilling may also have been used before Drake's time in Baku, Azerbaijan, though it is difficult to confirm this. But there is clear evidence that natural gas and oil seeps have been known since ancient times. Noah is said to have used asphalt to caulk the seams of the ark. Moses' mother used it to waterproof his cradle. The ancient Sumerians, Assyrians, and Babylonians used crude oil and asphalt from Hit on the Euphrates River as far back as five thousand years ago. The Greeks used oil to set an enemy fleet on fire, pouring it on the sea and setting it alight. The Persians used incendiary, oil-soaked arrows in the siege of Athens in 480 BC. American Indians used oil for medicinal purposes, as well as to waterproof their canoes.

These and other more recent uses involved the use of oil collected on or near the surface of the ground. Since 1859, when "Colonel" Drake drilled a well that ushered in the beginning of the age of petroleum, we have surveyed every corner of the land and the oceans and pumped out of the ground all the oil and gas we could get our hands on. Transportation, manufacturing, heating, lighting, and whole new industries (petrochemical) and materials (from drugs to dyes, from plastics to paint, from synthetic fibers to fertilizers) have transformed our societies and revolutionized our lives. Edwin Drake's quiet revolution continues. But oil is a finite resource, and the revolution cannot last forever.

Petroleum (oil) is a liquid mixture of about 15 percent hydrogen by mass and 85 percent carbon—a hydrocarbon—that occurs naturally in some rock formations. Its name "petroleum"—from the Latin words *petra*, "rock," and *oleum*, "oil"—reflects its origin. Oil is usually found in association with at least some natural gas, which consists predominantly of methane (CH_4), with 1 percent or less of other hydrocarbon gases of light molecular weight, such as ethane (C_2H_6)

and propane (C_3H_8). In contrast to this, petroleum consists of dozens of different hydrocarbons and comes in a variety of different liquid mixtures. Petroleum from a single source typically includes over a hundred different compounds.

Oil and natural gas, like coal, are formed from the remains of ancient organisms buried and preserved within the Earth's crust. Most organic material decomposes quickly at death by oxidation and anaerobic (oxygen-free) bacterial decay—whether buried or not—and the products are lost. Marsh gas (methane) bubbling up from stagnant swamp water is an example of decaying vegetation, where the decay products are lost to the atmosphere. But under certain rare conditions of burial, especially in oxygen-free environments, the products of decomposition of animals and plants can be preserved. This can happen with burial under both marine and terrestrial conditions, though most petroleum comes from marine rocks. Animals and plants typically produce different decay products. When marine organisms decay they yield mostly lipids (fatty acids), hydrocarbons, and proteins, which, on burial, yield oil and "wet" gas. In contrast, most plant material contains cellulose and lignin, which typically decompose to form coal and dry gas.

But it is not only the composition of the original material that determines the end products of this decay. The depth of burial also influences the products of decomposition and breakdown. Burial of organic matter in accumulating sediments and decay at shallow depths produces a compound called kerogen, which is rich in carbon and hydrogen. As depth, temperature, and pressure increase, this kerogen is, over time, converted to liquid petroleum, accompanied by the creation of gas, which increases in amount with increasing depth and temperature. Most oil is formed at depths of 7,500–15,000 feet, within the rocks bearing the organic debris. These are known as the *source rocks*. Source rocks typically have relatively low concentrations of petroleum, but the petroleum becomes concentrated by upward movement. Some 98 percent of all petroleum is formed in sedimentary source rocks and most of these are of marine origin. The organisms that have produced most of the world's petroleum are microscopic plankton, plants and animals that exist in countless numbers in the surface waters of the oceans, and whose remains rain down in a steady "snowfall" on the ocean floor. Other lacustrine (lake) deposits contain oil shales in some areas, as in Colorado and Utah. Elsewhere, as we have seen, burial of plant material under terrestrial conditions produces coal and natural gas rather than oil.

Compaction, with the consequent heat and pressure it produces, and its own buoyancy, has almost everywhere caused the petroleum to migrate from the source rock in which it was formed, and has thus allowed its concentration. Much is also lost over geological time by seepage, but commercial quantities of oil and gas have typically accumulated by being trapped in a *reservoir* of permeable rock, where they fill the pore spaces between the grains, their escape being prevented by various kinds of traps, typically consisting of a *caprock* or some other impermeable structure. About half of all productive reservoir rocks are sandstone and about half limestone or dolomite ($CaMgCO_3$).

Reservoir rocks have to be both porous (having significant pore space) and permeable (allowing petroleum to flow). Petroleum migrates upward because it is lighter than water, which accompanies it. Gas, in turn, is lighter than liquid petroleum and rests above it in the reservoir rock.

Oil and gas are formed in sedimentary rocks of all ages, but they become increasingly abundant in rocks of younger age, reflecting, perhaps, both the leakage or destruction of older accumulations and also the increasing abundance of the marine plankton and other organic material that have been major contributors to their formation.

Oil and gas are widely distributed geographically. Over 150 years of exploration have led to the discovery of oil and gas on all the continents except Antarctica—where international treaties prohibit drilling—as well as on the continental shelves of the oceans. The largest reservoirs, accounting for more than half of known reserves, are in the Middle East, but there are oil-producing sedimentary basins and major oil and gas fields in countries of the former Soviet Union, Venezuela, Mexico, Canada, the United States, North Africa, Nigeria, the North Sea, China, and Indonesia. Smaller fields exist in many other areas. Five countries (Saudi Arabia, Iraq, the United Arab Emirates, Kuwait, and Iran) account for almost two-thirds of the world's oil reserves.

Because oil and gas lie deeply buried, the search for petroleum involves not only careful geologic mapping over large surface areas but also intensive studies of subsurface structures and rock types. Geophysical methods of studying these conditions are now routinely used, involving especially seismic studies in which artificial shock waves are used to trace subsurface geologic structure. These methods are used both on land and offshore, and the search for future oil supplies increasingly involves offshore exploration.

Once discovered by this long and arduous pattern of exploration, the study begins as to whether a particular discovery will provide oil in commercial quantities. Exploratory boreholes are drilled over the indicated extent of the potential field; trial pumpings, flow patterns, and variability studies are recorded and analyzed; and a decision reached as to whether or not to exploit the structure. This involves a major financial investment.

Production

Early wells were drilled using cable-tool technique, in which a chisel-like bit is repeatedly hoisted and dropped from an overlying rig, shattering the underlying rocks. The chips and cuttings had then to be removed and such drilling could

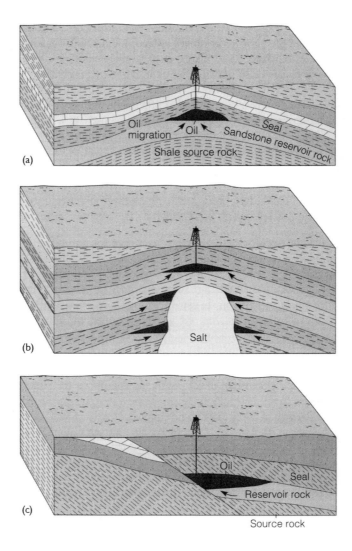

generally penetrate no more than a few feet a day. Though still used in some areas, this slow, limited method was replaced in the late 1800s by rotary drilling. A modern derrick may be 175 feet high and thus allows sixty- to ninety-foot sections of drill pipe to be assembled and handled. These drill pipes end in a drill bit—a heavy assembly of interlocking, coglike, diamond-tipped blades—which rotates within the drill hole. Water mixed with clay (drilling mud) is circulated inside the drill pipe and back up through the hole, cooling and lubricating the bit and carrying rock cuttings up to the surface.

These cuttings, together with electric logging of the hole, record characteristics of the rocks through which the drill passes. Essentially similar rotary drilling rigs

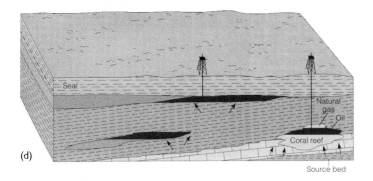

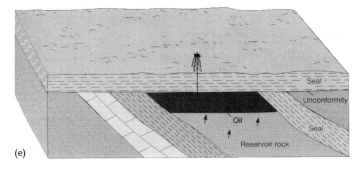

21.1 Oil and gas accumulation in various kinds of geologic traps: (a) anticline, (b) salt dome, (c) fault, (d) pinch out and reef, and (e) unconformity. From Brendan Murphy and Damian Nance, *Earth Science Today* (Pacific Grove, CA: Brooks/Cole, 1999). Used with permission.

are mounted on giant drilling platforms for offshore use. The most recent model can operate in water up to ten thousand feet in depth. Elaborate precautions are taken to avoid the danger of a sudden surge and blowout of oil or gas if they are encountered. Until recently, the most famous blowout and gusher in history was at Spindletop, near Beaumont, Texas, in 1901, which blew seven hundred feet of drill pipe over one hundred feet into the air and then gushed a hundred thousand barrels of oil a day to a height of 175 feet for nine days before it could be brought under control. The 2010 explosion the BP Deepwater Horizon drilling rig off the coast of Louisiana is a more recent and far more destructive example.

The production and management of producing wells is a specialized task, as is the transportation and refining of oil and gas. Gas is typically piped from wells directly to its point of usage. For international shipment over distances longer than about one thousand miles, it is generally liquefied to form LNG (liquefied natural gas) and shipped in tankers. Crude oil from the well is typically piped to the refinery.

Even in mature oil fields, as much as 60 percent of the oil may be left in the ground after years of pumping, so that extensive methods of secondary and tertiary recovery are used to prolong production.

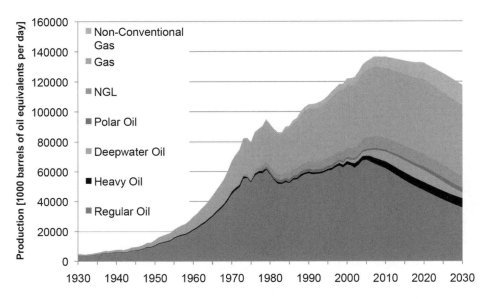

21.2 ASPO: Oil and gas production profiles, 2008 base case. Redrawn with permission from the Association for the Study of Peak Oil and Gas.

Natural Gas

Gas provided the most common lighting and cooking fuel for Europe and the United States for most of the nineteenth and early twentieth centuries. The gas, in this case, was coal gas, made by distilling coal. Gasworks were a common feature of the landscape.

But coal gas (largely carbon monoxide, CO) is toxic and has been replaced by electricity for lighting and cooking and by natural gas (mostly methane, CH_4) for more general use, especially heating and power generation. Natural gas is also widely used in manufacturing a range of products, from plastics to fertilizers. Its present widespread use in the United States developed in the years following World War II, as major pipeline networks were built across the country. Giant tankers were developed in the 1970s to transport liquefied natural gas (LNG), cooled to below –263.2°F, to allow intercontinental supply. In the last thirty years, production from the North Sea fields has provided a plentiful gas supply for Europe. Natural gas now provides about a quarter of the total primary energy consumption of both Europe and the United States.

Most gas occurs in association with oil, where it frequently lies at the top of oil reservoirs. It is also present in oil, giving it a frothy texture, and occurs below the "oil window" at fifteen thousand feet and deeper, where increasing temperature has cracked liquid oil, with its large, multicarbon molecules, into a one carbon molecule, methane. Until about fifty years ago, the absence of pipelines made natural gas a waste product of oil fields, and it was flared by controlled burning.

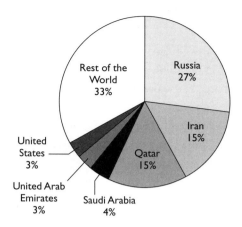

21.3 World natural gas reserves. Courtesy of CIA World Factbook.

Unattractive as that may appear, releasing it unflared would have been worse, as methane is an even more potent greenhouse gas than the carbon dioxide emitted by flaring. Today some gas is still flared—especially in Africa and the former Soviet Union—but most of it is carefully harvested and used. In the United States, some 31 percent is used for domestic and commercial heating, some 35 percent for industry, and some 24 percent for power generation. Some is also pumped back into oil reservoirs to assist in oil recovery.

What are the prospects for continuing supplies of natural gas? It is not clear that peak production has yet been reached or is imminent, but—as with petroleum—it seems probable that most of the major conventional (as opposed to tar sands or oil shale) gas fields have already been discovered and that future supplies will come from newly developed, smaller, and costlier fields. The outlook for continuing supplies is uncertain, but less critical than that for petroleum. Published figures suggest world conventional reserves that would last some sixty to seventy years at present rates of production (fig. 21.2).

The expansion of shale gas production, on the other hand, could greatly extend that estimate. Shale gas comes from dark shales, which are rich in organic material, and is classified as an "unconventional" source of natural gas. These black shales have low permeability, and this makes it difficult to extract their gas content. This extraction can be carried out, however, either by horizontal drilling or by fracturing the "tight" shale. The pressurized injection of water ("hydrofracking") breaks up the shale and allows the gas to be pumped out. This extraction process has created lively controversy, with claims that local groundwater may be polluted, wells contaminated, and air quality endangered by the chemical additives used in hydrofracking. There is also some debate as to whether or not the increased use of shale gas will reduce or increase greenhouse gas emissions.

Increased use of shale gas in the United States, Canada, Europe, Asia, and Australia could greatly extend the available lifetime of natural gas.

Prospects for Oil

Sharp increases in gasoline prices reflect a variety of causes including changes in both supply—caused by artificial restriction or manipulation of past supplies by OPEC, or the more recent lack of new refinery capacity, political unrest in the Middle East, or damage to producing and refining operations, such as that produced by Hurricane Katrina—and growing demand—created by the growing needs of developing nations, especially China and India. China, for example, has risen in the last decade to become the second-largest importer of petroleum in the world. It's tempting to assume that these shortages are short-term inconveniences, but all the indications suggest that the great age of cheap conventional oil, which has powered the expansion of the modern world and sustained the huge improvement in living standards of most of the world's growing population, is coming to an end. This won't happen overnight, of course; conventional oil production will peak, probably in the next decade or two, then level out for a while, and then slowly decline. That, combined with growing worldwide energy demand, will leave a huge energy gap, which will involve a fundamental and probably wrenching redesign and reallocation of the world's energy supplies.

In addition to the general question of petroleum availability, there is a policy question that must be addressed. Petroleum is used not only as a fuel but also as the source for several thousand petrochemical products. A major policy challenge raised by the limited life of petroleum resources is the relative balance between the use of remaining reserves for fuel, on the one hand, and as a source of petrochemical products on the other.

Petroleum Reserves

In spite of our dependence on petroleum, in spite of its ubiquity, in spite of a gas station at every major intersection, the days of cheap petroleum are numbered. Some well-informed petroleum geologists believe we have already reached peak production; others judge it to be a decade or two away. It is petroleum that has allowed us to enjoy the American way of life, providing, as it has, convenient, cheap transportation, a multitude of inexpensive products, and low-cost, unlimited domestic and commercial energy supplies.

How can we predict this peaking in oil supplies?

Consider, first, our past experience and changing demand. Drake and his backers were looking for a source of kerosene for lighting oil. For most of the nineteenth century, gasoline was merely a by-product of kerosene production, with little commercial value. By the turn of the century, Thomas Edison's lightbulb and the growing electricity-generating industry he created led to a decline in the market for lamp oil. But the invention at about the same time of the automobile and the subsequent development of the internal combustion engine led to an explosion in demand for gasoline and diesel fuels.

This demand continues unabated. At the midpoint of the twentieth century, for example, there were 50 million cars worldwide. By the end of the century that number had increased tenfold: there were 500 million, even though the global population had "only" tripled. As the developing countries continue to become more prosperous, automobile production and pollution will increase even more rapidly. In recent years world crude oil demand has been growing at more than 2 percent a year.

This growing demand has produced a series of reports and projections on future oil production. Though they varied in detail in their conclusions, virtually all suggest that peak conventional petroleum production has either already been reached or is likely to be reached within a limited (five- to twenty-five-year) period, with severe economic consequences. These studies are based on the assumption that virtually all Earth's major sedimentary basins are already known and more or less fully explored and that all or most of the largest conventional oil fields are already in production, with some of these already in decline.

It was M. King Hubbert, a longtime geologist on the staff of Shell and later the U.S. Geological Survey, who, in 1956, predicted that the peak in U.S. oil production would occur in 1970.[2] He proved to be right. In a later paper he predicted that world conventional oil production would peak in 2000.[3] A few authors agree, though most predict a longer life. Hubbert predicted that oil production would represent a bell-shaped curve (fig. 21.4), so that after peak production, there would be a prolonged period of declining—and thus more expensive—production.

The most authoritative projection of future supply was published in 2000 by the U.S. Geological Survey. It shows global oil production peaking in 2026, 2037, or 2047, depending on alternative assumptions about the volume of resources, ranging from 95 percent confidence (nineteen chances out of twenty that these reserves will be available) for 2026, the statistical mean (2037), to 5 percent confidence (only one chance in twenty of this much or more petroleum being available) for 2047. The projected rate of decline assumes a constant ratio of reserves to production of ten, an arbitrary figure, based in part on the production experience of the United States as a global model.

The Energy Information Administration has prepared a range of more detailed scenarios in which energy demand varies, while postpeak production remains constant at resources/production = 10.

---

2. M. King Hubbert, *Nuclear Energy and the Fossil Fuels: American Petroleum Institute Drilling and Production Practice Proceedings* (Houston: Shell Development Company, 1956), 5–75.

3. M. King Hubbert, "Energy Resources," in *Resources and Man*, ed. National Research Council, Committee on Resources and Man (San Francisco: W. H. Freeman, 1969), 196.

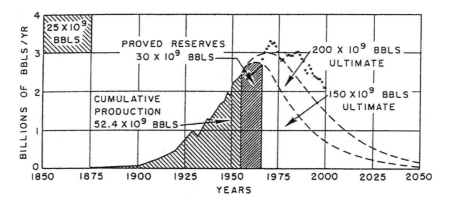

21.4 Hubbert's curve describing world oil production. On Hubbert's original 1956 graph, the lower dashed curve on the right gives Hubbert's estimate of U.S. oil production rates if the ultimate discoverable oil beneath the curve is 150 billion barrels. The upper dashed line, for 200 billion barrels, was his famous prediction that U.S. oil production would peak in the early 1970s. The actual U.S. oil production for 1956 through 2000 is superimposed as small circles. Since 1985, the United States has produced slightly more oil that Hubbert's prediction, largely because of successes in Alaska and in the far off-shore Gulf Coast. Kenneth S. Deffeyes, *Hubbert's Peak: The Impending World Oil Shortage* (Princeton: Princeton University Press, 2001); reprinted by permission of Princeton University Press.

These projections—and others like them—do not imply that we shall run out of crude oil. Unconventional oil (tar sands, heavy oils, oil shales) will probably play a more prominent role in the next hundred years. It is much more likely that oil will simply become too expensive to compete with cheaper and cleaner forms of energy. That's what has happened, in part, to coal in some countries. But that offers no comfort. The cheaper, cleaner fuels are not yet readily available; they will require the development and adoption of new energy technologies, including hydrogen fuel cells, nuclear, renewable, and other sources, as well, perhaps, as the extension of "oil" use, from unconventional resources, such as tar sands and oil shales.

Whether one accepts the earlier or the later estimate of peak production, the future of conventional petroleum production is one of decline. It may well be that the "peak" will prove to be an irregular plateau, but there is no doubt that we need a long-term alternative to petroleum. Perhaps an even more sobering curve is one that shows world petroleum production per person (fig. 21.5). The peak there was around 1980, and we are now well into decline. Because population will continue to increase, this decline in petroleum production per person will become more dramatic in its effects, especially as the developing world's use of personal transportation continues to increase.

Against this somber projection, it might be argued that more exploration, or improved production, or simply higher gas prices will guarantee future supplies.

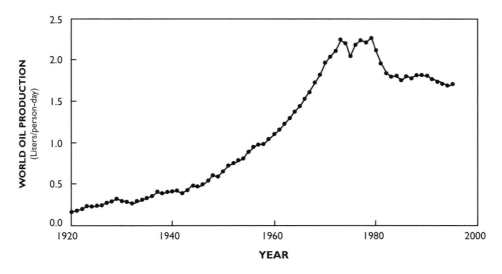

21.5 World petroleum production per person. From Albert Bartlett, "Thoughts on Long-Term Energy Supplies: Scientists and the Silent Lie," *Physics Today* 57, no. 53 (2004); reprinted with permission of the American Institute of Physics.

None of those offers us much hope for prolonging the age of conventional petroleum. Because we know that significant oil accumulations are formed only in sedimentary basins, the last century has seen the extensive survey of all the world's potential provinces. There are more than five hundred of these basins and all of them contain some petroleum. But only about 10 percent have proved to contain significant commercial reserves of oil and gas, and of these, seven account for some two-thirds of the world's reserves. Furthermore, the largest province—the Middle East—contains about two-thirds of the world's reserves. Improved recovery and production techniques are everywhere already being practiced but offer no prospect of significant long-term relief. Occasional discoveries of new offshore fields will almost certainly continue but are unlikely to make up for declining production in most major oil fields. As for price increases, we have them already and we shall certainly see more, in production costs, consumer demand and supply, and in taxes. But if present experience is any guide, their effect on gasoline use will be modest. Of course, petroleum is still being created in some modern sedimentary basins, but the rate at which it is created is far too slow to offer any prospect of relief for our present needs. Add to these constraints one more. The biggest proportion by far of all oil production comes from the Middle East, at present an area of political instability. Other producing areas may also experience supply disruptions, including Indonesia and Africa. (Figure 21.6.)

In limited and declining quantities, conventional oil will be with us for some decades yet. But the need to find an alternative fuel for transportation is urgent.

Two other factors will influence the timing of peak crude oil production. Improved efficiency could significantly reduce usage, though growing demand may counterbalance this. Exploitation of "nonconventional" oil from oil shale, tar sands, such as those of western Canada, and heavy oil deposits could extend crude life, perhaps even by as much as a century. These sources are already being developed but none is without economic or environmental problems. But the central concern remains: within the lifetime of most of today's consumers, conventional oil production will peak, its cost will increase, and alternative energy sources will be needed to replace it.

The production peak for natural gas will not be far behind that of conventional oil.

Nuclear Energy

Even to mention nuclear fuel as an alternative energy source to declining supplies of oil and gas is to invite the fury of some and the concern of many. Nuclear power is regarded by some as off-limits, largely because of the 1979 accident of Three Mile Island, the 1986 explosion of Chernobyl, and the tragic consequences of the earthquake and tsunami damage to the Fukushima Daiichi nuclear reactor complex in Japan in 2011. The public mood has been deeply disturbed by this tragedy, the full impact of which is still unclear.

Against these real concerns, one must also consider that generating electric power from nuclear energy releases no atmospheric CO_2, that reserves of nuclear fuel are substantial, and that there is now a formidable array of technical experience and expertise in designing and constructing newer nuclear reactors and in operating them safely. Some countries—France is one—have been generating most of their energy from nuclear power for many years without any accident. So let's describe the present situation, review the safety concerns, and then think about the future—if any—of nuclear energy.

Nuclear energy is generated either by splitting or combining the nuclei of radioactive elements. *Fission* involves the splitting of larger, heavy elements—thorium, uranium, and plutonium—to produce two new nuclei, with the release of vast amounts of energy. All present uses of nuclear energy—from nuclear submarines to nuclear power stations to nuclear weapons—involve the use of fission.

Fusion, in contrast, involves the combination of lighter elements, such as hydrogen and its isotopes deuterium and tritium, to produce a heavier element, with the release of energy. It is fusion energy that produces the heat and light

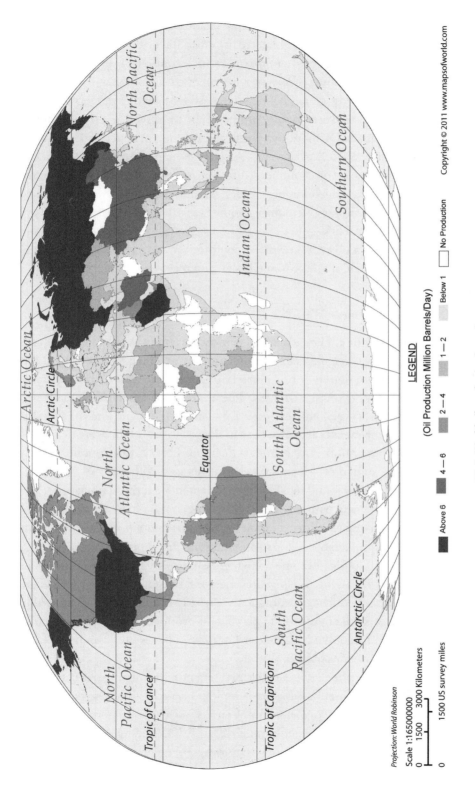

LEGEND

(Oil Production Million Barrels/Day)

Above 6   4 — 6   2 — 4   1 — 2   Below 1   No Production

21.6 World oil production by region.

Projection: World Robinson

Scale 1:165000000

0   1500   3000 Kilometers

0   1500 US survey miles

of the Sun and other stars. To generate this fusion reaction, hydrogen must be confined under high pressure and heated to more than 10–14 million degrees Kelvin, at which point significant numbers of atoms combine to form the heavier element helium and, in doing so, release great energy, far more than could ever be obtained by burning hydrogen as a fuel. The good news is that deuterium fuel is readily available from seawater, while tritium can be produced as a by-product of the fusion reaction itself. Furthermore, the fusion reaction produces no hazardous long-term radioactive waste, apart from the need to contain the reactor vessel itself. There is, however, as yet no commercial fusion reactor. Though experimental prototypes have yielded encouraging results, we have not yet attained sustained nuclear fusion, and it is likely to require another twenty to forty years of intensive development to create a first-generation commercial-scale fusion reactor. *If*—and it's a big *if*—nuclear power is to make a long-term contribution to our energy needs, the development of safe, commercial fusion reactors is a critical priority.

How Does Nuclear Power Work?

About 85 percent of the energy generated worldwide comes at present from burning fossil fuels. Now, burning is a chemical process in which electrons of atoms of fuel and oxygen are shared or transferred from one orbit around the nucleus to another, releasing energy in the process. The atomic nucleus remains unaffected in these chemical reactions. Nuclear power, in contrast, depends on the fact that some of the heavier elements are so large as to be unstable, and these are described as radioactive. These radioactive elements decay and break down spontaneously into one or more "daughter" elements, releasing energy as they do so. The same radioactive decay is also found in a few lighter elements—carbon, potassium, and rubidium—that have one naturally decaying isotope and a "loosely bound" nucleus. This radioactive breakdown is spontaneous, continuous, constant, measurable, and slow. So predictable is it in its natural state that it provides a method for measuring the age of rocks and some younger fossils in which radioactive minerals occur. Radioactive breakdown also yields energy: far more energy per unit mass, in fact, than that derived from burning fossil fuels. It is this process deep within the Earth that drives the engine of the planet and produces the sharp increase in temperature as we drill down into the crust or descend deep into a mine. It is the same process that provides nuclear power.

Now although the rate of this breakdown is constant in nature, it can be accelerated artificially by bombarding the atomic nucleus of a radioactive element with neutrons. The radioactive element of choice for generating nuclear energy is the uranium isotope with a mass number 235 (^{235}U). The mass number is the total number of neutrons and protons that form the nucleus, and an isotope is a variety having a distinctive number of neutrons, though the same number of protons, in the nucleus. When ^{235}U is bombarded by a neutron stream, it forms barium and krypton, releases neutrons, and produces energy, and so can activate adjacent nuclear material. When a *critical mass* of uranium is present, it can create a continuous chain reaction. Rapid release of energy in this chain reaction can be explosive if it is not controlled, but it can also be controlled and harnessed. The first is represented by a nuclear device, as atomic bombs are euphemistically called; the second is the basis of a nuclear power plant.

The energy released by such radioactive decay is immense. Every 0.03 ounces of ^{235}U can release energy equivalent to the burning of 5,950 pounds of coal; or, in other terms, every pound of ^{235}U can release the heat energy equivalent of more than 1,009,998 pounds of coal.

The uranium isotope ^{235}U is rare in nature, making up less than 1 percent of typical uranium ore. These ores are mined in open-pit operations and then concentrated to give a uranium oxide (U_3O_8, "yellowcake") that is enriched to produce small pellets of uranium dioxide with a high concentration of ^{235}U. It is these pellets that, contained in metal cylinders, form the fuel rods used in a nuclear reactor.

Early reactors were designed to employ "raw" untreated uranium ore as a fuel, but more modern reactors, which use enriched uranium, are simpler, safer, and more efficient in operation. Pebble reactors, for example, cannot melt down. Several designs exist, using either gas or boiling or pressurized water as a means of harnessing the heat of the nuclear disintegration to generate electricity. The speed of the nuclear reaction is controlled and maintained by setting or withdrawing carbon or steel rods into the reactor to absorb neutrons.

Fast breeder reactors are of more recent design, and allow a core of highly enriched uranium to be surrounded by containers of nonfissionable ^{238}U. The escaping neutrons from the core of the reactor convert this to a fissionable isotope of plutonium (^{239}Pu) at a rate greater than that of the depletion of the core. The heat is then conveyed—by either a molten metal mixture (potassium and sodium) or molten salt—to a system of heat exchangers, where it is used to generate electricity. The great benefit of this type of reactor is that it "breeds" its own nuclear fuel. If fission reactors are again to play a significant role in power in the next decade or two, they will most likely be of the fast-breeder type.

But is there a future for nuclear power? After all, nuclear power has been around for more than half a century, but it still has few advocates. The first commercial nuclear power station was opened by Queen Elizabeth in 1956 at Calder Hall in England. It was seen almost everywhere as a hopeful sign of human progress: of the turning of nuclear swords of Hiroshima and Nagasaki into electrical plowshares. But just over fifty years later, nuclear power has become, not a replacement, but rather a modest addition to our energy sources in many countries. In Britain, for example, it provides only some 13 percent of the total national energy supply. Six years after Calder Hall opened it was followed in a single year by the U.S. and the USSR's detonation of "test" hydrogen bombs equal in power to more than twenty thousand of the bomb dropped on Hiroshima, and the accompanying release of serious radiation. That, and the hazards of three serious accidents at nuclear power plants, have dashed hopes of a bright energy future and changed the terms of the nuclear debate. So any discussion of a nuclear future has to begin with a discussion of safety. If nuclear plants are not safe, there can be no nuclear-powered future.

Safety

How safe are nuclear power plants? It is understood that they will reduce greenhouse gas emissions and extend the useful life of oil and gas, but can we be reassured about their safety? The danger from nuclear reactors comes not from the fear of an atomic-bomb-type nuclear explosion, but rather the possibility of a reactor failure, such as a loss of primary coolant, that may allow a normally controlled nuclear reaction to run out of control, creating a "meltdown" of the reactor vessel. While modern, gas-cooled pebble-bed-type reactors are inherently free of meltdown risks, concern remains over the safety of older water-cooled models.

The first significant nuclear power failure was at Three Mile Island on the Susquehanna River near Harrisburg, Pennsylvania, in 1979. There a mechanical failure in the cooling system of a pressurized water reactor was aggravated by a series of human errors made by the operators, allowing the fuel exposure of the reactor and the release of a small quantity of hazardous radioactive material (xenon, ^{133}X). No one was injured and the long-term hazards were judged to be slight.

The tragic accident at Chernobyl, near Kiev in the former Soviet Union, in April 1986, seems to have involved a meltdown of the reactor vessel that allowed the escape of hazardous radioactive gases. A series of tests was conducted, in which—unaccountably—the reactor operators manually overrode all the automatic fail-safe emergency controls. The reactor power output surged to several hundred times its normal operating level, the reactor exploded, fire broke out, fuel was ejected, and winds carried a plume of radioactive gas (mainly iodine, ^{131}I) as far away as Scandinavia and Germany.

At least eighteen people died from exposure in the area of Chernobyl in the next few weeks. Other estimates put the number of casualties rather higher. Estimates provided by the World Health Organization, based on studies fourteen and nineteen years after the accident, suggest between forty-five and seventy-five people died as a result of the accident, most of them firemen and workers involved in the cleanup. Countless others over a wide area were exposed to dangerous levels of radiation. The long-term effects of this radiation are still unclear and may never be known.

The failure of the Tokyo Electric Power Company—TEPCO—nuclear complex at Fukushima Daiichi in March 2011 was a devastating accident, far more serious than either of those two earlier events. It was the direct consequence of damage caused by two other tragedies—an enormously powerful earthquake of magnitude 9.0 and a huge related tsunami, each of which produced havoc, death, and destruction over a wide area. The exact death toll from these two events is still unclear, but it is known to include 19,811 dead, with 17,541 people still missing. The number of fatalities arising from radiation is unknown, but it is likely to be very much smaller. The additive effects of these three devastating tragedies represent a catastrophe for the people of Japan.

Japan and the rest of the world need time to assess the harsh lessons of Fukushima Daiichi. Reactor design, location, construction, operation, maintenance, and oversight will, no doubt, face intense public scrutiny, as they should. After March 11, 2011, no one can ever pretend that nuclear power is without hazards. The more significant issue is the seriousness of the risks posed by nuclear power in comparison with those of other methods of power generation. Pollution from coal-fired power plants, for example, is estimated to kill more than one hundred thousand people a year. The World Health Organization estimates that indoor air pollution from biomass and coal causes 5 million premature deaths a year.

There are two other major risks posed by nuclear power. There is, first, the danger of weapons-grade plutonium falling into "the wrong hands." Recent confrontations with Iran and North Korea illustrate the seriousness of this threat. But these confrontations exist even a quarter century after the United States and several other Western nations suspended the construction of new nuclear plants. It is widely believed that the technical expertise, and perhaps some of the equipment, was provided to North Korea and Iran by a scientific official from the government of a developing country. It is no longer just the "Big Four"—the United States, Russia, Britain, and France—who have nuclear weapons. The nuclear genie is now out of the bottle and it is hard to see how the development of nuclear-weapon threats will be influenced, one way or the other, by the construction of additional nuclear power plants. What is needed everywhere, presumably, is stringent security and accountability at all nuclear power plants.

There is a second issue of public concern over our present and future use of nuclear power: the disposal of spent fuel and nuclear waste is a significant, but not an insuperable, problem. And it's a concern even if no new nuclear plants are built. Some of the materials involved will remain hazardous for a long time: hundreds or thousands of years. We need safe disposal arrangements, not only for spent nuclear fuel, but also for high-level, high-heat waste, such as the nitric acid used in reprocessing nuclear fuel. Nuclear power produces wastes from uranium enrichment, the manufacture of nuclear fuel, and reactor operation, as well as from the processing of spent fuel. It is the largest, but not the only, source of nuclear waste. Other radioactive waste comes from disposal of nuclear weapons and from small users, such as hospitals and universities.

Much nuclear waste is of a low or intermediate level, and can be readily handled by existing incineration or containment until it "cools off," but some nuclear fuel waste is high level, containing sufficiently high concentrations to require more elaborate containment.

Nuclear waste can cause damage both by direct contact (eating contaminated food or breathing contaminated air) and by radiation. So it has not only to be "contained" to prevent it from entering the food chain or the atmosphere, but also "sealed" to prevent radiation escaping.

Each of these requirements is technically feasible, but their implementation still involves technical problems. For example, in addition to the questions of *how* waste should be contained and shielded, the question of *where* raises serious issues. Aboveground storage is convenient—it keeps all future options open—but has yet to be designed for long-term (fifty years and more) storage. Underground storage in sealed facilities is attractive and is widely used, though still controversial because of groundwater movements. Partitioning of nuclear wastes by chemical separation into more-manageable residues and transmutation into shorter-lived or stable nuclides is attractive. Less conventional "solutions"—launching nuclear waste into outer space, dumping it at sea, disposing of it in down-turning crustal subduction zones or under ice sheets—all pose serious hazards. The most promising method of disposal at present seems to be underground disposal in "monitorable and retrievable" storage sites. Radioactive materials could then be stored directly or reprocessed and stored in stable borosilicate glass (vitrification) but would remain accessible for future use or further treatment. The requirements for safe underground storage are a safe engineered containment facility, located in a stable geologic site, typically a few hundred meters below ground level, with both artificial and natural containment barriers.

The objective is to develop sites involving only minimal risk (defined as 1 in 10^6) over a life span of 10^6 years, in reducing spent fuel radiation to the background level of natural radiation. There is evidence that storage of vitrified waste in radiation-proof containers in a radiation-proof repository in a safe and stable geologic setting will provide this security. Nonetheless, public acceptance of these risks is still a major barrier, particularly when comparisons or trade-offs with conventional energy sources are not considered.

No method of energy generation is risk free. For example, coal mining accidents and dam failures have both brought loss of life on a tragic scale in the recent past. Over three thousand coal miners are reported to have died in China in mine accidents in 2007, for example. Toxic emissions from existing coal-fired power stations will, alas, bring still more deaths. It is against this background that the hazards of nuclear energy—real as they are—have to be evaluated and judged (fig. 00).

But is not oil a sensible alternative for power generation? Oil supplies have a limited life, and if we continue to burn it as we now do for transportation, the increasing atmospheric pollution may well be a far more significant health hazard than the risks involved in the use of nuclear energy. And "cheap" oil supplies will not outlast the present century. If, then, we turn to coal, the greenhouse gas emissions and atmospheric pollution will be far worse. Nuclear power is not without its problems, but they are fewer than those of other fuels.

Supplies of Nuclear Fuel

If oil, gas, and even coal are dwindling resources, with limited long-term reserves, what of nuclear fuel? The particular uranium isotope used as nuclear fuel—^{235}U—is a rare component of naturally occurring uranium, and naturally occurring uranium is itself rarer still, making up only about 2 parts per million in the rocks of the Earth's crust. The earliest uranium mineral used was uraninite (UO_2, or pitchblende, found in igneous rocks), but other minerals are more generally used, especially those in sedimentary deposits formed from erosion of older igneous and metamorphic rocks. These include substantial deposits in the western United States, Canada, Africa, Australia, and other areas. General estimates of present reserves suggest availability for some seventy-five years at present demand rates, but it seems likely that their use in fast breeder reactors could provide substantial fuel supplies for one or two hundred years at several times present rates of consumption. Thorium reserves are much greater.

Present Use of Nuclear Power

Many nuclear power plants in the Western world are now approaching the end of their useful lives, and several have already been decommissioned. There are at present some 440 nuclear reactors operating in thirty countries and they contribute some 16 percent of world electricity as base load power. This is as much electricity as was produced from all sources in 1960. Sixteen countries depend on nuclear power for at least a quarter of their electricity. Although no nuclear plants have been completed in the United States for the last thirty years, sixty new plants are now under construction by fifteen countries, especially in China and other parts of Asia, but also including the United States and Europe. Almost all these countries are now committed to review of existing nuclear plants and to future nuclear policy.

It is noteworthy that those countries producing the largest proportion of their energy needs from nuclear fuel, such as France, have adopted a centralized design, construction, and operating program, thereby increasing the quality of control and oversight of operations.

21.7 Nuclear electricity generation by country in 2009 (in terawatt-hours)

Country	Amount
United States	815.2
France	387.8
Japan	262.3
Russia	148.5
South Korea	143.5
Germany	131.0
Canada	86.8
Ukraine	76.9
China	64.2
United Kingdom	62.1
Spain	51.0
Sweden	50.9
Belgium	44.7
Taiwan	41.1
Switzerland	26.0
Czech Rep.	25.0
Finland	22.1
India	15.0
Hungary	14.4
Bulgaria	13.9
Slovakia	12.8
Brazil	12.0
South Africa	11.8
Romania	10.5
Lithuania	10.1
Mexico	9.9
Slovenia	5.6
Netherlands	3.9
Pakistan	2.7

Source: Data courtesy of Europe's Energy Portal.

One factor in the future role of nuclear power is cost. Although the operating costs of nuclear plants are low, capital construction costs are high, with up to one half of the cost going in interest charges on construction funds. Legal challenges can greatly increase that cost by delays in construction, even after construction is approved. Recent increases in the prices of oil and gas, together with the environmental damage created by fossil fuels, are, however, already reducing the concern about the relative cost of nuclear power.

Apart from costs, public concerns about the operational safety of nuclear reactors, the disposal of nuclear waste, and the threat of the proliferation of nuclear weapons—by either the manufacture or theft of weapons-grade material—need to be addressed before an effective program for future energy supplies can be developed. But once they are addressed, nuclear energy will almost certainly play a significant role in future energy supplies.

One great irony about the widespread opposition to the use of nuclear power is that we derive virtually all our energy secondhand from nuclear power. The

Sun is the source not only of renewable energy (wind, hydropower, solar energy) but also of the power we derive from oil, gas, and coal. These are fossilized sunshine: the legacy of stored nuclear power. Only the limited geothermal energy we use is not derived from the Sun, but that also arises from radioactivity deep within the Earth itself. How ironic, then, that some refuse even to debate the issue of nuclear power as an increasing component of our future energy needs.

The Energy Gap

Expansion of renewable technologies, whatever they may turn out to be—wind, solar, hydro, biomass, geothermal, and others—will need time to develop and deploy to scale. So, too, will more efficient conversion and utilization of the energy we already consume. We must pursue new technology and new efficiency with a sense of urgency that we now lack. But it will take a while to get significant results: Ten, twenty, even thirty years, perhaps.

In the meantime, in most of the world "traditional" methods of energy production will be needed to fill that gap, including carbon-based fuels and nuclear energy. And that's a problem. In the Western world, for example, nuclear plants are aging and are in need of renewal and serial replacement. But that replacement is not taking place: there is still strong grassroots opposition to new nuclear construction in many countries, such as the United States and Germany. In the UK, to take one instance, the contribution of nuclear generation will fall from 23 percent (about 14,000 MWe) of total energy usage in 2002 to zero in about 2025 unless proposed new plants are constructed.[4] Should we, then, replace nuclear plants with other means of producing power? Should we increase the use of carbon-based fuels, build new traditional generating plants, to fill the gap? Well, that means still more carbon emissions, still more global warming and still more reliance on oil and gas, which are increasingly in short supply, and still more reliance on Middle Eastern suppliers.

So nuclear energy will be needed, at least to help us fill the "sustainability gap," and probably to continue to provide power well beyond it.

Renewable Energy

Energy use is more or less directly tied to living standards, industrial productivity, and security. So global energy use is not going to decline, even though efficiency continues to improve. But potential reserves of many of our present fuels—especially oil and natural gas—are declining, though new discoveries, new

---

4. David J. C. MacKay, *Sustainable Energy: Without the Hot Air* (Cambridge: UIT Cambridge, 2008), 342.

technology, and cost will determine their future. The secret to sustainability is to devise a long-term solution that will provide safe, accessible, inexhaustible energy sources, and to develop new intermediate energy programs to tide us over until we can get to that point. That could well be decades away and our interim energy mix will still depend heavily on oil, natural gas, coal, and other fossil fuels, as well as on some use of nuclear fuels. To ignore the urgency of the need to develop this interim mix is to invite wrenching economic and social disruption down the road, and all the vulnerabilities that this could entail.

But alternative energy strategies and new technologies are not the only need: just as we must also encourage conservation of the energy we use and improved efficiency in our existing power generation, we must also encourage the development of renewable energy as part of our future energy mix. Even if there were no shortage of oil and natural gas, renewable energy would still make sense. Petroleum has hundreds of better uses as the basis for petrochemical products, from medicine to synthetics to fertilizers, than it does as a fuel. Nuclear power still faces widespread public concern and opposition. Coal, though still plentiful, is a significant source of pollution as we use it at present.

Renewables, in contrast, though often requiring substantial capital investments to develop, are in potentially endless supply and immune to price fluctuations or cartel intervention. They are nonpolluting and they can become widely and cheaply distributed. Most renewable energy is ultimately derived from the Sun. Solar radiation not only provides energy for the light and warmth on which we depend, but also drives the weather cycle, creating winds and waves, evaporating water, driving the hydrologic cycle, and providing the basis for plant growth. The only renewable energy sources that are not derived directly from the Sun are limited to geothermal energy from Earth's internal heat, and tidal energy, in which the Moon, as well as the Sun, plays a role.

It is likely that the long-term solution to our energy needs will involve some mixture of sources, including advanced nuclear fission reactors, nuclear fusion (if this can be proved to be workable and affordable), hydrogen, and a variety of renewable sources. Existing renewable sources—including wind, solar, biomass, geothermal, and hydroelectric energy—could be significantly improved and expanded.

The direct sources of most existing renewable energy are hydroelectric plants and biomass, including wood burning. Though both are likely to increase, each presents problems. Because hydroelectric power plants involve the flooding of large areas, they have become increasingly unpopular. The Three Gorges Dam in China, for example, which was completed in 2008, has displaced almost 2 million people, and the 370-mile-long reservoir has flooded over thirteen hundred archaeological sites and vast areas of valuable agricultural land. Furthermore, such schemes are very costly to build and dams tend to silt up, some in relatively few decades. Although some hydro projects—such as Niagara Falls, and plants in Switzerland,

Norway, and Iceland—are well established, many of the sites for big hydropower plants have already been used. In turn, though biomass will play a growing role in providing energy, especially from the harvesting of such fast-growing trees as willows and poplar, wood burning is unlikely to increase much, partly because of the ravages of deforestation that it involves in many parts of the world.

In spite of these limitations, both sources of power remain important. In contrast to the destructive effects of "traditional" wood burning for fuel, biogas is a renewable resource, that provides positive environmental benefits. Biogas is produced by the breakdown of organic matter—biomass, crop waste, sewage, animal manure, and so on—under anaerobic conditions. This breakdown, which takes place in a sealed "digester," produces gas, consisting largely of methane and carbon dioxide. In rural communities small digesters can provide families with cooking and lighting. Rural China and Africa have many million such facilities. Large-scale digesters can provide whole communities with combined heat and power (CHP), as they do in parts of Germany, Denmark and Sweden. Upgraded biogas is used in places as a vehicle fuel.

The attractive thing about all renewables is that, once the technology is in place, they depend on resources—wind, Sun, tides, geothermal—that are either undepleted by their harnessing and use or easily renewable. Of course, they have also to be competitive in cost with existing sources, and they have to be distributed and made accessible, usually by converting them to electricity. They are not, however, a panacea. Initial construction costs tend to be high, and environmental passions can be intense. Opposition by homeowners to wind farms on Cape Cod or offshore near Martha's Vineyard is but one example of local concerns.

Solar Energy

Solar energy powers the surface of the planet. It drives the water cycle, creates climate, and promotes the growth of the plant kingdom. It is also the ultimate source of most of the energy we use, derived as this energy is from fossil fuels. Oil, gas, and coal are repositories of ancient sunlight. But they are nonrenewable, at least over the time span of human existence.

How then can we harvest the daily bounty of Earth's sunlight? We capture some of it already through existing technology. We harness it, for example, through hydroelectric schemes, which are driven by sunlight, and these generate some 20 percent of all Earth's electrical power. We harness it through wind power, which comes through air movement that reflects the action of sunlight. Wood burning, still the fuel of many of the world's poorer people, also harnesses sunlight. Biomass fuels are also the product of solar energy. But all these uses are indirect. Can solar power be harnessed directly?

Sunlight falling on the Earth is filtered, scattered, and absorbed by gases and dust of the various layers of the atmosphere, so that on a clear day at midlatitude only some 70 percent of direct solar radiation reaches Earth's surface. On cloudy days this figure may fall as low as 1 or 2 percent, but some scattered sunshine still reaches the surface even on a cloudy day. The amount of solar energy reaching Earth's surface depends, not only on cloud cover, but also on time of day and on the latitude of the particular place.

The heating capacity of this solar energy is huge, as every homeowner, pulling the shades at noon across a large south-facing window, knows. Harnessing it to produce low-level heating—say, of a hot water system—is relatively easy. In areas of abundant sunshine, solar heaters are frequently used on rooftops to provide power for domestic water heating. The traditional type of solar collector typically consists of a black panel, protected from outward loss of heat by a glass panel mounted above it and insulation below it and on its sides. Heat from the collector panel is transferred by circulating water or a freeze-resistant fluid to a storage tank, which is used either for hot water supply or for space heating. Solar photovoltaic roof tiles are now being produced. Even in cooler temperate areas, this passive, low-level, solar heating can be beneficial, especially if installed in well-designed and well-insulated buildings.

The use of solar energy to achieve higher temperatures depends on either concentrating the Sun's rays or using them to create a chemical reaction or an electric current to transfer energy. The principle of the first is the same as the principle of the burning glass, beloved by generations of children, to focus sunlight and set fire to paper. In commercial versions, some of which are operational, large banks of computerized reflectors are used to focus the Sun's rays on a central receiver, where water is converted to steam, which drives a turbine and so produces electricity. Other concentrated solar power plants (CSPs) using smaller parabolic reflector plants are also in use in a few areas.

Two factors at present limit the number of such large solar power plants. One is location: both land costs and construction costs are high, and, as long as traditional energy sources supply cheaper power, capital costs are too high to justify widespread construction. Second, some areas are far more favorable than others: deserts, for example, can provide maximum land and maximum sunshine, but are often remote from sites of heavy power consumption. Storage factors are overcome by feeding the electricity generated directly into the power grid or by storing a portion of the thermal energy collected during the day for use in power generation at night. If solar power is to play a significant role in power supply it will be necessary to transmit power from high solar impact generating areas to regions of high power use. This will require construction of new high voltage DC grids, allowing source mixing of solar, wind, or other renewable sources to high-demand industrial regions.

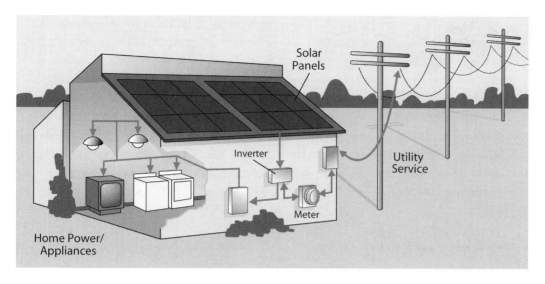

21.8 In this photovoltaic system, solar cells made of semiconductor materials within rooftop panels convert sunlight to DC current. An inverter converts the DC current to 240-volt alternating current suitable for household appliances; excess electricity feeds to the grid for credit or sale. A meter tracks production and consumption. Some systems include battery units for back up. Courtesy of DOE.

The second method of using solar energy is by using solar or photovoltaic cells, in which sunlight produces an electric current by direct interaction with the electrons of a semiconductor collector surface. At present, collectors, most frequently made of silica, are expensive to manufacture and have only limited efficiency of 15–20 percent. Other materials are cheaper, but even less efficient. Solar cells account at present for only some 1 percent of all the world's electrical energy production, though they are especially useful in such situations as supplying power to spacecraft, where cost is less important than access to power. They promise to be of particular value in rural areas, remote from grid supplies of electricity.

There are encouraging signs, however, that the efficiency of cells made of cheaper materials may soon be increased. Promising new collectors are being designed, using precise layering of certain materials within the structure of a plastic collector or by creating nanotubes. The extent of the use of photovoltaic cells in the future will depend on continuing refinements to reduce the costs of their manufacturing and installation and increase their efficiency.

Solar energy satisfies the essential energy needs of virtually all plants and animals except for our own species. In our case, it has been estimated to provide something less than 0.039 percent of our total energy needs. That proportion is likely to increase in the future. The difficult question is by how much.

Though solar energy is still not competitive with other sources of electrical generation, new technology is improving both concentrators and receivers used in direct solar heating. The alternative technology, by which solar cells convert sunlight directly into electricity, depends on the sensitivity of semiconductor materials to light. The photovoltaic (PV) solar cells used may be linked together to provide larger modules. At present, the costs and efficiency limit the wider use of photovoltaics. Passive solar heating and cooling is increasingly used for smaller buildings, by storing heat within the building mass. Though competitive with conventional fuels in some areas, such systems are not yet sufficiently well developed for wider use.

Hydroelectric Power

Hydropower is a particular form of solar power. Solar energy drives the water cycle and this, in turn, generates the streamflow that, acting under gravity, is the basis for hydropower.

The use of waterpower can be traced back to ancient Greece and beyond, and waterwheels were used extensively for grinding corn until the development of the steam engine. Some indication of their earlier importance is given by the records of the Domesday Book, a record of all property in England in the year 1086, which lists 5,624 waterwheels in southern England; that is, one for every four hundred people.

Each village of medieval Europe and early America had its millstream, where grain was ground by rotating millstones, driven by the paddles of a waterwheel. Toward the end of the nineteenth century, dams were constructed across rivers, creating a head of water designed to fall on turbine blades that drove early electrical generators.

Modern hydroelectric power involves using the force of running or falling water to drive turbine blades, which in turn drive electromagnets, whose operation generates electricity. The capacity of any hydroelectric plant depends on both the flow of the water and the height from which it falls: its "head." So the old waterwheels at stream level have now generally been replaced by high dams, which block the flow of a river and increase its potential energy by raising the height from which water falls into the turbines. By this means the potential energy of the reservoir water is transformed into the kinetic energy of the falling water, the mechanical energy of the turbine, and the electrical energy of the generator. Dams are also used to provide energy storage and they can also be of great value in flood control and agricultural development.

Hydroelectric power generation is remarkably efficient in comparison with other methods of power generation, up to twice that of plants using fossil fuels, for example. Against this, capital costs of construction are high, and the demand on land use is extensive; agricultural land and settlements are inundated, ecological disruption is frequent, and suitable sites are rare except in countries with substantial topographic relief, such as Switzerland and Norway. Any major dam construction is likely to involve intense disputes over the riparian rights of adjacent communities, states, or regions, as well as between the relative benefits of development versus conservation. Nor is hydroelectric power easily renewable over the longer term. Dam construction designed to disrupt the water flow also promotes the buildup behind the dam of sediment carried in the incoming water, leading to the silting up of reservoirs, typically over a century or less. The flooding of large areas of land has led in some places to the spread of disease, such as schistosomiasis, carried by waterborne parasites.

The capacity of some of the larger hydroelectric plants is, however, huge. One of the most recent plants—one of the largest in the world—was completed in 1991, at Itaipu, on the Paraná River, on the borders of Brazil and Paraguay. The enormous power plant generates 12,600 megawatts of electricity—enough to provide most of the needs of California—and enough to supply 25 percent of the energy needs of Brazil and 78 percent of those of Paraguay. Some 50 million tons of earth and rock had to be moved in diverting the river and building the dam for this vast system.

The particular contribution of hydroelectric power depends in large part on the landscape, topography, and climate of any given area. Canada, for example, which has abundant snow and rainfall and a wealth of suitable mountainous sites for tapping the energy of its river waters, produces almost two-thirds (350,000 gigawatt hours) of its total energy from hydroelectric sources. The United States, though it produces about 90 percent as much hydropower as Canada, now fills only about 10 percent of its total energy needs from this source, in contrast to the earlier part of the last century, when hydropower provided almost one-half of its energy needs.

Brazil and China, as well as Canada, Russia, Norway, Japan, Sweden, India, and France, also generate significant hydropower.

The relationship between topography, rainfall, and hydroelectric-generating capacity is illustrated by the contrast between Norway, which, with its mountainous topography and plentiful snow and rainfall, produces virtually all its electricity from hydropower, and the Netherlands, with its subdued topography, which produces relatively little. Iceland produces over 80 percent of its needs, and Austria some two-thirds of its needs, from hydropower.

In theory, the damming of all Earth's major river systems has the potential to supply some 30 percent of the world's present energy use. Promising as that sounds, it would involve a high price, as we saw in the case of the Three Gorges Dam in China, which is reported to have involved the displacement and relocation of about 2 million people. The human costs and environmental impact of any such major scheme are high, and the assessment of these costs in relation to the social benefits of the additional power generated is never easy and can never be quantified. The Vaiont disaster (discussed in chapter 6) is a stark reminder that, great as the benefits of a giant hydroelectric scheme to a region or country may be, the potential costs can outweigh them. Even when a scheme works perfectly, delivering power safely exactly as its design had predicted, the cost-benefit analysis of its construction will always be controversial.

There are, for example, concerns that have been raised about the effect of dams on fish migration and spawning, as well as their broader impact on downstream ecosystems. Though most hydroelectric plants have zero emissions during operation, those in tropical areas may produce methane and carbon dioxide from the decomposition of plant material in flooded areas. All reservoirs ultimately "silt up," and continuing sedimentation in the dam reservoir may add to the pressure on the dam walls or block the sluice gates, leading to failure. This is thought to have been a significant factor in the failure of the Banqiao Dam on the Ru River in China during a typhoon in 1975, which led to the failure of sixty-one other dams and many thousands of deaths.

In spite of the reality of these problems, the overall benefits of hydropower are substantial. The efficiency of hydroplants is high: about 90 percent, almost double that of coal-fired plants. Most do not produce greenhouse gases and most have longer lives than conventional coal-fired or nuclear plants. Several hydroelectric plants built at the turn of the twentieth century are still in use. Such plants can assist in flood control, irrigation, and erosion. These plants can also provide substantial storage facilities for surplus electricity produced during off-peak hours, by using it to pump water from a lower to a higher-level reservoir, where it is subsequently used to generate electricity. Hydroelectric plants are also less labor-intensive than other power plants and may shield consumers from escalations in fuel costs.

On balance, hydroelectric power is, in some respects, the most responsible and least damaging method of all those now available for power generation. But energy, like all essential human commodities, involves exchanging one thing for another. That's not unlike the simple biological exchange process on which all life depends. In a broad sense, it's what also defines our own lives and the way we choose to live them. Life involves endless transactions, with the environment, with our community, with our neighbors. Some of those transactions are effortless and automatic: we consume energy as we eat, or speak, or think, for example— even as we sleep. That's a basic transaction, an interchange. Other transactions are personal and far more complex: signing a contract, selecting a career, choosing a mate. And they are far more value laden.

And that is true of the particular methods we select to generate power. Hydroelectric power is clean and reliable, and is likely to continue to provide a significant and growing, though still limited, portion of our future total energy needs. It has been calculated, for example, that only about 13 percent of all the world's potential hydroelectric power has, as yet, been exploited, and that its full development could provide about a third of the world's total energy needs. In spite of environmental concerns about the flooding involved in dam construction, and the limited life of some plants because of reservoir silting, many of the world's developing areas, including China, South America, and Africa, have vast unused potential for hydropower.

Apart from large projects, small-scale (less than 30 megawatts) hydro projects will probably continue to be developed to serve local communities and power networks.

Tidal Power

Tidal power is a specific form of hydropower that is a potential source of energy. Tidal energy derives from the gravitational attraction and relative motions of the Earth, Moon, and Sun, and its magnitude is determined partly by coastal topography. The highest tides, for example, occur in restricted, funnel-like inlets, such as the Bay of Fundy and the Bristol Channel, while the highest velocity of tidal streams occurs in such sites as the Strait of Gibraltar and the Bosporus, where swift water currents are concentrated between adjacent headlands.

Tidal power is still in the developmental stage, with only a few prototypes. Only one prototype plant, located in Northern Ireland, is currently delivering power to the grid. Others are located in the Rance River in France, as well as Canada and Russia, and numbers of promising projects also exist in other parts of Europe, Australia, and North America. In contrast to the design of these plants, alternative barrage-type power plants use a barrage across a narrow tidal inlet to harness the "stored" energy of water driven through sluice gates at high tide, which is later released at low tide. Three small pilot barrage projects now exist, but their relatively high construction costs and concerns over their potential environmental impact may limit their future development.

Geothermal Energy

Within each of the broad groups of renewable energy sources there is room for substantial technological development and improvement. Geothermal energy is energy generated by and stored within the Earth. This energy arises from the breakdown of radioactive minerals. Geothermal heating plants tap the heated ground water below the surface by using heat pumps and use it to generate

electricity. The oldest geothermal power plant, for example, was built at Larder-
ello in Italy in 1904, but later plants have generally been developed only in areas
of recent volcanism—such as Iceland, the Philippines, Indonesia, New Zealand,
and California—where its cost is competitive with other energy sources. Iceland
meets 95 percent of its heating needs from geothermal energy, chiefly by using
district heating systems. In these areas of plentiful subsurface heat, this energy is
clean, competitive, reliable, and renewable, with base load dispatchability: an attractive combination. Such areas are, however, few in number.

But other promising possibilities exist. For example, any geothermal gradi-
ent can be tapped, including not only the traditional hydrothermal sources, but
also geopressure gradients, hot dry rocks, and, perhaps, the zones above magma
chambers. Temperature differences between warm surface oceanic water and
deeper, cooler water can be about 20°C and thus offer potential generating capacity, albeit with relatively low conversion efficiency. Cooler bottom water from
deep glacial lakes is already being used in air conditioning in some areas.

One ingenious example of renewable energy use in providing air conditioning
has been the development of lake source cooling at Cornell University in Ithaca,
New York. Cornell's campus is located on a hill above Cayuga Lake, a deep, gla-
cially formed body of water in New York's Finger Lakes region. Using the fact that
the deeper part of the lake remains cold at about 39°F all year, Cornell engineers
designed a closed circulating scheme to pump cold water from near the lake bottom at a depth of 250 feet, and pipe it into a shoreline heat exchanger, where it
lowers the temperature of circulating water in a separate closed loop that provides
campus cooling. The two water flows never mix. Water drawn from the lake is
returned to its surface in a shallow diffuser located five hundred feet offshore.
This nonpolluting, renewable energy source now supplies over 85 percent of the
energy needed for cooling the campus, not only reducing Cornell's electricity demand by 10 percent but also reducing annual greenhouse gas emissions by up to
thirty-seven tons of SO_2, sixteen tons of NO_x, and eleven thousand tons a year
of CO_2, as well as replacing the use of forty thousand pounds of harmful CFC
refrigerants.

One question that emerges is whether the use of geothermal heat could be extended beyond the "hot" areas in the western United States (California, Hawaii,
Utah, and Nevada) and elsewhere in the world where it is already produced. A
recent MIT-based study sponsored by the Department of Energy concluded that
sufficiently high thermal gradients, which exist almost everywhere in the United
States, might allow much larger production. The study concluded that "heat mining" could be economical in the short term using enhanced (or engineered) geothermal systems (EGS) technology. This technology drills several wells to reach
hot rock, connecting them with a fractured rock region through which water is
pumped. The steam produced is used to run electric generators at the surface.

Early demonstration projects show that, to reach adequate rock temperatures, the drilling needs to reach depths of 3–10 km. At 6 km depth the whole United States becomes a viable prospect. Although the highest-grade sites are in the western United States, where the proximity of hot rocks to the surface reduces drilling and production costs, production would also be feasible in other areas, particularly if district heating applications were feasible.

Research by the government into geothermal energy supply was active in the 1970s but lapsed as oil prices later declined. If the potential benefit of geothermal-based energy is to be developed, further research, development, and demonstration will be needed. Because geothermal energy is nonpolluting, has a small ecological footprint, provides a continuous energy supply, is sustainable, and can be developed in modular local units, it represents an important potential complement to other fuels.

Wind Energy

Wind is another of the several energy sources—hydropower, solar cells and collectors, biomass fuels, and ocean thermal units among them—created by the power of the Sun. Wind is air in motion, driven by unequal effects of solar radiation on different parts of the Earth. Local winds arise from the uneven heat absorption of land and water. Global wind patterns reflect the unequal heating of the equator and the poles.

The use of wind power goes back to ancient times. The sailing ships of ancient Greece and Phoenicia, the windmills of medieval Europe, the wind pumps of the Australian outback and the American West all testify to the capacity to convert the power of wind into other usable forms of energy, designed to achieve particular purposes.

The development of wind power brought prosperity to many regions. In the seventeenth century, for example, the development of advanced windmill designs in the Netherlands made that country one of the world's greatest industrial and naval powers. Across the Great Plains of the United States, wind pumps were a major means of opening up the region and were widely used to pump water before the advent of electricity.

Today most wind power is used to generate electricity by driving the blades of a turbine to power a generator. Wind turbines convert this air movement into mechanical and thus electrical energy. Recent improvements in design and materials have already led to marked improvements in efficiency, and use of wind energy is undergoing rapid expansion. Wind farms are being widely developed, with the electricity they produce being fed into utility power grids. Wind turbines range in size from small, 1–10 kw structures producing enough electricity for a single house, to huge turbines, thirty or more stories high, capable of generating 2 mw or more—enough electricity to supply the needs of five hundred average households. Each large wind turbine typically requires about two acres of land.

Not all areas have the capacity for useful wind generation, but the New England and Middle Atlantic coasts, parts of the Great Lakes, and the Great Plains, the Rockies, California, Hawaii, and Alaska have great potential. Storage capacity and sustainability are important factors in harnessing wind power. Wind strength of at least 14 mph is needed for economic power generation. In general, wind velocity increases logarithmically with height above the ground, so that the most effective wind turbines are likely to be larger and costlier than earlier models. Even in areas of sufficient wind strength, the intermittent occurrence of wind may limit the competitiveness of wind power.

Wind power has several clear advantages, however: it does not require the use of scarce water, as do other methods of power generation, it is endlessly renewable, widely available, clean, nonpolluting, and "free," at least as far as the fuel is concerned. But it also has some limitations: wind does not blow continuously; when it does blow, it blows at fluctuating speeds, so the capacity to store the power it produces is important. Though widely available, it is not uniformly distributed; it requires the construction of tall, "unsightly" towers and extensive areas for "wind farms," which may kill numbers of birds and bats. Some turbines may be noisy, and the relative cost of wind power is still a matter of debate.

It is difficult to prepare a balance sheet that weighs these issues against each other. The intermittent character of wind can be overcome by feeding the electricity it generates into a power grid; the unequal distribution of high winds can be overcome by bunching turbines into areas of high and frequent winds—cliffs, mountaintops, and mountain passes, for example—though these also happen to be, of course, some of the most beautiful places on the planet. That's why wind power divides the green supporters: it involves a trade-off between the benefits of clean energy and aesthetic values. Some of the aesthetic issues can be resolved in part by locating wind farms in unpopulated or offshore areas. So far as the hazards wind power poses to birds and other wildlife are concerned, it has been concluded that the number of birds killed by wind turbines is negligible in comparison with the number killed by other human activities and construction, including traffic, high-rise buildings, power lines, and environmental changes created by other means of generating power. Wind turbine designs can now accommodate variable fluctuations in wind speed and direction. Present studies indicate that to replace a traditional power plant, an area of some twenty square kilometers would be required to provide adequate space for a wind farm, but much of the land involved could also be used for agriculture and other purposes. One attractive consequence of this might be to allow farmers and other large landowners to make decisions on siting and construction, as well as providing a secondary source of income for farmers or landowners.

The economics of wind power are complex, involving subsidies for "green energy" in some states and the lack of any general policy that provides responsibility for payment of long-term environmental damage and cleanup costs from alternative "conventional" power generated by utilities and their users. But even discounting its environmental benefit, it appears that wind power is now becoming broadly competitive with other forms of power generation. New designs for wind turbines and new materials for construction are likely to continue to reduce this cost.

The *theoretical* potential of wind power to fill our energy needs is much greater than the present actual global use of wind power. Its actual use is now limited because its contribution depends on both the efficiency and capacity of wind generators. Present machines have an efficiency of 30–40 percent in converting wind energy to electricity, about equal to that of coal-fired generators. But their capacity—the capacity to produce electricity—is much lower. Wind turbines, because of daily and seasonal changes in wind velocity, operate only 65–80 percent of the time, and then mostly at less than full capacity. Their effective capacity is 30–35 percent, whereas a coal-fired plant typically operates at 75 percent capacity and a nuclear plant at 90 percent capacity. Because wind power is intermittent, wind turbines can never hope to match the capacity level of traditional power plants.

Yet for all their apparent benefits, wind farms are still highly controversial, especially among the residents of coastal and other areas where new sites are proposed. Moving them offshore, where wind speeds are typically higher than those on land, though it reduces aesthetic concerns, involves major additional costs, including those associated with corrosion and maintenance, though offshore farms in Europe have proved very effective.

Given the benefits of wind power, one might ask why it has supplied so little of our current power needs. Though many wind turbines exist, they satisfy only about 0.8 percent of the world's electricity. In the United States, that provision is about 0.4 percent, in Germany 6 percent, and in Denmark over 20 percent. Wind power installations are increasing more rapidly in percentage terms than any other type of power generation, quadrupling in a period of five years. That growth seems likely to continue; one recent projection estimates that wind power could supply 30 percent of the world's electricity by 2030. Other estimates suggest only one-tenth of that. But the variability of wind means that it will always be supplemental to "steady," backup, base-load power stations, however these are powered. Its more widespread use will, however, require improvements in the existing electricity grid.

Meanwhile, improvements in turbine engine design continue to increase the efficiency of wind power generation. Conspicuous among these is development of turbines mounted on a vertical axis, which greatly increases operating efficiency and reduces the danger to bird populations.

Wind energy continues to increase in use, partly because it is nonpolluting, and partly because of major reductions in its cost. The unit cost has fallen by 80 percent over the last twenty years. There are some states, California and Florida among them, that have adopted policies supporting renewable energy and where consumers, given the opportunity to pay a premium for purchasing "clean" electricity, have chosen in large numbers to support it. One power company—Florida Power and Light—now generates 40 percent of all the nation's wind energy. Wind power's future use will depend, in part, on whether states and the nation wish to pursue a renewable energy policy by sponsoring tax credits, as the federal government has done for wind and solar.

Wind energy, like other renewable systems, is unlikely to provide more than a partial solution to our overall energy needs, but its proportional contribution is likely to increase in the years ahead.

Future Energy Prospects

It is fashionable in some quarters to demonize fossil fuels and to campaign single-mindedly for their replacement and rejection, but it is worth at least a moment's reflection to acknowledge our indebtedness to them. Settled human societies have existed for the last ten thousand years or so, but the huge strides humanity has recorded in the last two hundred years in improved health, expanded food supplies, growing prosperity, burgeoning population, and increased longevity all rest in large part on a foundation of the abundant and cheap energy that fossil fuels have provided. The nineteenth century was powered by coal. The twentieth was powered by oil and gas. In our consideration of their direct ill effects—real as these are—we should at least acknowledge the enormous benefits that fossil fuels have provided, and still provide, for our society.

Furthermore, however intense and sincere the campaign to "eliminate oil and coal," or to "go green," we have, as yet, no adequate alternative energy source available to replace fossil fuels in the two vital areas—electricity generation and transportation—in which we are most dependent on them, but which are also the two greatest sources of pollution. For all their benefits, existing "clean energy" sources have substantial limitations as alternative fuels and all have some liabilities.

Nor does the use of ethanol or other biofuels provide an obvious alternative to petroleum. Already, the policy to replace 20 percent of the nation's gasoline by ethanol has resulted in a switch in land use away from food production, and a consequent steep rise in food prices worldwide.

Although both wind and tidal power will provide a growing proportion of future energy requirements, they provide only modest prospects as major replacements for existing energy sources, both because of the problems of storage and transmission in the case of wind power and the high capital costs and limited geographic availability of suitable sites in the case of tidal power.

The best prospects for alternative energy sources to fossil fuels are solar, wind, geothermal, biomass, and nuclear power. Each of these, as we have seen, provides both opportunities and challenges that will probably require several decades, not only of research and development, but also of prolonged public debate and political fortitude to resolve. In the meantime, we should and must make more efficient and clean use of the maligned fossil fuels, which, for all their ills, have "brought us hither." Without the ancient energy source they have entombed, preserved and made available to us, the social communities and lifestyles of our 7.0 billion fellow humans could not survive. Our very existence in these numbers requires the energy that fossil fuels provide.

Adequacy of Renewable Energy

Just how far short of our current energy consumption the sum total of available renewables falls has been illustrated by David MacKay,[5] who has calculated both the theoretical and the practical availability of renewable energy for the UK, where average energy consumption is 125 kilowatt hours (kWh) per day per person. This, incidentally, is about the same as the average use in other European countries, but considerably less than average U.S. consumption, which is some 250 kWh per day per person. MacKay calculated that, "after public consultation" (meaning weighing public opposition and considering relative costs), the maximum Britain could hope to derive from all renewables is about 18 kWh per day per person, in contrast to present average usage of 125 kWh per day per person. Other estimates reach much the same conclusion, though some regard them as unduly pessimistic. Renewables currently provide only some 1.05 kWh per day per person in the UK, considerably less than 1 percent of total usage.

Achieving a sustainable energy supply is possible either by reducing demand or by increasing supply, or by some combination of the two approaches. Those who argue that, because we were able to live on renewables before the Industrial Revolution, we should be able to do so today overlook the fact that the world's population was then only some one-sixth of what it now is, and that the average person probably then consumed only some 20 kWh per day and required a much larger land area to provide fuel than is now available in our crowded countries.

---

5. Ibid., 236.

The present average available land per person in the UK, for example, is only about one-twelfth what it was in preindustrial times.

In contrast, the same calculations show that North America could live on its renewable energy sources *if* (and it's a big *if*) it reduced its energy demand to that of the European average. The difference is explained largely by the possibility of a massive expansion of solar power, which could yield vast amounts of energy. But the area required for such solar collectors would be "Arizona-sized."[6]

So the potential longer-term contribution of renewables to our total energy supply will vary greatly with geographic location. A "portfolio" of renewable energy sources, including, say, solar, wind, low-head hydropower, geothermal, and biomass, could be selected to serve different regions, depending on regional characteristics and resources. Tucson, Arizona, would thus select a different portfolio mix from, say, Ithaca, New York. To make renewables more widely available will require not only the development of new technologies but also the construction of adequate transmission facilities and storage systems.

Energy Efficiency

Whatever mix of energy sources we adopt, one essential component of any broad policy will be greater efficiency of energy use. Energy efficiency reflects a mixture of costs, technology, and national capacity. Consider this: the United States used less oil and no more energy in 1986 than it did in 1973, even though its population substantially increased and its economic output grew at an average rate of about 2.5 percent per year. Had the 1972 energy use trends continued, its 1984 energy consumption would have been almost 40 percent higher. OECD countries also achieved comparable gains in energy efficiency, though the developing countries did not.

This improvement in energy efficiency was triggered by the 1973–1974 OPEC oil embargo. Energy efficiency improved as the cost of oil skyrocketed, consumers became energized, and technical improvements in vehicles, equipment, and buildings were reinforced by changes in energy sources and supply. Thus, although OECD countries' GDP grew by 29 percent from 1970 to 1985, they consumed less oil in 1985 than in 1975, partly by improving efficiency and partly by substituting other energy sources for petroleum. The combination of circumstances that produced that remarkable change involved both the widespread public response to increasing energy costs and technical improvements that led to improved efficiency. In contrast to this improvement, the developing countries' use of oil grew at roughly the same rate as their GDPs.

---

6. Ibid., 238.

We should not think of the goal of improvement in energy efficiency as being only a few percentage points at the margins, though every little bit helps. One example of major improvement is Shell's development of newly designed production—"monotowers" in the North Sea. These are powered by solar panels and wind turbines and reduce the power required for oil and gas extraction and monitoring from as much as 30 kilowatts to as little as 1.2 kilowatts—less power than it takes to boil water in a kettle.[7] Opportunity for comparable improvements in efficiency exists in many other areas.

Prospects and Policies

We are all children of the age of petroleum. It is to petroleum that we and all the peoples of the developed world largely owe our present levels of comfort, mobility, food supplies, and economic strength, as well as the availability of sources of useful petrochemical products. Petroleum has revolutionized our lives. But, as we have seen, we are now approaching or, as some would argue, have already reached, peak oil production from conventional sources. And our steadily growing global population will not only increase the demands for oil, gas, and coal, but will also lead to a marked increase in carbon emissions, with consequences for the global climate.

All this implies that we need a long-term solution to our energy needs if we are to remain viable as a human race. This will involve a true long-term solution for our energy needs, as well as a shorter-term, intermediate, transitional solution, perhaps covering the next few decades, until a more sustainable long-term energy solution is developed. This transitional phase is going to require us to continue to use the traditional fossil fuels on which we depend, but to use them far more efficiently and responsibly than we now do. That is especially true of the world's largest countries and fastest-growing economies, including China and India, which have huge reserves of coal. This transitional phase of energy invention is just as critical as the "long-term" solution to the world's energy problem.

Our future energy supply will be heavily influenced by the convergence of the challenges we have already identified and, in particular, on five trends that collectively represent the equivalent of "a perfect storm":

- If, as now seems probable, the Earth's population increases from its present 7.0 billion to level off at about 9 billion in 2075, we face the need to provide food, energy, and materials for almost 50 percent more people than we now feed.

---

7. "Monotowers," Shell Company, http://www.shell.com/home/content/innovation/bright_ideas/monotower/.

- Second, continuation of the present rising standard of living in the developing world will place substantial added demands on all resources, some of them already in short supply.

- Third, rising use of fossil fuels will add to the growing concentration of greenhouse gases in the atmosphere, and will increase the present trend of global climate change and disruption. Wally Broecker, a respected Earth scientist from Columbia University, has estimated that if "we do continue as we are, it is quite likely that we will triple the amount of CO_2 in the atmosphere by the end of this century." And that is a somber prospect.

- Fourth, some energy sources—especially conventional oil and gas—are already at or near their peak production levels.

- And finally, present levels of soil depletion, water table depression, and groundwater contamination will present growing problems for agriculture, especially in the poorer parts of the world.

The convergence of these five trends represents a major challenge in developing a strategy for future energy supplies.

It could be, of course, that one or more of those projections will be proved wrong: that some trends will be reversed. In the light of present evidence, it seems unlikely, but even if that turns out to be the case, we cannot delay our response for thirty or forty years. Any solutions we develop will probably involve significant lag times (ten to twenty years, perhaps) in their refinement, acceptance, and implementation, and will require still longer times to yield significant environmental benefits. To simply "wait and see what turns up" will increase both the difficulties and the hazards that we now face.

The implementation of any new energy policy will probably involve significant short-term costs, as well as a degree of inconvenience and adaptation, long before it provides any public benefits. That is not a popular proposition in any society, and it is especially difficult to persuade voters in democratic countries to embrace such a program. Nor is it easy to persuade corporate leaders in the market economies to adopt such policies if they involve significant added costs and few direct, short-term corporate benefits.

Any effective energy policy will win public support only to the extent that there is public understanding of the issues we face, of the relative costs and benefits of each of the various proposed solutions, and of the longer-term benefits of adopting—as well as the hazards of rejecting—comprehensive and responsible policies. That is the task both of public education and of public policy, and the essence of this understanding must be that energy is the essential foundation for every part of modern life. The huge differences in health, material well-being, and comfort between our lives and those of our ancestors two hundred or two thousand years ago lie in our use of energy that is both available and affordable. We have moved from unassisted human labor (manpower and womanpower) to animal power to water and wind power to steam power to electric power to oil power and gas power to nuclear power; and the positive impact on lifestyle, agriculture, industry, commerce, and freedom of movement has been enormous. To run out of gas—to find that petroleum and natural gas are becoming so scarce as to be too expensive for all their current uses—will have a comparable impact on every aspect of life; and it will not be a positive one unless we can develop secure alternatives for supplying our energy needs.

The Present Energy Budget

The contribution of various sources to United States energy consumption in 2009 was as follows:[8]

- Petroleum: 35.3 percent
- Natural gas: 23.4 percent
- Coal: 19.7 percent
- Nuclear: 8.3 percent
- Renewable energy: 7.7 percent

The convergence of the five trends we have reviewed above makes it unlikely that this will be the same mix in fifty years' time. The factors determining the mix will be reserves, availability, suitability, cost, and environmental impact.

Consider the simple classification in table 21.1. Though all these energy sources will probably play a role in our future energy supply, their relative contribution will reflect a combination of their reserves, distribution, cost, and impact.

There are major differences, not only in the total contribution of different sources to our overall energy budget, but also in the relative use of each source for particular purposes. Transportation, for example, is highly dependent on petroleum, which accounts for 97 percent of all fuel used for transportation in the United States. In an increasingly global economy, everything—from food to raw materials and manufactured products—depends on cheap and efficient transportation. Even some energy supplies and fuels—petroleum, natural gas, and coal, for example—require effective transportation systems for their delivery. The other critical needs in power supplies are versatility and availability to various applications and uses.

The most adaptable and transportable form of energy is electricity, and the most critical need for the future will be to develop clean fuel for transportation and electric power generation. That depends not only on cost and cleanliness but also on availability. Oil, for example, though a cleaner fuel than coal, is increasingly

---

8. U.S. Energy Information Administration, "Annual Energy Review 2009," Figure 2.0 Primary Energy Flow by Source and Sector, 2009 (Washington, DC: U.S. Department of Energy, August 2010), 37, ftp://ftp.eia.doe.gov/multifuel/038409.pdf.

21.9 Relative availability, distribution, cost, and environmental impact of major energy sources

Power Source	Reserves	Distribution/ Availability	Cost Without Impact Costs	Impact
oil	limited	limited	moderate at present	significant
coal	plentiful	widespread	low	severe
natural gas	limited	limited	moderate at present	benign (almost)
oil shale	plentiful	limited	low	significant
shale gas	plentiful	widespread	low	low if adequate fracking safeguards
geothermal	unlimited	limited	reasonable for high-grade prospects	low
nuclear	unlimited	widespread	high capital/low generating	potentially severe if design and storage problems unresolved
solar	plentiful	unlimited	high capital/low generating	benign
wind	unlimited	unlimited	competitive	benign
tidal	unlimited	limited	high capital/low generating	benign
hydroelectric	unlimited	limited	high capital/ low generating	benign

Sources: Author's adaptation from Jonathan Cowie and other sources.

imported from areas of political instability. Coal is plentiful, cheap, and available in large quantities within the United States and Europe, but its unrestricted use in power plants creates a level of atmospheric pollution that contributes both to global warming and to problems of human health. The trade-off between availability, adaptability, cost, and impact is never constant: the balance changes from time to time and from region to region. The trade-off is not, however, a contest between issues of equal significance. Environmental impact, for example, is likely to outweigh cost—at least when the issue is openly debated—and availability, when it approaches zero for some fuels, may give added significance to strategic and political issues. What is needed, then, is a framework within which policy can be developed and around which the complex and shifting issues of reserves, availability, adaptability, cost, and impact can be addressed.

Surrounding all these issues, and contributing to the changing pattern of energy use, are changes that will inevitably emerge from new technologies. That is why, alongside national and international energy strategies and policies, there must also be national and international programs of research and development, aimed at new methods of extraction and production, new techniques of power generation, and new levels of both efficiency and safety in energy use and environmental protection. These may be as varied as improving the efficiency of hydrocarbon production from oil shales, sequestering atmospheric carbon dioxide, and designing and developing safe fusion reactors. The ultimate purpose of all such research will be to increase energy supplies and thus to safeguard our future.

CHAPTER 22

Materials as Sustenance

The Occurrence of Useful Materials

Human history, as we have seen, is often divided on the basis of the materials that have characterized successive cultures: the Stone Age, Copper Age, Bronze Age, Iron Age, and so on. These substances and others have not only provided the material basis for civilization but have been so valuable that empires have been established, wars have been fought, and nations have been both created and destroyed in competition for their supply. The age of petroleum, in which we still live, provides striking examples.

We've already discussed some of the materials on which we are most dependent: petroleum, coal, air, soil, and water among them. But our daily lives also depend on the use of dozens of other materials, all of which are either refined from Earth's materials (from salt to silicon for computer chips, for example) or manufactured from them (plastics, for example). Everything else we use, from food to fabrics, also depends ultimately on the use of the Earth's resources. Some resources have many uses: diamonds can be used either as costly decorations or as inexpensive abrasives, for example. Others have more limited use.

All our mineral resources come, not from the "whole Earth," but from Earth's crust, whose relative thickness has been compared to that of a peach skin on a peach. And even within that thin skin, most useful minerals are rare. Ninety-eight percent of the crust is composed of only eight elements. These eight common elements include some that are also useful: iron makes up about 5 percent of the crust, for example; aluminum about 8 percent. Many other useful elements, however, occur in much lower concentrations and are relatively rare. Tin, for example, has an abundance in the crust of only some 0.00015 percent by weight; copper, 0.0058. So a copper ore that can be profitably mined has to have a natural concentration of fifty or more times this average amount. Other less abundant elements—gold, silver, and mercury, for example—must have natural concentrations of 1,000 to 100,000 times to provide economic ore deposits (fig. 3.7).

Some of the most important metals—including gold, silver, lead, and some copper and nickel—are formed around *igneous intrusions*. A few heavy metals and minerals—including chromium, platinum, and magnetite—crystallize early from molten magma and settle in the lower part of the molten material of these intrusions, where they solidify to form layered "deposits." Many other minerals—including the sulfides of lead (galena), zinc (sphalerite), iron (pyrite), and mercury (cinnabar)—are formed by the later cooling of hot, metal-rich, hydrothermal fluids, which form mineral-bearing veins in the rock surrounding an intrusive igneous body. Other hydrothermal deposits around an igneous intrusion may be more dispersed but still valuable. These disseminated deposits tend to be of lower concentration but of great extent. The great copper deposits of Chile and Utah are of this kind. Some hydrothermal deposits are also formed along

22.1 Per-capita consumption of minerals, 2010 (pounds per person)

Bauxite (Aluminum)	65
Cement	496
Clays	164
Coal	6,792
Copper	12
Iron Ore	357
Lead	11
Manganese	5
Natural Gas	8,091
Petroleum Products	6,792
Phosphate Rock	217
Potash	37
Salt	421
Sand, Gravel, Stone	14,108
Soda Ash	36
Sulfur	86
Uranium	0.22
Zinc	6
Other Metals	24
Other Nonmetals	332
Total	38,052

Sources: USGS, Energy Information Administration, U.S. Census Bureau.

mid-ocean ridges. Metallic sulfide deposits of copper, iron, and other metals have been observed in emissions from black smokers, where seawater, circulating in the heated crust, dissolves metallic sulfides, which are then precipitated to form sulfide deposits.

Still other important metals are the products of *metamorphism* around an igneous intrusion. Contact metamorphism in limestone frequently produces important ore minerals, while high-pressure regional metamorphism produces such "layered" minerals as graphite, talc, and asbestos, for example.

Weathering sometimes concentrates metals that once existed in dispersed form in minor amounts, to produce valuable commercial ores, either by leaching out less valuable elements—bauxite, the chief ore of aluminum, is an example—or by concentrating what had been more dispersed elements—some major copper deposits are an example.

Some of the most celebrated mineral deposits are those of heavy metals originally formed elsewhere that have been winnowed and concentrated by streams and wave action to form *placer deposits*. The gold of the 1849 California gold rush is an example. Commercial grade placer deposits of cassiterite ore (tin) are mined in Malaysia and Indonesia.

Other important minerals are of direct *sedimentary* origin. Salt and potash deposits are the result of the evaporation of ancient saline waters, and manganese modules are formed in quantity on the ocean floor. The great iron ore deposits of North America, Australia, and elsewhere are sedimentary rocks. Phosphates are extracted from apatite, formed as marine sedimentary deposits. Our more important fossil fuels—oil, gas, coal, oil shale, and tar sands—are of biological origin, formed by deposition and preservation of the remains of microorganisms within sedimentary rocks.

Not all the materials on which we depend are fuels or metals. A host of industrial materials—from fertilizers to fireclay, from ceramic clays to limestone and many other Earth materials—provide building and industrial materials that play a role in every aspect of our daily lives. Over 90 percent of all the nonfuel resources we use are nonmetallic, and many of these are of sedimentary origin. Building, engineering, and highway construction, for example, annually consume some 2 billion tons of aggregate, formed from crushed stone, sand, and gravel. Cement (formed from limestone and shale), clay tiles and bricks, gypsum wallboard and plaster are used in enormous quantities in house building and other construction projects. In contrast to most other minerals, these industrial minerals are wide in their distribution and plentiful in their supply. They are generally quarried and produced locally.

These and other mineral deposits are extracted by either underground mining, surface quarrying, opencut or open-pit mining, drilling, and, in the case of some potash and salt deposits, dissolving and evaporating. The particular method of extraction depends on the location, shape, grade, extent, value, and current market price of the commodity. Higher-grade ores, for example, may justify the high costs of underground mining, while lower-grade ores may not, though the latter may still be profitably extracted by surface mining. The high value of many placer mineral deposits—gold, diamonds, and titanium, for example—justifies the high costs of placer mining, where vast quantities of sand and gravel have to be excavated and sorted.

In almost all cases, the ore must be separated from the waste rock with which it is associated. This usually involves crushing and flotation to separate minerals from waste. The complete mining cycle also includes reclamation of the land that has been mined or quarried. This involves not only using waste to backfill mines and quarries but also restoring topsoil, drainage, and wildlife. Carefully done, this can be very successful, though it is frequently costly and sometimes controversial.

Segregation, Sources, and Supplies

Although minerals are widely distributed in crustal rocks, useful mineral deposits occur only when they have been segregated and concentrated by natural processes into workable deposits. That means that with one or two exceptions—limestone, coal, and bauxite among them—most Earth materials on which we depend are rare in their distribution and localized in their occurrence. Perhaps half the world's oil reserves are located in Saudi Arabia, for example. The United States, like other developed nations, relies heavily on a handful of countries for its supply of each of the many essential materials. One hundred percent of the bauxite and aluminum we use, for example, comes from just four countries: Australia, Guinea, Jamaica, and Brazil. One hundred percent of the manganese we use comes from South Africa, France, and Gabon. One hundred percent of the mica from India, Belgium, Brazil, and Japan. In fact, for fifteen critical nonfuel minerals, 85 percent or more of our supplies come from imports from a handful of countries.

So, although in theory we could extract useful minerals from almost any rock, the costs would be prohibitive, and we rely on ore deposits that are both localized and rare. That means, in turn, that we, and all industrialized nations, are dependent on a system of open trade that allows access to useful minerals at the most competitive costs. Some of these, though used in relatively small quantities, are critical to our national needs. A small number of metals—cobalt, manganese, chromium, titanium, and platinum—are of strategic importance. And for these, too, we are heavily reliant on imports. Over 90 percent of the platinum group metals we use, for example, come from South Africa, the UK, and the former Soviet Union.

22.2 2010 U.S. net import reliance for selected nonfuel mineral materials

Commodity	Percent	Major Import Sources (2006-09)[1]
ARSENIC (trioxide)	100	Morocco, China, Belgium
ASBESTOS	100	Canada
BAUXITE and ALUMINA	100	Jamaica, Brazil, Guinea, Australia
CESIUM	100	Canada
FLUORSPAR	100	Mexico, China, South Africa, Mongolia
GRAPHITE (natural)	100	China, Mexico, Canada, Brazil
INDIUM	100	China, Canada, Japan, Belgium
MANGANESE	100	South Africa, Gabon, China, Australia
MICA, sheet (natural)	100	China, Brazil, Belgium, India
NIOBIUM (columbium)	100	Brazil, Canada, Germany, Estonia
QUARTZ CRYSTAL (industrial)	100	China, Japan, Russia
RARE EARTHS	100	China, France, Japan, Austria
RUBIDIUM	100	Canada
STRONTIUM	100	Mexico, Germany
TANTALUM	100	Australia, China, Kazakhstan, Germany
THALLIUM	100	Russia, Germany, Netherlands
THORIUM	100	United Kingdom, France, India, Canada
YTTRIUM	100	China, Japan, France
GALLIUM	99	Germany, Canada, China, Ukraine
GEMSTONES	99	Israel, India, Belgium, South Africa
BISMUTH	94	Belgium, China, United Kingdom, Mexico
PLATINUM	94	South Africa, Germany, United Kingdom, Canada
ANTIMONY	93	China, Mexico, Belgium
GERMANIUM	90	Belgium, China, Russia, Germany
IODINE	88	Chile, Japan
RHENIUM	86	Chile, Netherlands
DIAMOND (dust, grit and powder)	85	China, Ireland, Russia, Republic of Korea
STONE (dimension)	85	Brazil, China, Italy, Turkey
POTASH	83	Canada, Belarus, Russia
COBALT	81	Norway, Russia, China, Canada
TITANIUM MINERAL CONCENTRATES	81	South Africa, Australia, Canada, Mozambique
SILICON CARBIDE	77	China, Venezuela, Netherlands, Romania
ZINC	77	Canada, Peru, Mexico, Ireland
BARITE	76	China, India
TIN	69	Peru, Bolivia, China, Indonesia
VANADIUM	69	Rep. of Korea, Czech Republic, Canada, Austria
TUNGSTEN	68	China, Canada, Germany, Bolivia
SILVER	65	Mexico, Canada, Peru, Chile
TITANIUM (sponge)	64	Kazakhstan, Japan, Ukraine, Russia
PEAT	59	Canada
PALLADIUM	58	Russia, South Africa, United Kingdom, Belgium
CHROMIUM	56	South Africa, Kazakhstan, Russia, China
MAGNESIUM COMPOUNDS	53	China, Austria, Canada, Brazil
BERYLLIUM	47	Kazakhstan, Kenya, Germany, Ireland
SILICON (ferrosilicon)	44	China, Russia, Venezuela, Canada
LITHIUM	43	Chile, Argentina, China
NICKEL	43	Canada, Russia, Australia, Norway
NITROGEN (fixed), AMMONIA	43	Trinidad and Tobago, Russia, Canada, Ukraine
ALUMINUM	38	Canada, Russia, China, Mexico
MAGNESIUM METAL	34	Canada, Israel, China, Russia
GOLD	33	Canada, Mexico, Peru, Chile
COPPER	30	Chile, Canada, Peru, Mexico
MICA, scrap and flake (natural)	27	Canada, China, India, Finland
GARNET (industrial)	25	India, Australia, China, Canada
PERLITE	25	Greece

22.2—cont.

SALT	24	Canada, Chile, Mexico, The Bahamas
VERMICULITE	22	China, South Africa
SULFUR	17	Canada, Mexico, Venezuela
GYPSUM	15	Canada, Mexico, Spain
PHOSPHATE ROCK	15	Morocco
IRON and STEEL SLAG	10	Japan, Canada, Italy, South Africa
CEMENT	8	China, Canada, Republic of Korea, Taiwan
IRON and STEEL	7	Canada, European Union, China, Mexico
PUMICE	7	Greece, Turkey, Iceland, Mexico
DIAMOND (natural industrial stone)	3	Botswana, South Africa, Namibia, India
LIME	2	Canada, Mexico
STONE (crushed)	1	Canada, Mexico, The Bahamas

Source: USGS.

[1]In descending order of import share.

Future Material Supplies

All Earth's resources are finite, but some are more limited than others. Some essential materials—iron, aluminum, clays, sand, and gravel, for example—have immense reserves. A few others are available in virtually unlimited quantity: magnesium and other materials can be extracted from seawater, for example, as long as energy is readily available. But most have more limited reserves. That does not mean that we are likely to run out of them, but rather that the problems and costs of discovering, extracting, and refining them will become more acute.

Consider the reasons for this:

- Earth's population is projected to increase from 7.0 billion to over 9 billion by 2050.
- Per-capita demand for everything—from meat-rich diets to automobiles—continues to increase, so the demand for mineral resources continues to increase. Thus, even if our population stabilizes, we can expect to continue to increase our consumption of mineral resources.
- Recycling is seen by some as the answer to our mineral problems. But, important as recycling is for some materials—metals, for example—it is not cost free. No material is ever lost, even after it is manufactured, used, and discarded. Discarding a soda can, for example, does not diminish the world's total supply of aluminum. After all, all materials are ultimately recycled by Earth's natural processes. But it does diminish the available, usable supply of aluminum, because Earth's recycling times are very long and we are discarding a metal that Earth has concentrated over aeons of geologic time. The costs of replacing that *natural* concentration are immense.
- Beyond that, recycling requires energy and is not possible for some of our critical mineral supplies. Oil and potash fertilizer, for example, decompose even as they are used.

- Some observers argue for the use of more substitute or replacement materials, and this is constantly being done. Aluminum, for example, is used in place of less abundant copper, and plastics are substituted for metals. But substitution simply shifts the supply burden, as well as often increasing the energy costs of manufacturing.

- Others argue that we should free up companies to increase the exploration and development of new sources of economic minerals. This involves not only the argument to expand drilling on the continental shelves but also a call to intensify every aspect of our search for new mineral deposits in the Arctic and elsewhere. Increased investment in geologic survey and exploration will certainly bring new discoveries, but apart from the environmental concerns, the costs of exploration are high and the costs of production are becoming still higher as the most productive sites have already been explored and exploited.

- One vast untapped resource is the oceans, but we need new techniques to reduce the extraction costs of the abundant materials they contain.

- Some have argued that mineral supplies are so vital that they should be in the hands of government rather than private corporations. A variety of economic factors influence present patterns of mineral exploitation and availability. Changes in price, demand, technological developments, and governmental policies can all have dramatic influence on mineral supply. Government taxes, royalty payments, depletion revenues, and a variety of controls, from export limits to operating restrictions, can have profound effects on private operating companies, some of whom have experienced expropriation and nationalization of their property and leases, with or without compensation. Africa, the Middle East, South America, and Asia provide examples of just this kind of expropriation. Apart from the legality of some of these actions, nationalization has, with some notable exceptions, produced operating results that compare unfavorably with those achieved by independent companies.

- There are always environmental concerns in any mining, quarrying, or drilling operation, and these are not always predictable. The Jeffrey asbestos mine in eastern Quebec, for example, was opened in the 1880s, when the town of Asbestos was developed alongside it. Few could then have predicted that the mine would ultimately involve the removal of 120,000 tons of ore and waste every day, becoming the largest asbestos mine in existence and supplying some 13 percent of total world production. In this case, a constructive partnership between the company and the community led to the relocation of the town and a remodeled city. Other environmental impacts have had less happy endings, and, while a degree of regulation is clearly needed to minimize the effects of extraction, it has to be balanced by the ability of the company to operate a profitable business venture.

- Beyond these challenges, there is another important consideration. All the issues we have just reviewed—population growth, rising per-capita consumption, recycling, and substitution—will involve substantial increases in demand for both energy and water. But our present energy resources are themselves limited. And water shortages already mark those parts of the world with the most rapid population growth.
- Not only that, but our current energy sources, especially wood, coal, and petroleum, are already having a deleterious effect on the atmosphere and there is an urgent need to reduce greenhouse gas emissions.

The Economics of Natural Resources

Modern industrialized societies exist on a foundation of Earth resources. At the base of the world's financial systems, gold provides the basic commodity on which monetary exchange rates are based. Monetary policy and financial stability—or lack of it—play a major role in resource exploration and exploitation.

All but a handful of the natural resources we use are the products of long-term geological concentration and cycles. We, in turn, influence those cycles: for example, by either creating soil erosion or promoting soil conservation. We also influence them by mining, quarrying, drilling, and otherwise extracting the host of material resources we need, and though the natural cycles ultimately accommodate this exploitation, replenishment is generally slow, with natural replacement cycles that far exceed the timescales of human use and activity. That means that most materials we use are effectively nonrenewable. Only construction materials exist in virtually unlimited supplies. Thus, while we speak of resources of a given mineral being substantial, its *reserves*—the quantity that may be extracted and used at a profitable price—are significantly smaller. The distinction between resources and reserves changes with economic and technological changes.

If, as is certain, population continues to increase, becoming 50 percent larger than its present level, and if the rapid industrialization and urbanization of the last two or three decades continues, natural-resource use will also continue to increase. The search for new deposits is only part of the problem of this growing demand. Land use, ownership questions, the scattered distribution and often remote location of known reserves, energy consumption, mining waste, and environmental degradation add to the problems of increasing demand. A mine producing ten thousand tons of copper a year, for example, may well produce 12 million tons of waste.

This means that future resources will require increasingly intensive and costly exploration, development, and production, and also that new sources and new

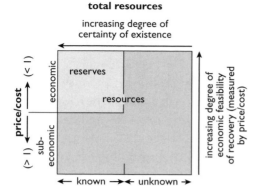

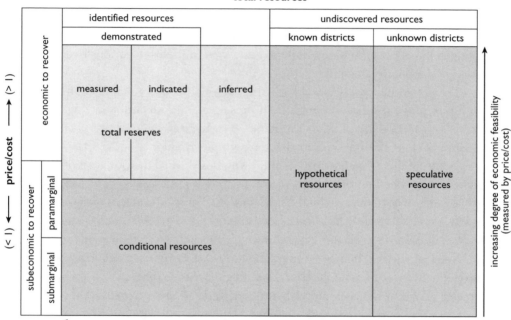

22.3 Resources, price, costs, and reserves. Smaller graph: the way in which the degree of certainty about the existence of a particular resource and the profitability of a deposit separate reserves from the whole field of resources. This is amplified in larger graph by the USGS's widely used classification of mineral raw materials. *Cambridge Encyclopedia of Earth Sciences* (New York: Crown Publishers and Cambridge University Press, 1981). Used with permission.

technologies will be required if we are to meet our growing needs. This is unlikely to be achieved by mining ever-lower-grade resources by existing means, not least because of the energy costs and environmental impact.

So in addition to reducing demand and recycling and reusing our various commodities, we need to find substantial new sources of supply and alternative means of production. A number of radical new procedures seem to offer encouragement. Studies are now under way on the use of microorganisms in the separation and concentration of useful minerals. These and other techniques may provide wholly new methods of extraction.

Two major submarine areas of exploration also hold great promise but at present lack the technology needed for anything other than limited development. Above constructive plate margins, such as the one that forms the Red Sea, large quantities of metal-rich brines and muds have been located that contain a variety of metals. There are some 80,000 km (49,700 miles) of spreading ocean ridges and many volcanic arcs, as well as some 350 known hydrothermal deposits. The potential of these sites is substantial.

On the deep ocean floors, great areas are covered by manganese nodules, which exist in such abundance as to provide a major complementary source of supply, not only for manganese, but also for nickel, copper, and cobalt. Encouraging commercial efforts now exist to mine these nodules in various deep trench areas of the South Pacific.

These two prospects alone have the potential to make major contributions to future mineral supply, though both present significant problems, not only in extraction technology, but also in national ownership and access, as well as in their impact on existing traditional supplies and suppliers of some materials. Steven Scott, of the University of Toronto, who has studied the potential of oceanic mineral resources, points out that ocean mining, though costly, has several major advantages over traditional land-based mining. These advantages include worker safety, small footprint, little social disturbance, and reusable infrastructure, as well as the lack of associated roads, towns, mineshafts, and excavations.

None of these schemes will replace the need for greater economy and improved efficiency in material extraction, manufacturing, and use, or the need for greater efforts at replacement, substitution, and recycling. Nor should we overlook the problems of either the limited geographic distribution and ownership of land-based reserves or the economic and political consequences of the present imbalance between material producer nations and consumer nations.

As in the case of other commodities we have discussed—water, food, and energy among them—major challenges exist if we are to provide resources for a still rapidly growing population, with an increasing appetite for almost everything.

CHAPTER 23

Prospects for Sustenance

Any discussion of the prospects for our long-term tenancy requires a survey of the general state of the property and an understanding of the impact of our occupancy. Earth is an ancient and resilient planet, constantly undergoing change in both its inhabitants and its surface geography. And though it derives most of its energy from its parent star within the solar system, it is otherwise, in broad terms, self-contained and self-renewing, at least over long cycles of time. Perhaps 98 or 99 percent of all Earth's species that have ever lived are now extinct. It is as idle to pretend that we can prevent extinction as it is to presume that we can prevent global climate change. Both will continue, whatever we may choose to do. Extinction and climate change are inescapable and they are interconnected. Past changes in climate and environment have been major influences on the distribution and survival of Earth's creatures, just as organisms, in turn, have always influenced their environment to some degree. That is how life works; that is the system. And it has worked well for over 3 billion years. Had it not, we should not be here.

But there is one new wrinkle. One species—ours—now has a global impact on both the fate of other species and on Earth's systems. And the scale and speed of that impact are troubling because our burgeoning population growth has created, not only an unprecedented demand for food and other resources, but also such a rapidly growing impact on our environment that it could exceed our planet's capacity to adapt and provide our future support.

The extent of this impact is troubling because it is no longer local or regional but global: rapid industrialization in India and China, for example, affects not only Delhi and Shanghai, but also Denver and Stockholm. The scale is troubling because our depletion of some vital resources—including oil and gas, as we have seen—far exceeds Earth's capacity to renew them. The speed is troubling because time, in this case, is not on our side. This does not mean we are doomed, but it does mean that—as never before—we must think and act globally, that we must develop new technologies, new sources, and new substitutes for many of the supplies and materials on which we depend.

So we need to get on with these tasks. The lead time required to develop commercially viable, renewable energy sources, for example, is probably ten to twenty years. Significant additions to our nuclear generating capacity will require five to ten years. But each of these raises social, environmental, economic, security, and geopolitical issues, and no nation can go it alone.

The key challenges of the twenty-first century are an increase of some 50 percent in our global population; our consequent need for resources, energy, food, and water; and our environmental impact. Like it or not, it is these issues that will determine the character, and perhaps the fate, of Earth's peoples, as well as many of its other inhabitants. Earth is resilient: it will survive. But we may not. It is time to rethink our tenancy.

So where are we—23 chapters, 320 pages, and 4.6 billion years since we started? Here's where we are. We are one of about 2 million recorded species that share the Earth. And one of those species—ours—now threatens so to modify the character of the benevolent planet that gave us birth that we may impact its ability to sustain us and our fellow creatures at anything like our present numbers.

But we are certainly not doomed. A trend, however threatening, is not destiny. We are free creatures, adaptable by nature, talented and creative, capable of extraordinary dedication and achievement when we are confronted by a common challenge and inspired to act responsibly. And this is just such a challenge. The trials we face come from the several broad changes we have just described, so let us be clear as to just what the challenge is. It is this: we have to find a way to deal humanely with the consequences of six troubling trends:

- a "built-in" increase in global population of some 50 percent, which will impose added strains on our planet's carrying capacity;
- energy constraints, with nuclear power as the only adequate alternative to fossil fuels yet available and alternative renewable energy sources not yet sufficiently developed to replace fossil fuels;
- widespread soil erosion and continuing deforestation;
- severe water table depletion in many densely populated areas;

- global climate change, which will have varying geographic effects; and
- growing urbanization, industrialization, and environmental transformation.

The impact of each of these trends will vary from region to region. Moderate climate change, with modest increases in temperature, for example, will benefit some temperate areas but may impose severe strains on agriculture in tropical areas. Growing population will affect Africa far more than Europe, which is now losing population, for example.

But none of these trends will act in isolation. Each will be influenced by and will reinforce all the others. Rising population will continue to create rising energy demands and will lead to increasing deforestation and biosphere degradation, for example, each of which will contribute to climate change. The consequences of these six trends—most the result of human influence—already impose severe stress on the stability of Earth's present congenial environment. And they will also place added stress on social, economic, and political systems and on the health and well-being of the world's poorest people. Therefore, each presents a degree of urgency.

Serious as these trends are, each can be stabilized or reduced, and the consequent problems also reduced, if we have the will to act. Any effective program will have to be based on a high degree of public acceptance and a wide variety of public initiatives. Nor is there any silver bullet, single-shot solution to our problems or any time-tested approach that will guarantee its effectiveness. Each of the trends we face is complex: there are various degrees of uncertainty in our own projections, and most of the hazards we face are global in their impact. So our solutions must ultimately be global, even though based on the cumulative effects of local efforts.

One question that arises in confronting any such issue as those we are now discussing is how best to encourage cooperation. Take climate change, for example. How can we get a grip on CO_2 emissions? Well, start with cars, the polluters of convenience and the choice for most of us for all transportation, from a trip to the drugstore to a three-hundred-mile journey. Automobiles are responsible for producing about a quarter of all anthropogenic (man-made) CO_2 emissions. And the number of automobiles is increasing rapidly in China, India, and other wealthier parts of the developing world. Sales of vehicles in China, for example, tripled between 1993 and 2003, while sales of passenger cars more than doubled. These increases will add to the challenge of restricting automobile emissions.

There are two basic approaches to curtailing carbon dioxide emissions: regulation on the one hand, and incentives on the other—or, of course, some combination of the two. "Add more gasoline taxes," urge the proponents. Well, taxes are already $4 or more a gallon in much of Europe, in contrast to about 47¢ in most

U.S. states. "Subsidize biofuels," urge others, but biofuels already enjoy a government subsidy of about one euro ($1.40) a gallon in Europe. Others argue that we should "require biofuels to make up a larger percentage of all retail gasoline sales." But some biofuels produce more CO_2 than gasoline if one also considers the polluting effects of the coal used in producing both. And biofuels are expensive; even with their government backing of $1 or more a gallon. What about adopting a new tax scheme that would give more substantial subsidies to hybrid electric-gas cars? Fashionable as hybrids are, according to one estimate hybrid cars will cost over $1,250 (1,062 euros) for the reduction of every metric ton (tonne) of CO_2. To achieve the same reduction by using biofuels, by contrast, would cost about 20–25 percent of that figure. But biofuels are still expensive, require vast acreage for their production, and so contribute to soil degradation.

None of the tax schemes yet envisioned or employed seems appropriate for this larger task. Higher taxes on fossil fuel use, production, or imports, for example, or a flat tax on CO_2 emissions, are all rather indiscriminate in their user effects and would fall disproportionately on particular segments of industry or society, such as lower-income groups, who tend to have the older and less fuel-efficient vehicles.

None of these solutions appears cheap, even if they prove to be effective. But suppose we think about all the ways in which CO_2 emissions could be reduced, calculate the costs of each, encourage national governments to set appropriate emission caps, and then allow manufacturers to decide how to achieve them, either by themselves reducing emissions, or by developing carbon sequestration, or by trading carbon credits. This, too, of course, would involve regulation, but it could also be combined with incentives.

This is the basis of the cap-and-trade policies, now being widely promoted, that set emission caps, allocate quotas for CO_2 emissions, and then allow trading in emissions permits by those industries that produce less carbon. This is similar to the Kyoto Protocol approach, but it has not yet proved successful in reducing emissions, and such schemes are also expensive. Although a recent estimate by the Congressional Budget Office projected that such legislation would cost the average U.S. household only $175 a year by 2020, a similar scheme introduced only a few years ago in the UK is already costing the average family $1,300 a year, over seven times the U.S. projected cost.[1]

Cap-and-trade schemes may be able to fix the level of carbon emissions, but they cannot control the trading price of permits, and fluctuations in that price could have severe effects on the world's economies. The prototype European Union Emissions Trading Scheme introduced in January 2005 has so far proved

---

1. "The Cap and Tax Fiction," *Wall Street Journal*, June 26, 2009, http://online.wsj.com/article/SB124588837560750781.html.

to be expensive but has yet to prove successful in reducing emissions to targeted levels. Even the setting of CO_2 emission targets involves policy questions that inevitably favor some producers over others. If emission targets are established on how much electricity a plant produces as opposed to how much CO_2 is emitted, for example, they would favor gas-fired generating plants and nuclear plants.

The recent European Union Emissions Trading Scheme failed, in part, because it created targets that were too high and thus too easy to achieve, but with refinement it is clear that consumer costs will become even greater, so that fundamental concerns over allocation policy and effectiveness remain with cap-and-trade schemes. It is this question of the costs of reducing emissions that troubles those who argue that any substantial increase in costs of fuel, whether from taxes or market price, will cripple the economy. Because increased energy costs will drive up the costs of everything else, they will multiply the hardships of half the world's population who already spend more than half their income on food.

It seems likely that, whatever overall strategies are adopted, they will benefit from requiring efficiency standards for vehicles, buildings, and appliances; eliminating subsidies for fuels, of whatever kind; and then allowing industry and the consumer to devise their own solutions.

The limited experience to date with the failure of policies to cap emissions has led at least one group at the LSE's Mackinder Programme and the University of Oxford's Institute for Science, Innovation and Society to argue that any long-term strategy based on emission cuts will continue to fail and should be replaced by a new approach based on improving efficiency and decarbonizing energy supplies.

Such alternative plans, while not perfect, would allow manufacturers to choose their own business strategies and so would make the overall price of greenhouse gas reduction less costly and more efficient than it might otherwise be.

Difficult as all these schemes are, they become even more difficult to devise and implement at a global level, as the 2009 G20 meeting demonstrated. The European nations may have been able to set common targets, but they have, as yet, been unable to reach them. Add developing nations to the list, and any agreement becomes difficult. Leaders of rapidly industrializing nations, such as India, have argued that they would accept a limit on GGEs only if it is the same per-person amount now enjoyed by citizens of the developed countries. That would mean a huge increase in overall emissions. China, which some Western leaders argue is subsidizing its exports at the cost of ignoring environmental emissions, has objected to any trade tariff on importation of its goods.

The key to any global program is the development of an effective strategy that does not cripple the world's economies and increase the burdens of the world's poorest peoples. Industry is the goose that laid the golden egg in this respect. It is not only the source of most greenhouse gas emissions from its processes and products—power plants, automobiles, and all the rest—it is also the source of the world's prosperity and provides the basis for whatever success we may have in improving the world's standards of living, including reducing the burdens of the neediest. Simply penalizing industry, or incapacitating manufacturing, will do little to help us. But comprehensive analysis, enlightened planning, and balanced standards and incentives can do much.[2]

Nor can we overlook the possibility of the lower comparative costs and greater benefits of adaptation, as opposed to remediation, as a means of addressing the future impacts of climate change.

So what are the prospects for sustenance? Is there any hope that we can develop a broad agreement that will sustain our future collective well-being?

Perhaps our best prospect is to draft a statement of general principles—a tenancy contract, if you like—and then encourage individual nations to consider and endorse or amend these principles so that each might decide how best to implement them. Certainly, given the difficulty of any one nation reaching agreement on an environmental action plan, the hope that we could impose one single plan on all the world would seem to be unrealistic. But the difficulties in even drafting the broad outlines of such a contract, great as they are, are exceeded only by the problems we may invite by failure to develop any meaningful collective strategies.

---

2. "Wanted: Fresh Air," *Economist,* July 9, 2009, http://www.economist.com/world/international/displaystory. cfm?story_id=14009113.

CHAPTER 24

Policies for Sustenance

There is, in the pursuit of sustainable development, the need for some modesty. Earth's ecosystems have undergone great change in the past, and they will continue to undergo change—most gradual, but some rapid—with or without any human interaction. But on that pattern of natural change we have imposed our own human influence. Much of our economic and social progress over the last two centuries has been made at the expense of these natural ecosystems. We need to recognize how critical a role these systems play in maintaining and stabilizing the resources and environment on which our continuing well-being depends. We must also recognize that any collective agreements to modify social behavior or instigate technological change to influence Earth's systems will involve significant time lags between agreement, implementation, and effect. For example, even if we agreed to stabilize the worldwide release of CO_2 at today's level, the atmosphere could well continue warming for several decades. Nor can we presume to guarantee human survival. We can no more guarantee the indefinite survival of the human species than we can guarantee that we as individuals will live forever. We are all mortal. No species lasts forever.

Nor are the results of future actions or particular programs entirely predictable. Different programs and priorities may compete or reinforce or sometimes be neutral with respect to one another. For example, economic growth, environmental quality, and social progress may either compete with or reinforce one another, depending on the particular context in which they occur. The gap between the *concept* of sustainability and the *achievement* of sustainability is immense. Add to that the challenge that, though the rate of population growth is decreasing, human population is projected to continue to grow to some 10 billion in 2050, until it later stabilizes at something around 8.5 billion. If you include the fact that hunger and poverty are still widespread, the task is even greater.

That is why responsible tenancy matters. That's why we have to understand the terms of our lease. That's why we have to work together, to hammer out some kind of social compact between us, that will allow us, not only to continue, ourselves, to enjoy the bounty of this rare and beautiful planet, but also to hand on our lease, unrestricted, unamended, and unencumbered, to those who follow us. Developing any such compact will be slow and difficult work; it will also be controversial and divisive with many uncertainties and ambiguities, and with few clear outcomes or guaranteed certainties. Failure to develop that compact is, by contrast, easy and relatively noncontroversial: it simply means following the path of least resistance, doing what we are doing now. Its outcome is less uncertain, less ambiguous. It will create damage to the lives of all Earth's peoples and creatures, especially the most vulnerable, and it will inflict damage, too, on the air and waters of the planet whose children we are. It's not easy to assess the scale or the degree of that damage, but it's clear that to do nothing, to ignore the indications of danger, would involve a global gamble, a whole-Earth experiment, whose outcome is far from clear, and whose results would take years to record, but whose impact is unlikely to be benign and whose results are unlikely to be short-lived.

What is needed, perhaps, is a road map that will show us the way toward not only human but also humane sustainability, a route with mileposts and direction signs to guide us, keep track of our progress, and if needed, help adjust our way.

So what should we do to recognize Earth's limits? How can we live responsibly on a finite planet? There is, alas, no existing recipe, no single solution. If we can develop even the outline of a solution, it must be nationally developed, internationally negotiated, globally applied, and individually observed. That, of course, is a very tall order and it will be possible only if there is a national basis for policy development. Without pretending to draft a policy, I think it is useful to consider the elements on which any strategy should rest and which any policy should recognize.

Moses issued Ten Commandments. In the clamor of the age in which we now live, there is no prospect of commandments carrying much weight or declarations receiving much attention. So let me instead presume to offer ten tentative proposals for our crowded planet—ten outlines for sustainable policies—including a footnote with a draft "contract clause" of tenant's responsibilities and obligations. Of course, I am not sure that proposals will fare any better than commandments, but proposals are at least open to debate and amendment, adoption or rejection. So here goes: What are the items of a tenancy agreement—a lease—that would allow us to sustain our own and other species on this rare, beautiful, and benevolent, but finite, planet?

1. Reduce Consumption of Materials in Short Supply

Survival requires sustenance: it does not require satiation. If humanity is to survive, we must reduce consumption of raw materials and energy to the point where we can maintain a steady level of supply as well as production. A few materials are already in short supply. Others will be in a few decades, fossil fuels among them. Can we reduce consumption, even as the world's population continues to increase? Can it be done, even as the industrialization of places such as China and India increases demand in those vast countries? We do not know, but like it or not, we are about to conduct the experiment. In many cases this will involve reducing consumption, not by deprivation, but by replacement, substitution, reuse, recycling, or redesign. But even as we do all these things—as we must—our growing population will increase the challenge of reaching a sustainable energy and raw material economy. We need to reduce consumption both of materials in short supply and of all materials whose production, distribution, and disposal increase energy consumption.

2. Redesign Buildings, Communities, Processes, Products, and Systems to Increase Efficiency and Reduce Emissions

Whatever mix of energy sources we adopt, one essential component of any broad policy will be greater efficiency of energy use. Energy efficiency reflects a mixture of costs, technology, and national capacity. As we have already seen, in the year 1986, the United States used no more energy and less oil than it did in 1973, even though its population substantially increased, and its economic output grew at an average rate of about 2.5 percent per year. Had the 1972 energy use trends continued, U.S. 1984 energy consumption would have been almost 40 percent higher. OECD countries also achieved comparable gains in energy efficiency, though the developing countries did not.

This improvement in energy efficiency was triggered by the 1973–1974 OPEC oil embargo. Energy efficiency improved as the cost of oil skyrocketed; consumers became energized and technical improvements in vehicles, equipment, and buildings were reinforced by changes in energy sources and supply.

Improving energy efficiency in both industrial processes and product design is a critical part of any energy policy. Many such technologies and designs already exist, but there have been few incentives for their widespread adoption.

For example, both domestic and commercial building designs are available that reduce energy consumption per square foot to just one-third to one-tenth of present average usage.

Petroleum is a nonrenewable resource. The largest SUV gives 18 miles per gallon on the highway, 14 in city driving. If some people must drive such large vehicles, we need to improve their mileage performance. But we need to do it also for every car. Unless we are to legislate that the world's people are limited to a choice between a hybrid vehicle and a bicycle, all cars—as well as all trucks, trains, buses, airplanes, and ships—have to be redesigned to achieve greatly improved efficiency and reduce harmful greenhouse gas emissions. Significant improvement will be relatively easy to achieve. The gas "crisis" of the 1970s demonstrated that. Major improvement will be more difficult, but there are encouraging precedents. Jet engine efficiency has increased by an order of magnitude, and jet engine emissions have been reduced by several orders of magnitude over the last thirty years. There is no reason we cannot redesign automobiles and trucks to produce much greater fuel efficiency as well as improve electric cars to reduce the use of fossil fuels.

But it's not just transportation vehicles that will need to be redesigned: it's transportation systems. That's a much bigger challenge. And it's a still bigger challenge to decide how to encourage the use of public transport and sustainable highway use. No one driving into or out of central L.A. or London at rush hour can doubt the need for rethinking the whole system of road use and vehicle charges.

But transport accounts for only some 20 percent of our total energy use. Buildings of all kinds account for much of our power use. Domestic energy use, industrial and manufacturing plants, and power generation all consume vast quantities of energy. And here too—from kitchen toasters to giant turbines, from power generation to power conversion and transmission—redesign will be critical, just as it will to farming methods and food production, and scores of other everyday processes.

Improvements will come, not only from major changes in technology, but also from modest improvements in handling and production. Ford, for example, recently developed a new way of painting its automobiles. The earlier process involved the application and separate drying of three coats of paint—the undercoat, color coat, and gloss coat. This was replaced by a system that supplies all three coats at once, then dries them and collects the fumes produced, using them to generate electricity, and so saving energy, as well as lowering costs and emissions.

An excellent overview of both the significance and the means of energy conservation is given by Amory Lovins and the work of the Rocky Mountain Institute (www.rmi.org).

Nor is energy use the only area in which we must increase efficiency. Water is another commodity where greater efficiency of use will become increasingly important. Present processes, from mining to agriculture, consume enormous quantities of water, whose treatment typically requires over nine times more energy than that of pristine water does. With 1 billion people already lacking clean drinking water and 2 billion lacking adequate sanitation, water is a precious commodity whose efficient use is becoming more urgent.

Increased efficiency and low-impact energy systems will bring benefits to the producer, the consumer, and the public in lower costs, conserved resources, and cleaner air, but it is far from clear that this will happen under its own steam. The decision to achieve increased efficiency will reflect not only the economic benefits but also public perception and public policy concerning existing hazards and the benefits and costs of improved schemes.

3. Reuse and Recycle Materials and Products

The Western world is a throw-away society. Until a quarter century ago, everything an affluent nation discarded as trash was dumped—either buried in landfills or mounded up and grassed over. Thus, Mount Trashmore in Virginia Beach. We now recycle everything from plastic, glass, and metal containers, to newspapers, cartons, and tires, and the material savings are substantial. To be sure, energy is required for the recycling process, but the reduction in use of petroleum-based plastics and forest-based paper and fiber is substantial. Reusing and recycling can contribute to a sustainable future and can be far more widely applied.

From refrigerators to ballpoint pens, from clothing to computers, we still throw things out. In earlier years, milk was delivered in bottles that were rinsed and returned. Groceries were carried home (on foot) from the neighborhood grocery store in a string bag, which lasted for years. Now milk cartons and plastic shopping bags offer great convenience and the family SUV increases carrying capacity, but we have scarcely begun to reap the larger benefits of recycling and reuse. From cartridges for ink-jet printers to engine components for automobiles, reusing and recycling are key components of resource sustainability.

Recycling provides significant benefits. Steel produced by recycling scrap, for example, requires only one-third as much energy as that produced directly from iron ore, reduces air pollution by 85 percent, cuts water pollution by 76 percent, and eliminates mining wastes. Newsprint from recycled paper requires 25–60 percent less energy to manufacture than that from wood pulp, reduces air pollution by 74 percent and water pollution by 35 percent, and preserves forests.

4. Replace Materials That Are in Short Supply with More Plentiful Alternatives

Replacement of materials has existed as long as humankind. In fact, we mark the course of human history by the changing use of materials: the Stone Age, the Bronze Age, the Iron Age, and so on. So also has replacement of one way of doing things (living by hunting and gathering, for example) by another (agricultural settlement and domestication). Until our own times, most of this replacement has been driven by durability, cost, and convenience: wood was frequently replaced by metal, which was later, in turn, replaced by plastic. But the problem is that supplies of many materials, as well as some of the replacement alternatives, are becoming limited.

Our challenge is to develop sustainable replacements, and that involves questions, not only of cost and convenience, but also of long-term availability. Plastics, for all their benefits, are petroleum-based, and future petroleum resources are in limited supply. Our best hope here is that direct replacement of use will be supplemented by direct replacement of means: think of the replacement of copper electricity cables and copper phone lines for fixed phones, first by using fiber-optic cable, made from quartz, Earth's most abundant mineral, and then their subsequent replacement by mobile phones using satellite relays.

One of our most urgent needs in this respect is to find alternative sources for energy. Now, energy is not in short supply: the Earth is bathed in energy. But one of our present major energy sources—petroleum—is in short supply, and the other—coal—is heavily polluting as we presently use it. The first provides the basis for most of our transport systems; the second is the basis for most of our electric power generation. Furthermore, the long-term value of petroleum as a basis for petrochemicals, and our heavy dependence on imported supplies, add to the urgency of developing alternative domestic sources. Wind, natural gas, clean coal, geothermal, solar, and nuclear sources will all need to be developed to increase the availability of these supplies across the country.

And we must continue to nurture research into additional new sources: biofuels, fusion, hydrogen, and the rest. Energy is the foundation of sustainability. It is the most basic of all our resource needs, the critical requirement for the quality of life of future generations.

5. Research New Technologies, New Materials, New Approaches

We need to evaluate every aspect of modern mechanized societies and research and develop new approaches and new technologies for many aspects of our industrial, agricultural, domestic, health, and commercial activities. Let me give one example: existing separation and purification of major commodities (water, fuels, and chemicals, for example) now consume some 15 percent of global energy use. The development of alternative separation techniques, such as greater use of membranes, bioagents, or solvents, could make a huge contribution to energy savings.

This is perhaps the most promising prospect of all for sustainability. We need to harness the best creative talents with the search for new approaches and new technologies. The advantages of cell phones over earlier fixed copper lines represents only one of the more obvious examples of the host of new technologies—from health care to agriculture to manufacturing—that have improved the prospects of sustainability. Fluorescent light bulbs are three to four times more efficient than standard incandescent bulbs and last appreciably longer, fluorine in drinking water reduces dental decay, bioengineering produces crops that are drought resistant, titanium replaces natural knee joints, a pacemaker controls heart irregularities, while uranium becomes a major source of energy, and water becomes a fuel. These are but a few of the newer technologies that have improved our daily lives and can promote sustainability.

6. Resource New Resources

One hundred and fifty years ago petroleum was a substance of no great significance and of no great value. It was its discovery and production in commercial quantities, together with the development of the internal combustion engine, that changed the world of the twentieth century. But one or two shallow wells in oil-soaked ground in rural Pennsylvania could not, of themselves, have brought about this revolution. That required plentiful petroleum, and that, in turn, required an understanding of its origin and occurrence and a method of discovering commercial reservoirs thousands of feet below the surface of the ground. The application of basic geology and applied geophysics has led to the discovery of huge petroleum reserves, many of them in remote and uninhabited places. And, as known supplies are depleted, further exploration, using new techniques, allows searches for new fields below the seas, as well as on the land, although the discovery of new reserves involves deeper drilling, more costly and often more hazardous conditions, smaller potential supplies, and more costly development, production, and transportation.

And what is true for petroleum is increasingly true for metals and other natural resources. Lower concentration ores, inaccessible sites, and more difficult geological conditions add to the cost and problems of production, and will continue to do so as scarcities increase. Reopening of old mines, refining of old waste heaps, reprocessing of used materials will all be part of this new resourcing. Even our conventional views of mining will be expanded, as scarce minerals are "mined" from the ocean floors and extracted from brines and seawater at a cost that becomes economically attractive.

Because many, perhaps most, of the Earth's largest and most conspicuous ore bodies and oil reservoirs have already been discovered, this new resourcing will be more costly and more speculative than that of earlier times. It will also be more controversial, as the tension between mineral extraction, scenic areas, and other alternative land uses becomes more hotly debated. Public policy will have to engage this issue: land and ocean use in the twenty-first century will become an increasingly lively public debate.

But resourcing new resources, valuable as it will be, will be inadequate to solve our problems of sustainability unless it is harnessed with the research and inventiveness we described under item 5: the search for new knowledge and the development of new technologies.

The development of most of these new technologies defied the conventional wisdom and even the wildest public imagination of the times in which they were introduced. Today, they are commonplace. Imagination, discovery, creativity: these are the drivers of both technological progress and social development. They are humanity's most distinctive and most scarce resource. Nurturing, supporting, and harnessing them to human needs is one of the most vital and most challenging aspects of survival. That is the business of schools, universities, business, government, and all societies. It is also the most important component of sustainability.

All six strategies we have so far considered—reducing consumption; redesigning processes, appliances, and products; reusing and recycling materials; replacing materials; researching new technologies; and resourcing new resources—involve the discovery, manufacturing, and use of materials and appliances. Their implementation rests largely in the intersection of the corporate, research, financial, and consumer worlds. But no business, no finance, and no end use takes place outside the political, social, and economic structures that support them. So also with the more general issue of sustainability, which involves not only personal, economic, business, and technological issues, but also major issues of public policy.

Of the many aspects of public policy, we here consider four that are particularly significant.

7. Rebalance Incentives, Taxes, and International Trade

Consumer conscience is a fine thing, but consumer costs and customer charges have repeatedly proved to be more effective in shaping public behavior. No government-imposed policies can succeed in a democracy without widespread public support and cooperation. But no public initiative in an area as large and complex as sustainability can hope to succeed without effective political leader-ship and judicious government policy. There are, perhaps, three broad areas where government leadership is essential. First, in incentives: encouragement is far more effective than compulsion, so the design of imaginative incentives to encourage reduction, redesign, recycling, replacement, research, and resourcing will be of prime importance. So also, second, will be charges, especially in the form of taxes and subsidies. Already governments tax vehicles and fuel, charge tolls on highways and bridges, license drivers, and regulate utility companies. Wise policy and effective charges can do much to encourage sustainable practices.

Third, the delicate balance between incentives, charges, and international trade is critical: burdens without incentives discourage cooperation, but an informed public will become partners in a fair and balanced policy that they understand.

8. Restore Damaged Ecosystems

Overfishing, industrial pollution, urban infringement, and global climate change have already had serious effects on many of the world's most precious ecosystems. From coral reefs to tropical forests, from mangrove swamps to the high Arctic tundra, many of Earth's most noble sites and valuable habitats are threatened, as are the species they support. But it is not simply the spectacular vistas and *Condé Nast Traveler* wild areas that require conservation. We depend on Earth's whole framework of interlocking ecosystems for our daily sustenance and well-being. It would be naive to expect that we could restore all such places to their pristine conditions, but we can limit further extensive damage, seek to restore some areas, and rehabilitate some populations. That, too, will require local enterprise as well as government initiative. The scale of this effort will range from the local trout stream to the North Atlantic fishing grounds. This restoration must also involve protection of existing ecosystems and reduction in the impact of contamination, salination, erosion, desertification, and urbanization. Soil loss and remediation will also have to be an important component of this initiative, as will a concern for the atmosphere. The public increasingly understands that the unregulated dump-ing of CO_2 into the atmosphere or pollutants into the soil is as serious a public concern as the unregulated dumping of sewage and effluents into rivers and lakes.

9. Respond as Communities, but Rekindle National Resolve

Although effective remediation and adaptation programs ultimately will need to be regional and even global, the framework for their discussion is likely to be shaped to a large extent by the particular economies, interests, and priorities of local communities. Deadlock and stalemate are likely to prevail in such deliberations unless broader, long-term interests are weighed against local and short-term interests, as well as the climatic convictions of particular constituencies. One effective means of approaching the debate had been outlined by Jonathan Cowie in Table 24.1.

So where do we start? "All politics," we are told, "are local." So, also, are all environmental issues. However public-spirited, I am likely to be far more influenced by a flood that swirls into my basement than by droughts in distant Africa

24.1 Two-by-two matrix for hypothetical global warming and global warming counter- (both adaptive and mitigating) measures

	Business as Usual	*Undertake greenhouse measures*
Human-induced global warming is real	Cheap energy prices Fossil fuel runs out more quickly Population rises faster (by comparison with greenhouse measures) Lower energy efficiency/higher international energy trade Less use of renewables/higher international energy trade Greater shifts in climatic belts with health and agriculture costs Greater rise in sea level, flooding, salination of groundwater, etc. No economic gains through manufacture and no implementation of energy efficiency and energy sustainability measures	More expensive energy prices Fossil fuels last longer Population rises slower (by comparison) Higher energy efficiency/lower international energy trade More use of renewables/lower international energy trade Slower/lesser shifts in climatic belts/lower impact costs Slower rise in sea level. Protection of low-lying coastland Economic gains through manufacture and implementation of energy efficiency and energy sustainablify measures
Global warming is fictitious	Cheap energy price Fossil fuel runs out more quickly Population rises faster (by comparison) Lower energy efficiency/higher international energy trade Less use of renewables/higher international energy trade No shifts in climatic belts, no health and agriculture costs No rise in sea level No sea defense costs No economic gains through manufacture and no implementation of energy efficiency and energy sustainability measures	More expensive energy prices Fossil fuels last longer Population rises slower (by comparison) Higher energy efficiency/lower international energy trade More use of renewables/lower international energy trade No shifts in climatic belts, no health and agriculture costs Agriculture contingency plans costs (but not implementation) No rise in sea level Sea defense contingency and some implementation costs Economic gains through manufacture and implementation of energy efficiency and energy sustainability measures

Source: Jonathan Cowie, *Climate and Human Change: Disaster or Opportunity?* (New York: Parthenon, 1998), used with permission.

or a typhoon in the faraway Pacific. Though all these events are linked in a complex web of interrelationships, any policy development is best nurtured at home. Concerns about the quality of neighborhood drinking water or local power plant emissions are real and sometimes urgent. Sustainability originates in a community approach. Public policy starts here; but it cannot end here. Any policy, to be effective, must be regional, and most need also to be national and international. And that's the difficult part, of course: witness the initial reluctance of some few industrialized nations—the United States and, until recently, Australia—to sign the Kyoto Protocol, and the failure of the many who did—all of Europe, except the UK and Finland—to live up to their commitments. If we are ever to achieve new international agreements, the debates and agreements at the local level will be the foundation.

10. Review Relative Benefits of Both Adaptation and Remediation in Responding to Climate Change

We have as yet no unanimity on the probable extent or duration of present trends in climate change, nor do we have a quantitative understanding of the impact they will make. Because of the huge potential costs of reduction in GGEs, as well as uncertainty over its effectiveness, we need a comprehensive review of the relative costs/benefits of *adaptation* to its impact in comparison with *remediation* of its sources. It is likely that the most effective overall program will be a judicious combination of the two approaches.

One estimate concludes, for example, that the European Union's proposed 20 percent reduction in emissions by 2020 would postpone the impact of global warming by only two years by the end of century, delaying until 2102 the temperature that would otherwise have been reached in 2100. And the cost of achieving that reduction is estimated at $90 billion annually. That sum invested in mitigation schemes and clean energy research could create enormous benefits and these could far exceed the benefits of CO_2 reduction.

Another estimate concludes that the funds needed to achieve the modest benefits that would arise from the reduction of GGEs could be much better used by investing them in simple health-related areas, which could drastically reduce the 4 million deaths a year from HIV/AIDS, for example, the 2.5 million from indoor and outdoor air pollution, the almost 2 million from lack of clean drinking water, and the million from malaria.[1] It has been estimated that the investment in

---

1. Bjorn Lomborg, *Cool It: The Skeptical Environmentalist's Guide to Global Warming* (New York: Random House, 2007).

greenhouse gas reduction needed to save one person from hunger would save five thousand if invested directly into programs directed to reduce hunger.

Addendum

One more thing: that draft addendum clause in the tenancy agreement. This is, perhaps, the most difficult and controversial section of the whole lease. It would state simply:

> The tenant shall keep the property clean and in good order, making good any damage for which he or she may be responsible, disposing of garbage only in the approved manner, not allowing more people than are appropriate in the landlord's opinion to occupy the property and conducting himself or herself so as not to jeopardize or harm other tenants.

Many leases have some such provision as this. Both parties—whether tenants or landlords—stand to benefit from it. The problem is not our disagreement in principle; it is our own collective inability to deliver in practice. No society "keeps the property in good order." No generation "makes good the damage" for which it is responsible. Most communities, in a host of different ways, "jeopardize or harm" others by the excessive demands they make on the resources of the planet. And it is the developed nations that create the most damage and make the most excessive demands. How, then, can we begin to develop a social contract that recognizes our impact and embodies our obligations, both to our neighbors and to future generations?

All existing projections, as we have seen, show population peaking and then stabilizing at a reduced level of some 8.5 billion. Growing prosperity tends to reduce population growth. Assuming that is likely to happen, our challenge becomes one, not of sustainable development—that would be impossible at our present rate of population growth—but of a sustainable future population some 50 percent larger than the present. What we need are transitional policies to get us there and sustain future human populations at that, or perhaps some lower, level.

And until we achieve a new dynamic equilibrium for our huge population, we shall need public policies and private support that respond to the social hardships that these transitional years will involve.

If we were dependent only on traditional resources, existing technology, and present practices, meeting these needs would be an overwhelming challenge. But we are not. Human resourcefulness can free us from that dependence. We must do everything we can to support programs and institutions that nurture the creativity of the human species and kindle the unbounded resilience of the human spirit. That means support for all the familiar, worldwide, everyday institutions that nurture creativity: families, communities, social services, and especially schools and universities. These, to paraphrase Cardinal Newman, are the small and ordinary means to a great and extraordinary end: the sustainable future well-being of our species.

Epilogue

Four and a half billion years ago Earth came into being, as a spinning, swirling cloud of hydrogen, helium, and dust slowly condensed to give birth to the Sun and its family of protoplanets. For aeons, the newborn planet was "without form and void," desolate, devoid of life, convulsed as impact followed impact and collision imploded on collision. Bombarded from without by the continuing rain of cosmic debris, Earth was also heated from within by its own growing radioactivity, so that gradually its temperature rose. Later, its slowly hardening surface crust was repeatedly ruptured by violent volcanic outbursts and explosive eruptions of steam and other gases from its vents and fissures. It was from these emanations that the oceans slowly came into existence and the atmosphere gradually emerged. For millennia, ceaseless torrential rains drenched the surface of the planet, blanketed now in clouds so thick that its long darkness was broken only by the blinding light of violent electric storms that convulsed its nascent atmosphere.

From such inhospitable beginnings we have come to be, the descendants of the first lowly living cells that emerged some 3.5 billion years ago and of the long cavalcade of countless later forms—fish, amphibians, reptiles, and mammals—to which they eventually gave rise. Our pedigree is long; our ancestors many. We are the offspring of a multitude of creatures, many as remote from us in time as they are unlike us in form. And we are but one of these many. Nor are we linked only to these ancient ancestors. We share this rare and beautiful planet with more than 2 million other living species of animals and plants.

With each of them, we are co-tenants, alike in our formation from the dust of our planetary home and alike, too, in our utter dependence on its daily benevolence. Structure by structure, molecule by molecule, element by element, we reflect our affinity with them. And yet, for all our affinity, we are ourselves; like other creatures in many ways, and yet unlike them in other fundamental respects: unlike in language, unlike in range of skills, unlike in self-awareness, unlike in complexity of social structure, unlike in refinement of purpose and action. And unlike, too, in the heavy demands that we make and the growing injuries that we inflict on the planet that gave us birth.

How strange that is, given our unique ability to recognize its bounty, to illustrate its exquisite beauty, and to articulate its extraordinary fitness, to understand its ancient history and to know its inner secrets. We recognize that uniqueness and celebrate our skills in the very name we choose to give ourselves: *Homo sapiens,* wise man. Surely we are, of all species, the most knowing.

But knowledge is not wisdom. Wisdom is the way in which knowledge is held. Whether we have wisdom enough to recognize our condition and comprehend our present situation, we are about to discover. Whether we have will enough to amend our ways, we are now about to learn. Whether we have courage enough to honor our obligations to our parent planet, we are about to see.

"Speak to the earth and it shall teach thee," Job, a suffering and bewildered man, was told in the book that bears his name. In our generation we have spoken to the Earth as never before. We have assessed its health, charting its vital signs—its changing temperature, its uneven pulse, its spasmodic exhalations, its constant restlessness. With exquisite technology we have discovered its hidden secrets and recorded every contour of its surface; we have searched its deepest oceans, probed its restless interior, and uncovered its intricate history.

We have spoken to the Earth in all the comprehensive jargon of our science and all the elaborate subtlety of our art. But our speaking has been one long analytical interrogation, one uninterrupted inquisition. We have barely paused for breath. We have not had time enough to participate in a conversation. It is time now to listen; time now to learn. Have we now the patience and the humility to allow this ancient planet, this elderly parent, to teach us? That is the question on which our future tenancy depends. That is the question on which humanity's future hangs.

Related Reading

Chapter 1. The Third Planet

Conkin, Paul K. 2007. *The State of the Earth: Environmental Challenges on the Road to 2100.* Lexington: University Press of Kentucky.

Fukuyama, Francis. 1992. *The End of History and the Last Man.* New York: Free Press.

Gorst, Martin. 2001. *Measuring Eternity: The Search for the Beginning of Time.* New York: Broadway Books.

Goudie, Andrew. 2000. *The Human Impact on the Natural Environment.* Oxford: Blackwell.

Leslie, John. 1998. *The End of the World: The Science and Ethics of Human Extinction.* London: Routledge.

McKibben, Bill. 1989. *The End of Nature.* New York: Random House.

Pollard, Sidney. 1968. *The Idea of Progress: History and Society.* London: C. A. Watts.

Rhodes, Frank H. T., Richard O. Stone, and Bruce D. Malamud. 2008. *Language of the Earth.* Oxford: Blackwell.

Saul, John Ralston. 1995. *The Unconscious Civilization.* Toronto: Anansi.

Wright, Ronald. 2004. *A Short History of Progress.* New York: Carroll and Graf.

Chapter 2. The Home Planet

Adams, F., and G. Laughlin. 1999. *The Five Ages of the Universe.* New York: Touchstone.

Allegre, C. 1992. *From Stone to Star.* Cambridge, MA: Harvard University Press.

Asimov, Isaac, and R. Hantula. 2003. *The Sun: Isaac Asimov's 21st Century Library of the Universe and Solar System.* Milwaukee: Gareth Stevens.

Barrow, John D., Simon Conway Morris, Stephen J. Freeland, and Charles L. Harper Jr. 2008. *Fitness of the Cosmos for Life: Biochemistry and Fine Tuning.* Cambridge: Cambridge University Press.

Beatty, J. Kelly, Carolyn Collins Petersen, and Andrew L. Chaikin, eds. 1998. *The New Solar System.* 4th ed. Cambridge: Cambridge University Press and Sky Publishing.

Chown, M. 1993. *Afterglow of Creation.* New York: Arrow.

Clark, S. 1995. *Towards the Edge of the Universe.* New York: Springer Verlag.

Couper, H., and Nigel Henbest. 2006. *Universe: Studying Stunning Satellite Imagery from Outer Space.* San Diego: Thunder Bay Press.

Dinwiddie, Robert, and Martin Rees. 2005. *Universe.* New York: DK Publishing.

Gingerich, Owen. 1992. *The Great Copernicus Chase and Other Adventures in Astronomical History.* Cambridge, MA: Sky Publishing.

Hartman, J., and R. Miller. 1991. *The History of the Earth: An Illustrated Chronicle of an Evolving Planet.* New York: Workman.

Kuhn, K. F. 1994. *In Quest of the Universe.* St. Paul, MN: West.

Silk, J. 1988. *The Big Bang: The Creation and Evolution of the Universe.* New York: Freeman.

Smith, Alexander Gordon. 2006. *The Solar System.* New York: Barnes & Noble.

Vita-Finzi, C. 2008. *The Sun: A User's Manual.* New York: Springer.

Vollmann, William T. 2006. *Uncentering the Earth: Copernicus and the Revolutions of the Heavenly Spheres.* New York: W. W. Norton.

Weinberg, S. 1996. *The First Three Minutes.* New York: Flamingo.

Chapter 3. The Rocky Planet

Achenbach, Joel, and Peter Essick. 2000. "Life beyond Earth." *National Geographic,* January.

Allegre, C. 1998. *The Behavior of the Earth: Continental and Sea Floor Mobility.* Cambridge, MA: Harvard University Press.

Barrow, John D., Simon Conway Morris, Stephen J. Freeland, and Charles L. Harper Jr. 2008. *Fitness of the Cosmos for Life: Biochemistry and Fine Tuning.* Cambridge: Cambridge University Press.

Condie, K. C. 1997. *Plate Tectonics and Crustal Evolution.* 4th ed. Boston: Butterworth-Heinemann.

Cox, A., and R. B. Hart. 1986. *Plate Tectonics: How It Works.* Palo Alto: Blackwell.

Erickson, J. 1992. *Plate Tectonics: Unraveling the Mysteries of the Earth.* New York: Facts on File.

Glen, W. 1982. *The Road to Jaramillo: Critical Years of the Revolution in Earth Sciences.* Palo Alto: Stanford University Press.

Grotzinger, John, Thomas H. Jordan, Frank Press, and Raymond Siever. 2006. *Understanding the Earth.* 5th ed. San Francisco: W. H. Freeman.

Hamblin, W. Kenneth, and Eric H. Christiansen. 2004. *Earth's Dynamic Systems.* 10th ed. Upper Saddle River, NJ: Prentice Hall.

Keary, P., and F. J. Vine. 2009. *Global Tectonics.* 3rd ed. Cambridge, MA: Blackwell.

Lunine, J. 1999. *Earth: Evolution of a Habitable World.* Cambridge: Cambridge University Press.

Marshak, Stephen. 2001. *Earth: Portrait of a Planet.* New York: W. W. Norton.

McPhee, J. A. 1998. *Annals of the Former World.* New York: Farrar, Strauss & Giroux.

Moores, E., ed. 1990. *Plate Tectonics: Readings from "Scientific American."* New York: Freeman.

Sullivan, W. 1991. *Continents in Motion: The New Earth Debate.* 2nd ed. New York: American Institute of Physics.

Wiener, Jonathan. 1990. *The Next 100 Years.* New York: Bantam Books.

Chapter 4. The Blue Planet

Anderson, R. 1986. *Marine Geology.* New York: Wiley Press.

Cone, J. 1991. *Fire under the Sea.* New York: Morrow.

Denny, Mark. 2008. *How the Ocean Works: An Introduction to Oceanography.* Princeton, NJ: Princeton University Press.

Dinwiddie, Robert, Philip Eales, Sue Scott, Michael Scott, Kim Bryan, David Burnie, Frances Dipper, and Richard Beatty. 2008. *Ocean: The World's Last Wilderness Revealed*. New York: DK Publishing.

Ellis, Richard. 2003. *The Empty Ocean: Plundering the World's Marine Life*. Washington, DC: Island Press.

Glover, Linda K., and Sylvia A. Earle. 2004. *Defying Ocean's End: An Agenda for Action*. Washington, DC: Island Press.

Hollister, Charles D., and Bruce C. Heezen. 1972. *The Face of the Deep*. New York: Oxford University Press.

Kennett, J. P. 1982. *Marine Geology*. Englewood Cliffs, NJ: Prentice Hall.

Scrutton, R. A., and M. Talwani, eds. 1982. *The Ocean Floor*. New York: Wiley Press.

Chapter 5. The Veiled Planet

Ahrens, C. D. 1998. *Essentials of Meteorology: An Introduction to the Atmosphere*. New York: Wadsworth.

Burroughs, W. J. 1999. *The Climate Revealed*. Cambridge: Cambridge University Press.

Frior, John. 1992. *The Atmosphere: A Global Challenge*. New Haven, CT: Yale University Press.

——. 2002. *The Crowded Greenhouse: Population, Climate Change, and Creating a Sustainable World*. New Haven, CT: Yale University Press.

Lutgens, F. R., E. J. Tarbuck, and Denis Tarza. 2009. *The Atmosphere: An Introduction to Meteorology*. 11th ed. Englewood Cliffs, NJ: Prentice Hall.

Rumney, George R. 1970. *The Geosystem: Dynamic Integration of Land, Sea, and Air*. Dubuque, IA: Wm. C. Brown.

Schneider, Stephen H., and Randi Londer. 1984. *The Coevolution of Climate and Life*. San Francisco: Sierra Club Books.

Walker, J. C. G. 1977. *Evolution of the Atmosphere*. New York: Macmillan.

Chapter 6. The Hazardous Planet

Bradford, Marlene. 2001. *Natural Disasters*. Pasadena, CA: Salem Press.

Bryant, Edward. 2005. *Natural Hazards*. Cambridge: Cambridge University Press.

Burton, Ian, Robert William Kates, and Gilbert F. White. 1993. *The Environment as Hazard*. New York: Guilford Press.

Cullen, Heidi. 2010. *The Weather of the Future*. New York: Harper/Harper Collins.

Davis, Lee. 2002. *Natural Disasters*. New York: Checkmark Books.

deBoer, Yelle Zeilinga, and Donald Theodore Sanders. 2002. *Volcanoes in Human History: The Far-Reaching Effects of Major Eruptions*. Princeton: Princeton University Press.

Dudley, Walter C. 1998. *Tsunami*. Honolulu: University of Hawaii Press.

Erikson, Jon. 2001. *Quakes, Eruptions and Other Geological Cataclysms: Revealing the Earth's Hazards*. 2nd rev. ed. New York: Facts on File.

Gunn, Angus M. 2001. *The Impact of Geology on the United States: A Reference Guide to Benefits and Hazards*. Westport, CT: Greenwood Press.

Hyndman, Donald W., and David Hyndman. 2005. *Natural Hazards and Disasters*. Lexington, KY: Thomson Higher Education.

Keller, Edward A., and Robert H. Blodgett. 2007. *Natural Hazards*. Upper Saddle River, NJ: Prentice Hall.

Kusky, Timothy M. 2003. *Geological Hazards: A Sourcebook*. Westport, CT: Greenwood Press.

Lewis, John S. 1999. *Comet and Asteroid Impact Hazards on a Populated Earth*. San Diego: Academic Press.

McGuire, Bill. 1999. *Apocalypse: A Natural History of Global Disasters*. London: Cassell.

——. 2004. *World Atlas of Natural Hazards.* London: Hodder Arnold.

Murck, Barbara C., Brian J. Skinner, and Stephen C. Porter. 1996. *Dangerous Earth: An Introduction to Geologic Hazards.* Hoboken, NJ: John Wiley & Sons.

Palmer, Trevor. 2003. *Perilous Planet Earth: Catastrophes and Catastrophism through the Ages.* Cambridge: Cambridge University Press.

Schneiderbauer, Stefan. 2008. *Risk and Vulnerability to Natural Hazards.* Saarbrücken, Germany: VDM Verlag.

Smith, Keith. 2004. *Environmental Hazards: Assessing Risks of Reducing Disaster.* 5th ed. Florence, KY: Routledge.

Wilson, E. O. 2006. *The Creation: An Appeal to Save Life on the Earth.* New York: W. W. Norton.

Weise, Elizabeth. 2006. "Is There Anywhere Safe to Live?" *USA Today,* April 20.

Chapter 7. The Ancient Planet

Albitton, Claude C. Jr. 1986. *The Abyss of Time: Changing Conceptions of Earth's Antiquity after the Sixteenth Century.* Los Angeles: Jeremy P. Tarcher.

Brush, Stephen G. 1996. *Transmuted Past: The Age of the Earth and the Evolution of the Elements from Lyell to Patterson.* New York: Cambridge University Press.

Dalrymple, G. Brent. 1991. *The Age of the Earth.* Stanford: Stanford University Press.

Gillispie, Charles C. 1996. *Genesis and Geology: A Study of the Relations of Scientific Thought, Natural Theology and Social Opinion in Great Britain, 1790–1850.* Cambridge, MA: Harvard University Press.

Gorst, Martin. 2001. *Measuring Eternity.* New York: Broadway Books.

Haben, Francis C. 1959. *The Age of the World: Moses to Darwin.* Baltimore: Johns Hopkins University Press.

Holland, Charles H. 1999. *The Idea of Time.* Chichester: John Wiley Books.

Holmes, Arthur. 1913. *The Age of the Earth.* London: Harper and Brothers.

Lewis, Cheryl, and Simon J. Kuck, eds. 2001. *The Age of the Earth: From 4004 B.C. to AD 2002.* London: Geological Society of London.

Macdougal, Doug. 2008. *Nature's Clocks: How Scientists Measure the Age of Almost Everything.* Berkeley: University of California Press.

York, Derek. 1997. *In Search of Lost Time.* Bristol: Institute of Physics.

York, Derek, and Ronald M. Farquhar. 1972. *Earth's Age and Geochronology.* Oxford: Pergamon.

Chapter 8. The Bountiful Planet

Barrow, John D., Simon Conway Morris, Stephen J. Freeland, and Charles L. Harper Jr. 2008. *Fitness of the Cosmos for Life: Biochemistry and Fine Tuning.* Cambridge: Cambridge University Press.

Craig, James R., David J. Vaughn, and Brian J. Skinner. 2001. *Resources of the Earth: Origin, Use and Environmental Impact.* 3rd ed. New York: Prentice Hall.

Dorr, Anne. 1987. *Minerals: Foundations of Society.* 2nd ed. Alexandria, VA: American Geological Institute.

Durant, Will, and Ariel Durant. 1968. *The Lessons of History.* New York: Simon & Schuster.

Fodor, Eben. 1991. *Looking at Energy from the Human Perspective.* Eugene, OR: Southern Willamette Alliance.

Jackson, Patrick W. 2006. *The Chronologer's Quest: Episodes in the Search for the Age of the Earth.* Cambridge: Cambridge University Press.

Lovering, T. S. 1943. *Minerals in World Affairs.* New York: Prentice Hall.

McDivitt, J. F., and Gerald Manners. 1974. *Minerals and Men: Resources for the Future.* Washington, DC: Johns Hopkins University Press.

Park, C. F., Jr. 1968. *Affluence in Jeopardy: Minerals and the Political Economy.* San Francisco: Freeman, Cooper.

Poss, J. R. 1975. *Stones of Destiny: Keystones of Civilization.* Houghton: Michigan Technological University.

Raymond, Robert. 1986. *Out of the Fiery Furnace: The Impact of Metals on the History of Mankind.* University Park: Pennsylvania State University Press.

Ritchie-Calder, Lord. 1970. "Mortgaging the Old Homestead." *Foreign Affairs* 48 (2): 207–20.

Sass, Stephen L. 1998. *The Substance of Civilization: Materials and Human History from the Stone Age to the Age of Silicon.* New York: Arcade Publications.

York, Derek. 1997. *In Search of Lost Time.* Bristol: Institute of Physics.

Youngquist, Walter. 1997. *GeoDestinies.* Portland, OR: National Book Company.

Chapter 9. The Finite Planet

Brown, Lester R., and Hal Kane. 1994. *Full House: Reassessing the Earth's Population Carrying Capacity.* Worldwatch Environmental Alert Series. New York: W. W. Norton.

Cohen, Joel E. 1995. *How Many People Can Earth Support?* New York: W. W. Norton.

Daly, Herman E., and Kenneth N. Townsend, eds. 1993. *Valuing the Earth: Economics, Ecology and Ethics.* Cambridge, MA: MIT Press.

Ehrlich, Paul R., and Anne H. Ehrlich. 1990. *The Population Explosion.* New York: Simon & Schuster.

Grant, Lindsey. 1992. *Elephants in the Volkswagen: Facing Tough Questions about Our Overcrowded Country.* New York: W. H. Freeman.

Harrison, Paul. 1992. *The Third Revolution: Environment, Population and a Sustainable World.* London: I. B. Tauris.

Meadows, Donella H., Dennis L. Meadows, and Jorgen Randers. 1992. *Beyond the Limits: Counting the Global Collapse, Envisioning a Sustainable Future.* Toronto: McClelland Stewart.

Meadows, Donella H., Dennis L. Meadows, Jorgen Randers, and William W. Behrens. 1972. *The Limits to Growth: A Report for the Club of Rome's Project on the Predicament of Mankind.* New York: New American Library.

Pearce, David, and Jeremy J. Wenford. 1993. *World without End: Economics, Environment and Sustainable Development.* New York: Oxford University Press.

Pimentel, David, and Carl W. Hall, eds. 1989. *Food and Natural Resources.* Toronto: Academic Press.

Scientific American. 1990. *Managing Planet Earth: Readings from the "Scientific American."* New York: W. H. Freeman.

Taylor, Ronnie. 1992. *Poverty, Population and the Planet.* Luton: Friends of the Earth.

Turner, B., ed. 1990. *The Earth as Transformed by Human Action: Global and Regional Changes in the Biosphere over the Past 300 Years.* New York: Cambridge University Press.

Weiner, Jonathan. 1990. *The Next One Hundred Years: Shaping the Fate of Our Living Earth.* New York: Bantam Books.

Wilson, Edward O. 2006. *The Creation: An Appeal to Save Life on Earth.* New York: W. W. Norton.

Chapter 10. The Singular Planet

Barrow, John D., Simon Conway Morris, Stephen J. Freeland, and Charles L. Harper Jr. 2008. *Fitness of the Cosmos for Life: Biochemistry and Fine Tuning.* Cambridge: Cambridge University Press.

Bernal, J. D. 1967. *The Origin of Life.* London: Weiderfeld & Nicholson.

Davies, Paul. 1992. *The Mind of God: The Scientific Basis for a Rational World*. New York: Simon & Schuster.

Dawkins, Richard. 1986. *The Blind Watchmaker*. New York: W. W. Norton.

de Duve, Christian. 1995. *Vital Dust: Life as a Cosmic Imperative*. New York: Basic Books.

——. 2002. *Life Evolving: Molecules, Mind and Meaning*. Oxford. Oxford University Press.

Fortey, Richard. 1999. *Life: A Natural History of the First Four Billion Years of Life on Earth*. New York: Vintage Books.

Fry, Iris. 2000. *The Emergence of Life on Earth: A Historical and Scientific Overview*. Piscataway, NJ: Rutgers University Press.

Gore, Al. 2006. *An Inconvenient Truth*. Emmaus, PA: Rodale.

Gould, Steven Jay. 1989. *Wonderful Life: The Burgess Shale and the Nature of History*. New York: W. W. Norton.

Henderson, L. J. 1913. *The Fitness of the Environment: An Enquiry into the Biological Significance of the Properties of Matter*. Repr., Boston: Beacon Press, 1958.

Rees, Martin. 2000. *Just Six Numbers: Deep Forces That Shape the Universe*. New York: Basic Books.

Wilson, Edward O. 1998. *Consilience: The Unity of Knowledge*. New York: Knopf.

Chapter 11. The Uninhabitable Planet

Cloud, Preston. 1988. *Oasis in Space: Earth History from the Beginning*. New York: W. W. Norton.

Craig, James R., David J. Vaughn, and Brian J. Skinner. 2001. *Resources of the Earth: Origin, Use and Environmental Impact*. 3rd ed. New York: Prentice Hall.

Fowler, C. M. R., C. J. Ebinger, and C. J. Hawkesworth. 2002. *The Early Earth: Physical, Chemical and Biological Development*. London: Geological Society of London.

Taylor, S. R. 1992. *Solar System Evolution*. Cambridge: Cambridge University Press.

Chapter 12. The Living Planet

Bakker, R. T. 1986. *The Dinosaur Heresies*. New York: William Morrow.

Beerling, David. 2007. *The Emerald Planet: How Plants Changed Earth's History*. Oxford: Oxford University Press.

Benton, M. J. 1990. *Vertebrate Paleontology*. London: Unwin Hyman.

Diamond, Jared. 1997. *Guns, Germs, and Steel: The Fates of Human Societies*. New York: W. W. Norton.

Glaessner, Martin F. 1984. *The Dawn of Animal Life*. Cambridge: Cambridge University Press.

Gould, Stephen Jay. 1989. *Wonderful Life: The Burgess Shale and the Nature of History*. New York: W. W. Norton.

——, ed. 1993. *The Book of Life*. New York: W. W. Norton.

Halstead, L. Beverly. 1982. *The Search for the Past*. Garden City, NY: Doubleday.

Houghton, John. 2009. *Global Warming: The Complete Briefing*. 4th ed. Cambridge: Cambridge University Press.

Imbrie, John, and K. P. Imbrie. 1979. *Ice Ages: Solving the Mystery*. Short Hills, NJ: Enslow.

Leakey, Richard, and Roger Lewin. 1992. *Origins Reconsidered: In Search of What Makes Us Human*. New York: Doubleday.

Leslie, John. 1998. *The End of the World: The Science and Ethics of Human Extinction*. London: Routledge.

Lewin, R. 1984. *Human Evolution: An Illustrated Introduction*. New York: W. H. Freeman.

Margulis, Lynn. 1982. *Early Life*. Boston: Science Books International.

Margulis, Lynn, Clifford Matthews, and Aaron Haselton, eds. 2000. *Environmental Evolution: Effects of the Origin and Evolution of Life on Planet Earth*. 2nd ed. Cambridge, MA: MIT Press.

Martin, P. S., and R. G. Klein. 1984. *Pleistocene Extinction*. Tucson: University of Arizona Press.

Rhodes, Frank H. T. 1976. *The Evolution of Life*. Harmondsworth: Penguin Books.

Savage, R. J. G., and M. R. Long. 1986. *Mammal Evolution: An Illustrated Guide*. New York: Facts on File.

Stanley, Steven M. 1993. *Exploring Earth and Life through Time*. New York: W. H. Freeman.

Sutcliffe, A. J. 1985. *On the Track of Ice Age Mammals*. Cambridge, MA: Harvard University Press.

Tattersall, I. 1995. *The Last Neanderthal: The Rise, Success, and Mysterious Extinction of Our Closest Human Relatives*. New York: Macmillan.

Tattersall, I., E. Delson, and J. Van Couvering. 1988. *Encyclopedia of Human Evolution and Prehistory*. New York: Garland.

Thomas, B. A., and R. A. Spicer. 1987. *The Evolution and Palaeobiology of Land Plants*. London: Croon Helen.

Vidal, G. 1984. "The Oldest Eukaryotic Cells." *Scientific American*, February.

Chapter 13. The Warming Planet

Archer, David. 2007. *Global Warming: Understanding the Forecast*. Malden, MA: Blackwell.

Broecker, Wallace S., and Robert Kunzig. 2008. *Fixing Climate: What Past Climate Changes Reveal about the Current Threat—and How to Counter It*. New York: Hill & Wang.

Calvin, William. 2002. *A Brain for All Seasons: Human Evolution and Abrupt Climate Change*. Chicago: University of Chicago Press.

Cronin, Thomas M. 2009. *Paleoclimates: Understanding Climate Change Past and Present*. New York: Columbia University Press.

Cowie, Jonathan. 2007. *Climate Change: Biological and Human Aspects*. Cambridge: Cambridge University Press.

Dow, Kristin, and Thomas E. Downing. 2006. *The Atlas of Climate Change: Mapping the World's Greatest Challenge*. Berkeley: University of California.

Emanuel, Kerry. 2007. *What We Know about Climate Change*. Cambridge: Boston Review Books / MIT Press.

Fagan, Brian. 2000. *The Little Ice Age: How Climate Made History 1300–1850*. New York: Basic Books.

Fookes, Peter G., and E. Mark Lee. 2007. "Climate Variation: A Simple Geological Perspective." *Geology Today* 23 (2): 66–73.

Hoffman, Paul, and Daniel Schrag. 2000. "Snowball Earth." *Scientific American,* January.

Houghton, John. 2009. *Global Warming: The Complete Briefing*. 4th ed. Cambridge: Cambridge University Press.

Hulme, Mike. 2009. *Why We Disagree about Climate Change*. Cambridge: Cambridge University Press.

Imbrie, John, and K. P. Imbrie. 1979. *Ice Ages: Solving the Mystery*. Short Hills, NJ: Enslow.

Intergovernmental Panel on Climate Change (IPCC). 2007. *Climate Change 2007: The Physical Science Basis*. New York: Cambridge University Press.

Jones, P. D., A. E. J. Ogilvie, T. D. Davies, and K. R. Briffa. 2001. *History and Climate: Memories of the Future?* New York: Kluwer Academic / Plenum Publishers.

Kartén, Wibjörn, Eigil Früs-Christensen, and Bergt Dahlström. 1993. *The Earth's Climate: Natural Variations and Human Influence*. Stockholm: Elforsk, AB.

Lamb, H. H. 1995. *Climate, History, and the Modern World*. New York: Routledge.

Lomborg, Bjorn. 2001. *The Skeptical Environmentalist*. Cambridge: Cambridge University Press.

Lunine, Jonathan I. 1999. *Earth: Evolution of a Habitable World*. Cambridge: Cambridge University Press.

Macdougall, Doug. 2004. *Frozen Earth: The Once and Future Story of Ice Ages*. Berkeley: University of California Press.

Martin, P. S. 1963. *The Last 10,000 Years*. Tucson: University of Arizona Press.

Maslin, Mark. 2004. *Global Warming: A Very Short Introduction*. Oxford: Oxford University Press.

Michaels, Patrick J., and Robert C. Balling Jr. 2000. *The Satanic Gases: Clearing the Air about Global Warming*. Washington DC: Cato Institute.

Moore, T. G. 1995, *Global Warming: A Boon to Human and Other Animals*. Stanford: Hoover Institute, Stanford University.

Philander, S. George. 2003. "Why Global Warming Is a Controversial Issue." In *Global Climate: Current Research and Uncertainties in the Climate System*, edited by Xavier Rodó and Francisco A. Comín, 25–33. Berlin: Springer-Verlag.

Pielou, E. C. 1991. *After the Ice Age: The Return of Life to Glaciated North America*. Chicago: University of Chicago Press.

Rotberg, R. I., and T. K. Rabb, eds. 1981. *Climate and History: Studies in Interdisciplinary History*. Princeton, NJ: Princeton University Press

Singer, S. Fred. 1997. *Hot Talk, Cold Science: Global Warming's Unfinished Debate*. Oakland, CA: Independent Institute.

Stanley, Steven. 1996. *Children of the Ice Age: How a Global Catastrophe Allowed Humans to Evolve*. New York: Harmony Books.

Stern, Nicholas. 2007. *Intergovernmental Panel on Climate Change: The Scientific Basis*. Cambridge: Cambridge University Press.

Walker, Gabrielle. 2003. *Snowball Earth: The Story of the Great Global Catastrophe That Spawned Life as We Know It*. New York: Crown Publishers.

Walker, Gabrielle, and Sir David King. 2008. *The Hot Topic: What We Can Do about Global Warming*. Orlando: Harcourt.

Chapter 14. The Polluted Planet

Bily, Cynthia. 2006. *Pollution (Introducing Opposing Viewpoints)*. San Diego: Greenhaven.

Cherni, Judith A. 2002. *Economic Growth versus the Environment: The Politics of Wealth, Health and Air Pollution*. New York: Palgrave.

Clark, R. B. 1947. *Marine Pollution*. 4th ed. Oxford: Clarendon Press.

Crandall, Robert W. 1983. *Controlling Industrial Pollution: The Economics and Politics of Clean Air*. Washington, DC: Brookings Institution Press.

Elsom, Derek. 1996. *Smog Alert: Managing Urban Air Quality*. London: Earthscan.

Fenger, Jess, and Christian J. Tiell. 2009. *Air Pollution: From a Local to a Global Perspective*. London: Royal Society of Chemistry.

Goudie, Andrew. 2000. *The Human Impact on the Natural Environment*. 5th ed. Cambridge, MA; MIT Press.

Goudie, A. S., and H. Viles. 1997. *The Earth Transformed*. New York: Blackwell.

Hill, Margarita K. 2010. *Understanding Environmental Pollution*. 3rd ed. Cambridge: Cambridge University Press.

Johnson, Steven Beilin. 2006. *The Ghost Map: The Story of London's Most Terrifying Epidemic and How It Changed Science, Cities, and the Modern World*. New York: Riverhead Books.

Little, Jane Braxton. 2009. "The Carbon Equation." *Nature Conservancy*, Winter.

Meyer, W. B. 1996. *The Human Impact on the Earth*. Cambridge: Cambridge University Press.

Pickering, K. T., and L. A. Owen. 1997. *Global Environmental Issues*. 2nd ed. New York: Routledge.

Simmons, I. G. 1996. *Changing the Face of the Earth: Culture, Environment, History*. London: Blackwell.

Soros, Marvin S. 1997. *The Endangered Atmosphere: Preserving a Global Commons*. Columbia: University of South Carolina Press.

Theodore, Mary K., and Louis Theodore. 1997. *Major Environmental Issues Facing the 21st Century*. Upper Saddle Creek, NJ: Prentice Hall.

Turner, B. L., William C. Clark, Robert W. Kates, John F. Richards, Jessica C. Mathews, and William B. Meyer, eds. 1993. *The Earth as Transformed by Human Action: Global and Regional Changes in the Biosphere over the Past 300 Years*. Cambridge: Cambridge University Press with Clark University Press.

Viessman, Warren W., Mark I. Hammin, Elizabeth Perez, and Paul A. Chadik. 2008. *Water Supply and Pollution Control*. Upper Saddle River, NJ: Pearson Prentice Hall.

Chapter 15. The Crowded Planet

Archer, David. 2009. *The Long Thaw: How Humans Are Changing the Next 100,000 Years of Earth's Climate*. Princeton, NJ: Princeton University Press.

Bledsoe, Caroline, and J. A. Johnson-Kuhn. 1999. *Critical Perspectives on Schooling and Fertility in the Developing World*. Washington, DC: National Academy Press.

Brown, Lester. 2001. *Eco-Economy: Building an Economy for the Earth*. New York: Norton.

Cohen, Joel E. 1995. *How Many People Can Earth Support?* New York: W. W. Norton.

Conkin, Paul K. 2007. *The State of the Earth*. Lexington: University of Kentucky Press.

Ehrlich, Paul R., and Anne H. Ehrlich. 2008. *The Dominant Animal: Human Evolution and the Environment*. Washington, DC: Island Press.

Friedman, Thomas L. 2008. *Hot, Flat and Crowded: Why We Need a Green Revolution—and How It Can Renew America*. New York: Farrar, Strauss and Giroux.

Klein, David R. "The Introduction, Increase, and Crash of Reindeer on St. Matthew Island." http://dieoff.org/page80.htm.

Livi-Bacci, Massimo. 2001. *A Concise History of World Population*. 3rd ed. Malden, MA: Blackwell.

Meadows, Donella H., Jorgen Randers, and Dennis Meadows. 2004. *Limits to Growth: The 30-Year Update*. White River Junction, VT: Chelsea Green.

Meyer, W. B. 1996. *Human Impact on the Earth*. Cambridge: Cambridge University Press.

Pickering, K. T., and Lewis A. Owen. 1997. *Global Environmental Issues*. 2nd ed. London: Routledge.

Pimentel, David, and Marcia H. Pimentel. 2008. *Food, Energy and Society*. Boca Raton: CRC Press.

Pimentel, David, Michele Whitecraft, Zachary R. Scott, Leixin Zhao, Patricia Satkiewicz, Timothy J. Scott, Jennifer Phillips, et al. 2010. "Will Limited Land, Water, and Energy Control Human Population Numbers in the Future?" *Human Ecology* 38 (5): 599–611.

Ponting, Clive. 1991. *A Green History of the World: The Environment and the Collapse of Great Civilizations*. London: Sinclair-Stevenson.

Simmons, I. G. 1996. *Changing the Face of the Earth: Culture, Environment and History*. 2nd ed. Oxford: Blackwell.

Stanton, William. 2004. *The Rapid Growth of Human Populations 1750–2000: Histories, Consequences, Issues, Nation by Nation*. Essex: Multi-Science.

Tainter, Joseph A. 1988. *The Collapse of Complex Societies*. Cambridge: Cambridge University Press.

Weeks, John R. 2004. *Population: An Introduction to Concepts and Issues.* Florence, KY: Wadsworth.

Chapter 16. The Sustainable Planet

Broecker, Wallace S. 1987. *How to Build a Habitable Planet.* Palisades, NY: Eldigio Press.

Brown, Lester R. 2009. *Plan B 4.0: Mobilizing to Save Civilization.* New York: W. W. Norton.

Brown, Lester R., Christopher Flavin, and Sandra Postel. 1991. *Saving the Planet: How to Shape an Environmentally Sustainable Global Economy.* New York: W. W. Norton.

Conkin, Paul K. 2007. *The State of the Earth: Environmental Challenges on the Road to 2100.* Lexington: University of Kentucky Press.

deVilliers, Marq. 2001. *Water: The Fate of Our Most Precious Resource.* New York: Mariner Books.

Diamond, Jared. 2005. *Collapse: How Societies Choose to Fail or Succeed.* New York: Viking.

Esty, Daniel C., and Andre Winston. 2006. *Green to Gold: How Smart Companies Use Environmental Strategy to Innovate, Create Value and Build Competitive Advantage.* New Haven, CT: Yale University Press.

Friedman, Thomas L. 2008. *Hot, Flat and Crowded: Why We Need a Green Revolution and How it Can Renew America.* New York: Farrar, Straus and Giroux.

Gleick, Peter H. 1993. *Water in Crisis: A Guide to the World's Fresh Water Resources.* New York: Oxford University Press.

——. 2000. *The World's Water.* Washington, DC: Island Press.

Kennedy, Donald. 2006. *State of the Planet 2006–2007.* Washington, DC: Island Press.

Lomborg, Bjorn. 2001. *The Skeptical Environmentalist: Measuring the Real State of the World.* Cambridge: Cambridge University Press.

MacKay, David J. C. 2009. *Sustainable Energy without the Hot Air.* Cambridge: UIT.

Mackenzie, Fred T. *Our Changing Planet: An Introduction to Earth System Science and Global Environmental Change.* 4th ed. Upper Saddle River, NJ: Pearson Education.

Patel, Raj. 2007. *Stuffed and Starved: The Hidden Battle for the World Food System.* Brooklyn: Melville House.

Pearce, Fred. 2006. *When the Rivers Run Dry: Water; The Defining Crisis of the Twenty-First Century.* Boston, MA: Beacon Press.

Ponting, Clive. 1991. *A Green History of the World: The Environment and the Collapse of Great Civilizations.* London: Sinclair-Stevenson.

Postel, Sandra. 1997. *Last Oasis: Facing Water Scarcity.* New York: W. W. Norton.

Postel, S. L., G. C. Daly, and P. R. Ehrlich. 1996. "Human Appropriation of Renewable Fresh Water." *Science* 291:85.

Rees, Martin. 2003. *Our Final Hour.* New York: Basic Books.

Rosengrant, Mark A., Xining Cai, and Sarah A. Cline. 2002. *World Water and Food to 2025: Dealing with Scarcity.* Washington, DC: International Food Policy Research Institute.

Schneiderman, Jill S., ed. 2000. *The Earth around Us: Maintaining a Livable Planet.* New York: W. H. Freeman.

Shiklomanov, I. A., and John C. Rodda. 2003. *World Water Resources at the Beginning of the Twenty-First Century.* Cambridge: Cambridge University Press.

Tainter, Joseph A. 1988. *The Collapse of Complex Societies.* Cambridge: Cambridge University Press.

Van Beers, C., and A. de Moor. 2001. *Public Subsidies and Policy Failures: How Subsidies Distort the Natural Environment, Equity and Trade and How to Reform Them.* Northampton, MA: Edward Elgar.

Wilson, E. O. 2006. *The Creation: An Appeal to Save Life on Earth.* New York: W. W. Norton.

Youngquist, Walter B. 1997. *GeoDestinies: The Inevitable Control of Earth's Resources over Nations and Individuals.* Portland, OR: National Book Company

Chapter 17. Water as Sustenance

Barton, Maude. 2007. *Blue Covenant: The Global Water Crisis and the Coming Battle for the Right to Water.* New York: The New Press.

Cech, Thomas V. 2009. *Principles of Water Resources: History, Development, Management and Policy.* 3rd ed. Hoboken, NJ: John Wiley & Sons.

Chartres, Colin, and Samyuktha Varma. 2010. *Out of Water: From Abundance to Scarcity and How to Solve the World's Water Problems.* Upper Saddle River, NJ: FT Press.

Conkin, Paul K. 2007. *The State of the Earth: Environmental Challenges on the Road to 2100.* Lexington: University of Kentucky Press.

deVilliers, Marq. 2001. *Water: The Fate of Our Most Precious Resource.* New York: Mariner Books.

Eckholm, E. P. 1989. *Down to Earth: Environment and Human Needs.* London: Pluto Press.

Gleick, Peter H. 1993. *Water in Crisis: A Guide to the World's Fresh Water Resources.* New York: Oxford University Press.

Gleick, Peter H., and associates. 2009. *The World's Water: The Biennial Report on Freshwater Resources.* Washington DC: Island Press.

Hillel, D. J. 1992. *Out of the Earth: Civilization and the Life of the Soil.* New York: Free Press.

Hoffman, Stephen J. 2009, *Planet Earth: Investing in the World's Most Valuable Resource*, Hoboken, NJ: John Wiley & Sons.

Hyams, Edward. 1952. *Soil and Civilization.* London: Thames & Hudson.

Kennedy, Donald. 2006. *State of the Planet 2006–2007.* Washington, DC: Island Press.

Montgomery, David. 2007. *Dirt: The Erosion of Civilizations.* Berkeley: University of California Press.

Morgan, R. C. 1995. *Soil, Erosion and Conservation.* 2nd ed. Oxford: Wiley-Blackwell.

Murphy, B., and Damian Nance. 1998. *Earth Science Today.* Florence, KY: Brooks/Cole.

Patel, Raj. 2007. *Stuffed and Starved: The Hidden Battle for the World Food System.* Brooklyn: Melville House.

Pearce, Fred. 2006. *When the Rivers Run Dry: Water—The Defining Crisis of the Twenty-First Century.* Boston: Beacon Press.

Pimentel, David, ed. 1993. *World Soil Erosion and Conservation.* Cambridge: Cambridge University Press.

Postel, Sandra. 1997. *Last Oasis: Facing Water Scarcity.* New York: W. W. Norton.

Prud'homme, Alex. 2011. *Clean, Clear, and Cold: The Fate of Fresh Water in the Twenty-First Century.* Roseburg: Scribner.

Rogers, Peter, Susan Leal, and Edward J. Markey. 2010. *Running Out of Water: The Looming Crisis and Solutions to Conserve Our Most Precious Resource.* New York: Palgrave Macmillan.

Rosengrant, Mark A., Xining Cai, and Sarah A. Cline. 2002. *World Water and Food to 2025: Dealing with Scarcity.* Washington, DC: International Food Policy Research Institute.

Shiklomanov, I. A., and John C. Rodda. 2003. *World Water Resources at the Beginning of the Twenty-First Century.* Cambridge: Cambridge University Press.

Van Beers C., and A. de Moor. 2001. *Public Subsidies and Policy Failures: How Subsidies Distort the Natural Environment, Equity and Trade and How to Reform Them.* Northampton, MA: Edward Elgar.

Wild, Alan. 1993. *Soils and the Environment.* Cambridge: Cambridge University Press.

Chapter 18. Air as Sustenance

Barry, R. G., and R. J. Chorley. 2010. *Atmosphere, Weather and Climate*. 9th ed. London: Routledge.

Brewer, George. 1980. *Air in Danger: Ecological Perspectives on the Atmosphere*. New York: Cambridge University Press.

Brimblecombe, P. 1987. *The Big Smoke: A History of Air Pollution in London since Medieval Times*. London: Routledge.

Crandall, Robert W. 1983. *Controlling Industrial Pollution: The Economics and Politics of Clean Air*. Washington, DC: Brookings Institution Press.

Cowie, J. 1998. *Climate and Human Change: Disaster or Opportunity?* New York: Parthenon.

Desonie, Dana. 2007. *Atmosphere: Air Pollution and Its Effects*. New York: Chelsea House.

Elsom, D. 1996. *Smog Allergy: Managing Urban Air Quality*. London: Earthscan.

Fenger, Jes. 2009. *Air Pollution: From a Local to a Global Perspective*. New York: Springer-Verlag.

Frederick, John E. 2007. *Principles of Atmospheric Science*. Sudbury, MA: Jones & Bartlett.

Frior, John. 1990. *The Changing Atmosphere: A Global Challenge*. New Haven, CT: Yale University Press.

Graedel, T. E., and P. J. Crutzen. 1993. *Atmospheric Change: An Earth System Perspective*. San Francisco: Freeman.

Hayward, Steven F., and Joel M. Schwartz. 2008. *Air Quality in America: A Dose of Reality on Air Pollution Levels, Trends and Health*. Washington DC: AEI Press

Jacobson, Mark Z. 2002. *Atmospheric Pollution: History, Science and Regulation*. Cambridge: Cambridge University Press.

Kidd, J. S., and Renee A. Kidd. 2005. *Air Pollution: Problems and Solutions*. 2nd ed. New York: Facts on File.

Lutgens, Frederick K., and Edward J. Tarbuck. 2009. *Earth Science*. 12th ed. Upper Saddle River, NJ: Prentice Hall.

McElroy, Michael. 2002. *The Atmospheric Environment: Effects of Human Activity*. Princeton, NJ: Princeton University Press.

Michaels, Patrick J., and Robert C. Balling Jr. 2000. *The Satanic Gases: Clearing the Air about Global Warming*. Washington, DC: Cato Institute.

Park, C. C. 1987. *Acid Rain: Rhetoric and Reality*. London: Methuen.

Saha, Kshudiram. 2010. *The Earth's Atmosphere: Its Physics and Dynamics*. New York: Springer.

Soroos, Marvin S. 1997. *The Endangered Atmosphere: Preserving a Global Commons*. Columbia: University of South Carolina Press.

Stewart, T. Charles. 1979. *Air Pollution, Human Health and Public Policy*. New York: Lexington Books.

Vallero, David. 2008. *Fundamentals of Air Pollution*. 4th ed. Burlington, MA: Elsevier, Academic Press.

Van Veers, C., and A. de Moor. 2001. *Public Subsidies and Policy Failures: How Subsidies Distort the Natural Environment, Equity and Trade and How to Reform Them*. Northampton, MA: Edward Elgar.

Wark, Kenneth, Cecil F. Warner, and Wayne T. Davis. 1997. *Air Pollution: Its Origin and Control*. 3rd ed. Upper Saddle River, NJ: Prentice Hall.

Chapter 19. Soil as Sustenance

Adams, J. A. 1986. *Dirt*. College Station: Texas A&M Press.

Eckholm, E. P. 1989. *Down to Earth: Environment of Human Needs*. London: Pluto Press.

Granger, A. 1992. *Controlling Tropical Deforestation*. London: Earthscan.

Hillel, D. J. 1992. *Out of the Earth: Civilization and the Life of the Soil*. New York: Free Press.

Hyams, Edward. 1952. *Soil and Civilization*, London: Thames & Hudson.

Logan, W. B. 1995. *Dirt: The Ecstatic Skin of the Earth*. New York: W. W. Norton.

Montgomery, David. 2007. *Dirt: The Erosion of Civilizations*. Berkeley: University of California Press.

Morgan, R. P. C. 2005. *Soil, Erosion and Conservation*. 3rd ed. Malden, MA: Blackwell.

Owens, O. S. 1980. *Natural Resource Conservation: An Ecological Approach*. New York: Macmillan.

Pimentel, David, ed. 1993. *World Soil Erosion and Conservation*. Cambridge: Cambridge University Press.

Troeh, Frederick R., J. Arthur Hobbs, and Roy L. Donahue. 2003. *Soil and Water Conservation for Productivity and Environmental Protection*. 4th ed. Upper Saddle River, NJ: Prentice Hall.

Wild, Alan. 1993. *Soils and the Environment*. Cambridge: Cambridge University Press.

Williams, Michael. 2003. *Deforesting the Earth*. Chicago: University of Chicago Press.

Chapter 20. Food as Sustenance

Altieri, M. A. 1995. *Agrocology: The Science of Sustainable Agriculture*. Boulder, CO: Westview Press.

Dreze, Jean, and Amartya Sen. 1991. *Hunger and Public Action*. Oxford: Oxford University Press.

Islan, Nurul, ed. 1995. *Population and Food in the Early Twenty-First Century*. Washington, DC: International Food Policy Research Institute.

McConnell, Kathryn. 2004. "One-Seventh of World's People Suffer from Hunger, Experts Say." America.gov. http://www.america.gov/st/washfile-english/2004/November/20041129142905AKllennoCcM0.7834589.html.

Mitchell, Donald O., Melinda D. Ingco, and Ronald C. Duncan. 1997. *The World Food Outlook*. Cambridge: Cambridge University Press.

Morgan, Kevin, Terry Marsden, and Jonathan Murdoch. 2006. *Worlds of Food: Place, Power and Provenance in the Food Chain*. Oxford: Oxford University Press.

National Research Council. 2010. *Toward Sustainable Agricultural Systems in the Twenty-first Century*. Washington, DC: The National Academies Press.

Pimentel, David, and Marcia Pimentel, eds. 1996. *Food, Energy and Society*. Rev. ed. Niwot, CO: University of Colorado Press.

Roberts, W. 2008. *The No-Nonsense Guide to World Food*. Oxford: New Internationalist.

Chapter 21. Energy as Sustenance

Adams, Robert McCormick. 1996. *Paths of Fire: An Anthropologist's Inquiry into Western Technology*. Princeton, NJ: Princeton University Press.

Bartlett, Albert. 2004. "Thoughts on Long-Term Energy Supplies: Scientists and the Silent Lie." *Physics Today* 57:53–55.

Bryce, Robert. 2010. *Power Hungry: The Myths of "Green" Energy and the Real Fuels of the Future*. New York: PublicAffairs

Coley, David. 2008. *Energy and Climate Change: Creating a Sustainable Future*. Chichester: John Wiley & Sons.

Eerkens, Jeff. 2006. *The Nuclear Imperative: A Critical Look at the Approaching Energy Crisis*. New York: Springer.

Elliot, David. 2007. *Nuclear or Not? Does Nuclear Power Have a Place in a Sustainable Energy Future?* New York: Palgrave Macmillan.

Fodor, Eben. 1991. *Looking at Energy from the Human Perspective*. Eugene, OR: Southern Willamette Alliance.

Friedman, Thomas L. 2008. *Hot, Flat and Crowded: Why We Need a Green Revolution—and How It Can Renew America.* New York: Farrar, Straus and Giroux.

Grant, Lindsey. 2005. *The Collapsing Bubble: Growth and Fossil Energy.* Santa Ana, CA: Seven Locks Press.

Heintzman, Andre, and Evan Solom, eds. 2005. *Fueling the Future: How the Battle of Energy Is Changing Everything.* Toronto: House of Ananasi Press.

International Energy Agency. 2008. *Energy Technology Perspectives: Scenarios and Strategies to 2050.* OECD Publishing.

International Energy Agency and Organisation for Economic Co-operation and Development. 2001. *World Energy Outlook: Assessing Today's Supply to Fuel Tomorrow's Growth.* Paris: OECD / IEA.

Lovering, T. S. 1943. *Minerals in World Affairs.* New York: Prentice Hall.

MacKay, David J. C. 2008. *Sustainable Energy: Without the Hot Air.* Cambridge: UIT.

Manwell, J. F., J. G. McGowan, and A. L. Roger. 2002. *Wind Energy Explained: Theory, Design, and Application.* New York: John Wiley.

Rifkin, Jeremy. 2002. *The Hydrogen Economy: The Creation of the Worldwide Energy Web of the Redistribution of Power on Earth.* New York: Penguin.

Roberts, Paul. 2005. *The End of Oil: The Decline of the Petroleum Economy and the Rise of a New Energy Order.* London: Bloomsburg.

Salvador, Amos. 2005. *Energy: A Historical Perspective and 21st Century Forecast.* AAPG Studies in Geology 54. Tulsa: American Association of Petroleum Geologists.

Sass, Stephen L. 1998. *The Substance of Civilization: Materials and Human History from the Stone Age to the Age of Silicon.* New York: Arcade.

Scheer, Hermann. 2004. *The Solar Economy: Renewable Energy for a Sustainable Future.* London: Earthscan.

Simon, Christopher A. 2006. *Alternative Energy: Political, Economic, and Social Feasibility.* Lanham, MD: Rowman & Littlefield.

Smil, Vaclav. 2005. *Energy at the Crossroads: Global Perspectives and Uncertainties.* Cambridge, MA: MIT Press.

Youngquist, Walter. 1997. *GeoDestinies: The Inevitable Control of Earth Resources over Nations and Individuals.* Portland, OR: National Book Company.

Alternative Energy Sources

Baker, A. C. 1971. *Tidal Power.* London: Peter Peregrinus.

Boyle, Godfrey. 2004. *Renewable Energy: Power for a Sustainable Future.* 2nd ed. Oxford: Oxford University Press.

European Wind Energy Association. 2009. *Wind Energy—the Facts: A Guide to the Technology, Economics and Future of Wind Power.* London: Earthscan.

Kruger, Paul. 2006. *Alternative Energy Resource: The Quest for Sustainable Energy.* Hoboken, NJ: Wiley.

Scheer, Herman. 2007. *Energy Autonomy: The Economic, Social, and Technological Case for Renewable Energy.* London: Earthscan.

Coal

Thomas, Larry. 2002. *Coal Geology.* Chichester: John Wiley & Sons.

Hydropower/Geothermal

Baker, A. C. 1991. *Tidal Power.* London: Peter Peregrinus.

Gupta, Harsh K., and Sukanta Roy. 2006. *Geothermal Energy: An Alternative Resource for the 21st Century.* New York: Elsevier Science.

Massachusetts Institute of Technology. 2006. *The Future of Geothermal Energy: Impact of Enhanced Geothermal Systems (EGS) on the United States in the Twenty-first Century.* http:// geothermal.inel.gov.

Oil and Natural Gas

Ahebrandt, Thomas S., Ronald R. Charpentier, J. R. Klett, James W. Schoker, Christopher J. Schenk, and Gregory F. Vlimshetz. 2005. *Global Resource Estimates from Total Petroleum Systems.* AAPG Memoir 86. Tulsa: American Association of Petroleum Geologists.

Campbell, C. J. 1988. *The Coming Oil Crisis.* Multi-Science Publishing Company 7 and Petroconsultants SA.

Campbell, Colin, and Jean Laherrere. 1998. "The End of Cheap Oil." *Scientific American,* March.

Deffeyes, Kenneth. 2001. *Hubbert's Peak: The Impending World Oil Shortage.* Princeton, NJ: Princeton University Press.

Goodstein, David. 2004. *Out of Gas: The End of the Age of Oil.* New York: W. W. Norton.

Huber, Peter W., and Mark P. Mills. 2005. *The Bottomless Well: The Twilight of Fuel, the Virtue of Waste, and Why We Will Never Run Out of Energy.* New York: Basic Books.

USGS World Energy Assessment Team. 2000. *U.S. Geological Survey World Petroleum Assessment 2000.* Digital Data Series. Washington, DC: USGS.

Nuclear Energy

Bodansky, David. 2004. *Nuclear Energy: Principles, Practices, and Prospects.* New York: Springer-Verlag.

Kursunogammalu, Behram N., Arnold Perlmutter, and Stephan L. Mintz, eds. 2004. *The Challenges to Nuclear Power in the Twenty-First Century.* New York: Plenum.

Lovelock, James. 2005. *The Revenge of Gaia.* London: Penguin, Allen Lane.

Mercall, Gene, ed. 2006. *Nuclear Power.* Pontiac, MI: Cengage Gale.

Morris, Robert C. 2000. *The Environmental Case for Nuclear Power.* London: Continuum International.

Newton, David E. 2006. *Nuclear Power.* New York: Facts on File.

Tucker, William. 2008. *Terrestrial Energy: How Nuclear Power Will Lead the Green Revolution and End America's Energy Odyssey.* Savage, MD: Bartleby Press.

Chapter 22. Materials as Sustenance

Adams, Robert McCormick. 1996. *Paths of Fire: An Anthropologist's Inquiry into Western Technology.* Princeton, NJ: Princeton University Press.

Daly, H. E., and K. N. Townsend, eds. 1993. *Valuing the Earth: Economies, Ecology and Ethics.* Cambridge, MA: MIT Press.

Dorr, Anne. 1987. *Minerals: Foundations of Society.* 2nd ed. Alexandria, VA: American Geological Institute.

Fodor, Eben. 1991. *Looking at Energy from the Human Perspective.* Eugene, OR: Southern Willamette Alliance.

Hetherington, L. E., N. E. Idoine, T. J. Brown, P. A. Lusty, L. Hitchen, and T. Bide. 2007. *United Kingdom Minerals Yearbook 2007.* Keyworth, Nottinghamshire: British Geological Survey.

Kesler, Stephen E. 1994. *Mineral Resources: Economies and the Environment.* Upper Saddle River, NJ: Prentice Hall.

Kinstler, James Howard. 2005. *The Long Emergency.* New York: Atlantic Monthly Press.

Lovering, T. S. 1943. *Minerals in World Affairs.* New York: Prentice Hall.

McDevitt, J. F., and Gerald Manners. 1974. *Minerals and Men: Resources for the Future.* Washington, DC: Johns Hopkins University Press.

Park, C. F. Jr. 1968. *Affluence in Jeopardy: Minerals and the Political Economy.* San Francisco: Freeman Cooper.

——. 1975. *Earthbound: Minerals, Energy and Man's Future.* San Francisco: Freeman Cooper.

Poss, J. R. 1975. *Stones of Destiny: Keystones of Civilization.* Houghton: Michigan Technological Institute.

Raymond, Robert. 1986. *Out of the Fiery Furnace: The Impact of Metals on the History of Mankind.* University Park: Pennsylvania State University Press.

Sass, Stephen L. 1998. *The Substance of Civilization: Materials and Human History from the Stone Age to the Age of Silicon.* New York: Arcade Publishing.

Simon, J. L., ed. 1995. *The State of Humanity.* Cambridge: Blackwell.

Tickell, Sir Crispin. 1994. "The Future and Its Consequences." *Interdisciplinary Science Reviews* 19 (4): 273–79.

Treister, Michael Yu. 1995. *The Role of Metals in Ancient Greek History.* London: Brill.

U.S. Geological Survey. 2007. *Minerals Yearbook.* Washington, DC: U.S. Department of the Interior.

Youngquist, Walter. 1997. *GeoDestinies: The Inevitable Controversy of Earth's Resources over Nations and Individuals.* Portland, OR: National Book Company.

Chapter 23. Prospects for Sustenance

Ashby, E. 1978. *Reconciling Man with the Environment.* London: Oxford University Press.

Bailey, R., ed. 1995. *The True State of the Planet.* New York: Free Press.

Brown, Lester R. 2006. *Plan B 2.0: Rescuing a Planet under Stress and a Civilization in Trouble.* New York: W. W. Norton.

——. 2009. *Plan B 4.0: Mobilizing to Save Civilization.* New York: W. W. Norton.

Brown, Lester R., Christopher Flavin, and Sandra Postel. 1991. *Saving the Planet: How to Shape an Environmentally Sustainable Economy.* New York: W. W. Norton.

Diamond, Jared. 2005. *Collapse: How Societies Choose to Fail or Succeed.* New York: Penguin.

Mackenzie, Fred T. 2010. *Our Changing Planet: An Introduction to Earth System Science and Global Environmental Change.* 4th ed. Upper Saddle River, NJ: Pearson Education.

Ponting, Clive. 1991. *A Green History of the World: The Environment and the Collapse of Great Civilizations.* London: Sinclair-Stevenson.

Rees, Martin. 2003. *Our Final Hour.* New York: Basic Books.

Roberts, N., ed. 1994. *The Changing Global Environment.* Oxford: Blackwell.

Sachs, Jeffrey. 2005. *The End of Poverty: Economic Possibilities for Our Time.* New York: Penguin.

Simon, Julian L. 1998. *The Ultimate Resource 2.* Princeton, NJ: Princeton University Press.

Tainter, Joseph A. 1988. *The Collapse of Complex Societies.* Cambridge: Cambridge University Press.

Trainer, F. E. 1995. "Can Renewable Energy Sources Sustain Affluent Society?" *Energy Policy* 23 (12): 1009–26.

Ward, Peter. 2001. *Future Evolution.* San Francisco: W. H. Freeman.

Wilson, E. O. 2006. *The Creation: An Appeal to Save Life on Earth.* New York: W. W. Norton.

Chapter 24. Policies for Sustenance

Brown, Lester R. 2009. *Plan B 4.0: Mobilizing to Save Civilization.* New York: W. W. Norton.

Ponting, Clive. 1991. *A Green History of the World: The Environment and the Collapse of Great Civilizations.* London: Sinclair-Stevenson.

Rees, Martin. 2003. *Our Final Hour.* New York: Basic Books.

Sachs, Jeffrey. 2005. *The End of Poverty: Economic Possibilities for Our Time.* New York: Penguin.

Simon, Julian L. 1998. *The Ultimate Resource 2.* Princeton, NJ: Princeton University Press.

Tainter, Joseph A. 1988. *The Collapse of Complex Societies.* Cambridge: Cambridge University Press.

Tickell, Crispin. 1993. "The Human Species: A Suicidal Success?" *Geographical Journal* 159 (2): 219–26.

Wilson, E. O. 2006. *The Creation: An Appeal to Save Life on Earth.* New York: W. W. Norton.

Index